Applied Probability and Statistics

BAILEY · The Elements of Stochastic Processes with Applications to the Natural Sciences
BARTHOLOMEW · Stochastic Models for Social Processes, *Second Edition*
BENNETT and FRANKLIN · Statistical Analysis in Chemistry and the Chemical Industry
BHAT · Elements of Applied Stochastic Processes
BLOOMFIELD · Fourier Analysis of Time Series: An Introduction
BOX and DRAPER · Evolutionary Operation: A Statistical Method for Process Improvement
BROWNLEE · Statistical Theory and Methodology in Science an Engineering, *Second Edition*
BURY · Statistical Models in Applied Science
CHERNOFF and MOSES · Elementary De
CHOW · Analysis and Control of Dynam
CLELLAND, deCANI, BROWN, BURS
Statistics with Business Applications, *Se*
COCHRAN · Sampling Techniques, *Secon*
COCHRAN and COX · Experimental Desig
COX · Planning of Experiments
COX and MILLER · The Theory of Stochastic Processes, *Second Edition*
DANIEL and WOOD · Fitting Equations to Data
DAVID · Order Statistics
DEMING · Sample Design in Business Research
DODGE and ROMIG · Sampling Inspection Tables, *Second Edition*
DRAPER and SMITH · Applied Regression Analysis
DUNN and CLARK · Applied Statistics: Analysis of Variance and Regression
ELANDT-JOHNSON · Probability Models and Statistical Methods in Genetics
FLEISS · Statistical Methods for Rates and Proportions
GOLDBERGER · Econometric Theory
GROSS and CLARK · Survival Distributions
GROSS and HARRIS · Fundamentals of Queueing Theory
GUTTMAN, WILKS and HUNTER · Introductory Engineering Statistics, *Second Edition*
HAHN and SHAPIRO · Statistical Models in Engineering
HALD · Statistical Tables and Formulas
HALD · Statistical Theory with Engineering Applications
HARTIGAN · Clustering Algorithms
HOEL · Elementary Statistics, *Third Edition*
HOLLANDER and WOLFE · Nonparametric Statistical Methods
HUANG · Regression and Econometric Methods
JAGERS · Branching Processes with Biological Applications
JOHNSON and KOTZ · Distributions in Statistics
Discrete Distributions
Continuous Univariate Distributions-1
Continuous Univariate Distributions-2
Continuous Multivariate Distributions

continued on back

Taylor Series: 57-8.

Normal Approximation and Asymptotic Expansions

For non unif rates, see index.

R. N. Bhattacharya

Associate Professor of Mathematics
University of Arizona

R. Ranga Rao

Professor of Mathematics
University of Illinois

John Wiley & Sons, New York • London • Sydney • Toronto

TO GOURI AND SHANTHA

Library of Congress Cataloging in Publication Data

Bhattacharya, Rabindra Nath, 1937-
Normal approximation and asymptotic expansions.

(Wiley series in probability and mathematical statistics)
Bibliography: p.
Includes index.
1. Central limit theorem. 2. Convergence. 3. Asymptotic expansions. I. Rao, Ramaswamy Ranga, 1935- joint author. II. Title.

QA273.67.B48 519.2 75-35876
ISBN 0-471-07201-X

Printed in the United States of America

10 9 8 7 6 5 4 3 2 1

Preface

This monograph presents in a unified way various refinements of the classical central limit theorem for independent random vectors and includes recent research on the subject. Most of the multidimensional results in this area are fairly recent, and significant advances over the last 15 years have led to a fresh outlook. The increasing demands of application (e.g., to the large sample theory of statistics) indicate that the present generality is useful. It is rather fortunate that in our context precision and generality go hand in hand.

Apart from some material that most students in probability and statistics encounter during the first year of their graduate studies, this book is essentially self-contained. It is unavoidable that lengthy computations frequently appear in the text. We hope that in addition to making it easier for someone to check the veracity of a particular result of interest, the detailed computations will also be helpful in estimations of constants that appear in various error bounds in the text. To facilitate comprehension each chapter begins with a brief indication of the nature of the problem treated and its solution. Notes at the end of each chapter provide some history and references and, occasionally, additional facts. There is also an Appendix devoted partly to some elementary notions in probability and partly to some auxiliary results used in the book.

We have not discussed many topics closely related to the subject matter (not to mention applications). Some of these topics are "large deviation," extension of the results of this monograph to the dependence case, and rates of convergence for the invariance principle. It would take another book of comparable size to cover these topics adequately.

We take this opportunity to thank Professors Raghu Raj Bahadur and Patrick Billingsley for encouraging us to write this book and giving us advice. We owe a special debt of gratitude to Professor Billingsley for his many critical remarks, suggestions, and other help. We thank Professor John L. Denny for graciously reviewing the manuscript and pointing out a number of errors. We gratefully acknowledge partial support from the National Science Foundation (Grant. No. MPS 75-07549). Miss Kanda Kunze and Mrs. Sarah Oordt, who did an excellent job of typing the manuscript, have our sincere appreciation.

R. N. Bhattacharya
R. Ranga Rao

Tucson, Arizona
Urbana, Illinois
August 1975

Contents

List of Symbols

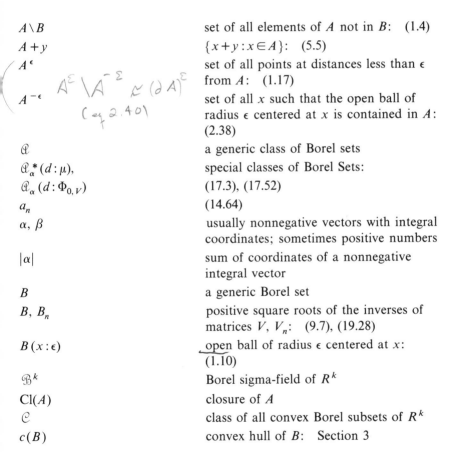

$A \setminus B$	set of all elements of A not in B: (1.4)
$A+y$	$\{x+y : x \in A\}$: (5.5)
A^{ϵ}	set of all points at distances less than ϵ from A: (1.17)
$A^{-\epsilon}$	set of all x such that the open ball of radius ϵ centered at x is contained in A: (2.38)
$\mathcal{A}$	a generic class of Borel sets
$\mathcal{A}_{\alpha}^{*}(d:\mu)$, $\mathcal{A}_{\alpha}(d:\Phi_{0,V})$	special classes of Borel Sets: (17.3), (17.52)
a_n	(14.64)
α, β	usually nonnegative vectors with integral coordinates; sometimes positive numbers
$\lvert\alpha\rvert$	sum of coordinates of a nonnegative integral vector
B	a generic Borel set
B, B_n	positive square roots of the inverses of matrices V, V_n: (9.7), (19.28)
$B(x:\epsilon)$	open ball of radius ϵ centered at x: (1.10)
$\mathcal{B}^k$	Borel sigma-field of R^k
$\mathrm{Cl}(A)$	closure of A
$\mathcal{C}$	class of all convex Borel subsets of R^k
$c(B)$	convex hull of B: Section 3

$\text{cov}(\mathbf{X}, \mathbf{Y})$	covariance between random variables $\mathbf{X}, \mathbf{Y}$: (A.1.5)
$\text{Cov}(\mathbf{X})$	matrix of covariances between coordinates of a random vector $\mathbf{X}$: Appendix A.1
D	average covariance matrix of centered truncated random vectors $\mathbf{Z}_1, \ldots, \mathbf{Z}_n$: (14.5)
D^α	αth derivative: (4.3)
$d(0, \partial A)$	euclidean distance between the origin and ∂A: Section 17
$d_0(G_1, G_2)$	(17.50)
d_p	Prokhorov distance: (1.16)
d_{BL}	bounded Lipschitzian distance: (2.51)
$d(P, \Phi)$	(12.47)
Det V, Det D	determinant of a matrix V or D
det L	absolute value of the determinant of the matrix of basis vectors of a lattice L: (21.20)
$\Delta(A, B)$	Hausdorff distance between sets A and B: (2.62)
$\Delta_{n,j,s}$, $\bar{\Delta}_{n,s}$	(14.4)
$\bar{\Delta}_{n,s}(\epsilon)$	(14.105), (14.106)
$\Delta^*_{n,s}$	(15.7)
$\tilde{\Delta}^*_{n,s}$	(17.55)
$\delta^*_{n,s}$	(18.4)
∂A	topological boundary of A: (1.15)
$E\mathbf{X}$, $E(\mathbf{X})$	expectation or mean of a random variable or random vector $\mathbf{X}$: (A.1.2), (A.1.3)
ϵ, ε	generic small numbers
$\in$	symbol for "belongs to"
$\hat{f}$	Fourier transform of a function f: (4.5)
$\tilde{f}$	(4.4)
$f_y(x)$	$f(x+y)$: (11.5)
$f * g$	convolution of functions f and g: (4.9)
f^{*n}	n-fold convolution of a function f: (4.11)

$\mathcal{F}$	a generic class of Borel-measurable functions
$\mathcal{F}^*$	fundamental domain for the dual lattice L^*: (21.22)
Φ	normal distribution on R^k with zero mean and identity covariance matrix
ϕ	density of Φ
$\Phi_{m,V}$	normal distribution with mean m and covariance V
$\phi_{m,V}$	density of $\Phi_{m,V}$: (6.31)
Φ_{r_0}	(15.5), (18.10)
$G_{a,m}$, $g_{a,m}$	a special probability measure and its density: (10.7)
g_T	(16.7)
$\gamma(f:\epsilon)$, $\gamma^*(f:\epsilon)$	(11.8), (11.18)
η_r, $\bar{\eta}_{r,n}$	(9.8), (19.32)
I	$k \times k$ identity matrix
I_A	indicator function of the set A
$\text{Int}(A)$	interior of A
K_ϵ	a smooth kernel probability measure assigning either all or more than half its mass to the sphere $B(x:\epsilon)$: (11.6), (11.16), (15.26)
χ_ν	νth cumulant, average of νth cumulants of $\mathbf{X}_1, \ldots, \mathbf{X}_n$: (6.9), (9.6), (14.1)
χ'_ν	average of νth cumulants of centered truncated random vectors $\mathbf{Z}_1, \ldots, \mathbf{Z}_n$: (14.3)
$\chi_{\nu,j}$, $\bar{\chi}_{\nu,n}$	νth cumulant of $\mathbf{X}_j$, $1 \leqslant j \leqslant n$, and their average: (9.6), Sections 19, 20
$\chi_s(z)$	(6.16)
L	a lattice: Section 21
L^*	lattice of periods of $\lvert f \rvert$, f being the characteristic function of a lattice random vector: (21.9), (21.19)
$L(c,d)$	a Lipschitzian class of functions: (2.50)
$l_{s,n}$	Liapounov coefficient: (8.10)
λ, Λ	smallest and largest eigenvalues of an average covariance matrix V: Section 16

λ_k	Lebesgue measure on R^k
$\Lambda_{r,n}(F)$	(23.8)
$M_r(f)$, $M_0(f)$	(15.4)
$M_f(x:\epsilon)$, $m_f(x:\epsilon)$	supremum and infimum of f in $B(x:\epsilon)$: (11.2)
$\mathfrak{M}$	set of all finite signed measures on a metric space
μ^+, μ^-, $\lvert\mu\rvert$	positive, negative, and total variations of a finite signed measure μ: (1.1)
$\lVert\mu\rVert$	variation norm of a signed measure μ: (1.5)
$\hat{\mu}$	Fourier–Stieltjes transform of μ: (5.2)
$\mu * \nu$	convolution of two finite signed measures μ, ν: (5.4)
μ^{*n}	n-fold convolution of μ: (5.6)
$\mu\circ T^{-1}$	signed measure induced by the map T: (5.7)
μ_α	αth moment, average of αth moments of $\mathbf{X}_1,\ldots,\mathbf{X}_n$: (6.1), (14.1)
μ'_α	average of αth moments of centered truncated random vectors $\mathbf{Z}_1,\ldots,\mathbf{Z}_n$: (14.3)
$\mu_r(t)$, $\beta_s(t)$	(8.4)
$\nu!$	$\nu_1!\nu_2!\ldots\nu_k!$ where $\nu=(\nu_1,\ldots,\nu_k)$ is a nonnegative integral vector
ν_r, ν_0	special signed measures: (15.5)
P	a probability measure, a polyhedron
$\mathcal{P}$	set of all probability measures on a metric space
$\hat{P}$	characteristic function of a probability measure P: (5.2)
$\tilde{P}_s(z:\{\chi_\nu\})$	a special polynomial in z: (7.3)
$P_r(-\phi_{0,V}:\{\chi_\nu\})$	a polynomial multiple of $\phi_{0,V}$: (7.11)
$P_r(-\Phi_{0,V}:\{\chi_\nu\})$	signed measure whose density is $P_r(-\phi_{0,V}:\{\chi_\nu\})$
P_a	a special polyhedron: (3.19)
$p_n(y_{\alpha,n})$	point masses of normalized lattice random vectors $\mathbf{X}_1,\ldots,\mathbf{X}_n$: (22.3)
$p'_n(y'_{\alpha,n})$	point masses of normalized truncated lattice random vectors: (22.3)

Q_n	distribution of $n^{-1/2}(\mathbf{X}_1+\cdots+\mathbf{X}_n)$, where $\mathbf{X}_1,\ldots,\mathbf{X}_n$ are independent random vectors having zero means and average covariance matrix V (or I)
Q_n''	distribution of $n^{-1/2}(\mathbf{Y}_1+\cdots+\mathbf{Y}_n)$, where $\mathbf{Y}_j$'s are truncations of $\mathbf{X}_j$'s: (14.2)
Q_n'	distribution of $n^{-1/2}(\mathbf{Z}_1+\cdots+\mathbf{Z}_n)$, where $\mathbf{Z}_j=\mathbf{Y}_j-E\mathbf{Y}_j$: (14.2)
$q_{n,m}$, $q'_{n,m}$	local expansions of point masses of Q_n, Q_n' in the lattice case: (22.3), (22.38), (23.2)
$\rho(x,A)$	distance between a point x and a set A: (1.18)
ρ_s	sth absolute moment, average of sth absolute moments of $\mathbf{X}_1,\ldots,\mathbf{X}_n$: (6.2), (9.6), (14.1)
ρ_s'	average of sth absolute moments of centered truncated tandom vectors $\mathbf{Z}_1,\ldots,\mathbf{Z}_n$: (14.3)
$\rho_{s,j}$, $\bar{\rho}_{s,n}$	sth absolute moment of $\mathbf{X}_j$, $1\leqslant j\leqslant n$, and their average: (14.1), (17.55)
S_j, S_α	special periodic functions: (A.4.2), (A.4.14)
$\mathcal{S}$	Schwartz space: (A.4.13)
σ_{k-1}	surface area measure on the unit sphere of R^k: Section 3
$\lVert T\rVert$	norm of a matrix T: (14.17)
τ_r	(16.6)
$\tau(f:2\epsilon)$, $\tau^*(f:2\epsilon)$	(11.8), (11.18)
V	average of covariance matrices of random vectors $\mathbf{X}_1,\ldots,\mathbf{X}_n$: (9.6), (14.5)
$\omega_f(A)$	oscillation of f on A: (2.7), (11.1)
$\omega_f(x:\epsilon)$	oscillation of f on $B(x:\epsilon)$: (2.7), (11.3)
$\bar{\omega}_f(\epsilon:\mu)$	average modulus of oscillation of f with respect to a measure μ: (11.23)
$\omega_f^*(\epsilon:\mu)$	$\sup_y \bar{\omega}_{f_y}(\epsilon:\mu)$: (11.24)
$\lvert x\rvert$	$\lvert x_1\rvert+\cdots+\lvert x_k\rvert$, where $x=(x_1,\ldots,x_k)$: (4.8)

$y_{\alpha,n}, y'_{\alpha,n}$	(22.3)
$\mathbf{Z}^+$	set of all nonnegative integers
$(\mathbf{Z}^+)^k$	set of all k-tuples of nonnegative integers
$\Vert\cdot\Vert, \langle,\rangle$	euclidean norm and inner product
$\Vert\cdot\Vert_p$	L^p-norm
$\mathbf{Z}$	set of all integers

CHAPTER 1

Weak Convergence of Probability Measures and Uniformity Classes

Let Q be a probability measure on a separable metric space S every open ball of which is connected (e.g., $S=R^k$). In the present chapter we characterize classes $\mathfrak{F}$ of bounded Borel-measurable functions such that

$$\sup_{f\in\mathfrak{F}}\left|\int_S f\,dQ_n-\int_S f\,dQ\right|\to 0 \qquad (n\to\infty), \tag{1}$$

for every sequence $\{Q_n:n\geqslant 1\}$ of probability measures converging weakly to Q. Such a class is called a Q-uniformity class. It turns out that $\mathfrak{F}$ is a Q uniformity class if and only if

$$\sup_{f\in\mathfrak{F}}\omega_f(S)<\infty, \qquad \lim_{\epsilon\downarrow 0}\left[\sup_{f\in\mathfrak{F}}\int_S \omega_f(x:\epsilon)Q(dx)\right]=0, \tag{2}$$

where $\omega_f(S)$ is the (total) oscillation of f on S, and $\omega_f(x:\epsilon)$ its oscillation on the open ball of radius ϵ centered at x. This suggests that the appropriate characteristics of f on which the rate of convergence $\int f\,dQ_n \to\int f\,dQ$ depends are (i) $\omega_f(S)$ and (ii) the average oscillation function $\epsilon\to\int\omega_f(x:\epsilon)Q(dx)$. Specialized to indicator functions of Borel sets A, this says that the rate of convergence $Q_n(A)\to Q(A)$ depends on the function $\epsilon\to Q((\partial A)^\epsilon)$, where ∂A is the boundary of A and $(\partial A)^\epsilon$ is the set of all points whose distances from ∂A are less than ϵ. We have pursued this line of thinking in Chapters 3 and 4 to obtain appropriate rates of convergence for the central limit theorem.

Section 1 contains a brief review of those aspects of weak convergence theory that are relevant for proving results on characterization of uniformity classes in Section 2. These two sections are not used in the sequel (except as motivation). In Section 3 we obtain estimates such as

$$\sup_{C \in \mathcal{C}} \Phi((\partial C)^{\epsilon}) \leq d(k)\epsilon \qquad (\epsilon > 0), \tag{3}$$

where Φ is the standard normal distribution on R^k, $\mathcal{C}$ is the class of all (Borel-measurable) convex subsets of R^k, and $d(k)$ is a positive number depending only on k. We have several occasions in Chapters 3 and 4 to use these estimates for deriving rates of convergence $Q_n(C) \to \Phi(C)$, $C \in \mathcal{C}$, where Q_n is the distribution of the normalized sum of n independent random vectors.

1. WEAK CONVERGENCE

In this section we briefly review some standard results in the theory of weak convergence of probability measures.

Throughout this section S denotes a *metric space* with a metric ρ. The *Borel sigma-field* $\mathcal{B}$ *of* S is the smallest sigma-field containing the class of all open subsets of S. We say μ is a *(signed) measure on* S if it is a (signed) measure defined on $\mathcal{B}$. The class of all finite signed measures on S is denoted by $\mathcal{M}$, and the subclass of $\mathcal{M}$ comprising all probability measures is denoted by $\mathcal{P}$. Given a finite signed measure μ on S, one defines three associated set functions μ^+, μ^-, $|\mu|$, called the *positive, negative*, and *total variations* of μ, respectively, by

$$\begin{aligned} \mu^+(B) &= \sup\{\mu(A) : A \subset B,\ A \in \mathcal{B}\}, \\ \mu^-(B) &= -\inf\{\mu(A) : A \subset B,\ A \in \mathcal{B}\}, \qquad (B \in \mathcal{B}) \\ |\mu| &= \mu^+ + \mu^-. \end{aligned} \tag{1.1}$$

The so-called *Jordan–Hahn decomposition*[†] asserts that μ^+ and μ^- (and, therefore $|\mu|$) are finite measures on S satisfying

$$\mu = \mu^+ - \mu^-. \tag{1.2}$$

For every finite signed measure μ on a *separable* metric space S, we define the *support of* μ as the smallest closed subset of S whose complement has $|\mu|$-measure zero: that is,

$$\text{support of } \mu = \cap\{F : F \text{ closed}, |\mu|(S \backslash F) = 0\}, \tag{1.3}$$

[†]See Halmos [1], pp. 121–123.

where for any two sets A, B we write

$$A \setminus B = \{x : x \in A, x \notin B\}. \tag{1.4}$$

Note that the separability of the metric space S ensures that the complement of the right side of (1.3) has zero $|\mu|$-measure.

The class $\mathfrak{M}$ of (set) functions on $\mathfrak{B}$ into R^1 is a real linear space with respect to pointwise addition and multiplication by real scalars. It is a Banach space when endowed with the *variation norm*

$$\|\mu\| = |\mu|(S) \qquad (\mu \in \mathfrak{M}). \tag{1.5}$$

Let $C(S)$ denote the *class of all complex-valued, bounded, continuous functions on* S. The *weak totopogy* on $\mathfrak{M}$ is the weakest topology (on $\mathfrak{M}$) that makes the maps

$$\mu \to \int f\,d\mu \qquad [f \in C(S)] \tag{1.6}$$

on $\mathfrak{M}$ into the complex field **C** continuous. The right side of (1.6) always stands for the *Lebesgue integral of* (a μ-integrable, complex-valued, Borel-measurable function) f *on* S. The *Lebesgue integral of* f *on a Borel set* B is denoted by

$$\int_B f\,d\mu. \tag{1.7}$$

When it becomes necessary to indicate the variable of integration, we also write

$$\int f(x)\,\mu(dx) \tag{1.8}$$

instead of $\int f\,d\mu$.

In this monograph we are particularly concerned with the relativized weak topology on the class $\mathfrak{P}$ of all probability measures on S. In this topology *convergence of a sequence* $\{Q_n\}$ *of probability measures to a probability measure* Q means

$$\lim_n \int f\,dQ_n = \int f\,dQ \tag{1.9}$$

for every f in $C(S)$. The following theorem gives several characterizations of weak convergence of a sequence of probability measures.

THEOREM 1.1. *Let* S *be a metric space. Let* Q_n $(n=1,2,\ldots)$, Q *be probability measures on* S. *The following statements are equivalent.*

(i) Q_n converges weakly to Q.

(ii) $\lim_n \int f\,dQ_n = \int f\,dQ$ for every uniformly continuous f in $C(S)$.

(iii) $\overline{\lim_n}\; Q_n(F) \leqslant Q(F)$ for every closed subset F of S.

(iv) $\underline{\lim_n}\; Q_n(G) \geqslant Q(G)$ for every open subset G of S.

For a proof of this theorem we refer to Billingsley [1] (Theorem 2.1, pp. 11–14) or Parthasarathy [1] (Theorem 6.1, pp. 40–42).

Let $B(x:\epsilon)$ denote the *open ball with center* x *and radius* ϵ,

$$B(x:\epsilon)=\{y:y\in S,\ \rho(x,y)<\epsilon\} \qquad (x\in S,\quad \epsilon>0). \tag{1.10}$$

For an arbitrary real-valued function f on S we define, for each positive number ϵ, the *oscillation function* $\omega_f(\cdot:\epsilon)$,

$$\omega_f(x:\epsilon)=\sup\{|f(z)-f(y)|:y,z\in B(x:\epsilon)\} \qquad (x\in S). \tag{1.11}$$

For a complex-valued function $f=g+ih$ (g, h real-valued), define

$$\omega_f(x:\epsilon)=\omega_g(x:\epsilon)+\omega_h(x:\epsilon) \qquad (x\in S,\quad \epsilon>0). \tag{1.12}$$

The oscillation function is Borel-measurable on the (Borel-measurable) set on which it is finite.† The set of points of discontinuity of f is Borel-measurable and can be expressed as

$$\left\{x:\omega_f\left(x:\frac{1}{n}\right)\not\to 0 \quad \text{as} \quad n\to\infty\right\}. \tag{1.13}$$

Now let Q be a probability measure on S. A complex-valued function f on S is said to be Q-*continuous* if its points of discontinuity comprise a set of Q-measure zero. In particular, if the *indicator function* I_A of a set A, taking values one on A and zero on the complement of A, is Q-continuous, we say A is a Q-*continuity set*. Since the set of points of discontinuity of I_A is precisely the boundary ∂A of A, A is a Q-*continuity set* if and only if

$$Q(\partial A)=0. \tag{1.14}$$

Recall that the (topological) *boundary* ∂A *of a set* A is defined by

$$\partial A=\mathrm{Cl}(A)\backslash \mathrm{Int}(A), \tag{1.15}$$

†See relations (11.1)–(11.4) and the discussion following them.

where $\mathrm{Cl}(A)$, $\mathrm{Int}(A)$ are the *closure* and *interior of* A, respectively.

LEMMA 1.2. *Let* Q *be a probability measure and* f *a complex-valued, bounded, Borel-measurable function on a metric space* S. *The following statements are equivalent.*

(i) f is Q-continuous.

(ii) $\lim_{\epsilon \downarrow 0} Q(\{x : \omega_f(x:\epsilon) > \delta\}) = 0$ for every positive δ.

(iii) $\lim_{\epsilon \downarrow 0} \int \omega_f(x:\epsilon)\, Q(dx) = 0$.

Proof. Let D be the set of discontinuities of f. As $\epsilon \downarrow 0$ the sets $\{x : \omega_f(x:\epsilon) > \delta\}$ decrease to a set D_δ. Now (i) means $Q(D)=0$ and (ii) means $Q(D_\delta)=0$ for all $\delta > 0$. Since $D = \bigcup_{n \geqslant 1} D_{1/n}$, (i) and (ii) are equivalent. Since, as $\epsilon \downarrow 0$, the functions $\omega_f(\cdot : \epsilon)$ are uniformly bounded and decrease to a function that is strictly positive on D and zero outside, (iii) is equivalent to $Q(D)=0$. Q.E.D.

The next theorem provides two further characterizations of weak convergence of a sequence of probability measures.

THEOREM 1.3. *Let* $\mathrm{Q_n}$ $(\mathrm{n}=1,2,\dots)$, Q *be probability measures on a metric space* S. *The following statements are equivalent.*

(i) $\{Q_n\}$ converges weakly to Q.

(ii) $\lim_n Q_n(A) = Q(A)$ for every Borel set A that is a Q-continuity set.

(iii) $\lim_n \int f\, dQ_n = \int f\, dQ$ for every complex-valued, bounded, Borel-measurable Q-continuous function f.

Although it is not difficult to prove this theorem directly, we note that it follows as a very special case of Theorem 2.5.

We conclude this section by recalling that if the metric space S is separable, then the weak topology on $\mathcal{P}$ is metrizable and separable, and that a metrization is provided by the *Prokhorov distance* $\mathrm{d_p}$ *between two probability measures* $\mathrm{Q_1}$ *and* $\mathrm{Q_2}$, which is defined by

$$d_p(Q_1,Q_2) = \inf\{\epsilon : \epsilon > 0,\ Q_1(A) \leqslant Q_2(A^\epsilon) + \epsilon \text{ and } Q_2(A) \leqslant Q_1(A^\epsilon) + \epsilon \text{ for all Borel sets } A\}, \tag{1.16}$$

where

$$A^\epsilon = \{x : x \in S,\ \rho(x,A) < \epsilon\}, \tag{1.17}$$

$\phi^{\varepsilon} := \phi$

and $\rho(x,\emptyset):=\infty$

$$\rho(x,A)=\inf\{\rho(x,y):y\in A\}. \tag{1.18}$$

For these and other details we refer the reader to Billingsley [1], (Appendix III, pp. 233–242) or Parathasarathy [1] (Chapter II, pp. 39–55).

2. UNIFORMITY CLASSES

Unless otherwise specified, throughout this section S denotes a *separable metric space.*

Let Q be a given probability measure on S. Our main task in this section is to characterize those classes $\mathcal{F}$ of complex-valued, bounded, Borel-measurable functions on S for which

$$\lim_n \sup_{f\in\mathcal{F}}\left|\int f\,dQ_n-\int f\,dQ\right|=0 \tag{2.1}$$

for every sequence of probability measures $\{Q_n\}$ converging weakly to Q. Such a class $\mathcal{F}$ is called a Q-*uniformity class of functions.* A class $\mathcal{A}$ of Borel subsets of S is called a Q-*uniformity class of sets* if

$$\lim_n \sup_{A\in\mathcal{A}}|Q_n(A)-Q(A)|=0 \tag{2.2}$$

for every sequence of probability measures $\{Q_n\}$ converging weakly to Q. Thus $\mathcal{A}$ is a Q-uniformity class of sets if and only if $\mathcal{F}=\{I_A:A\in\mathcal{A}\}$ is a Q-uniformity class of functions. We need some preparation before we can characterize uniformity classes.

In the following lemma we deal with (signed) measures on a sigma-field $\mathcal{E}$ of subsets of an abstract space Ω. For a finite signed measure μ on this space the variation norm $\|\mu\|$ is again defined by (1.5) with Ω replacing S (and $\mathcal{E}$ replacing $\mathcal{B}$).

LEMMA 2.1. (Scheffé's theorem) *Let* $(\Omega,\mathcal{E},\lambda)$ *be a measure space. Let* $\mathrm{Q_n}$ $(\mathrm{n}=1,2,\ldots)$, Q *be probability measures on* $(\Omega,\mathcal{E})$ *that are absolutely continuous with respect to* λ *and have densities* (*i.e., Radon-Nikodym derivatives*) $\mathrm{q_n}(\mathrm{n}=1,2,\ldots)$, q, *respectively, with respect to* λ. *If* $\{\mathrm{q_n}\}$ *converges to* q *almost everywhere* (λ), *then*

$$\lim_n\|Q_n-Q\|=0.$$

Proof. Let $h_n = q - q_n$. Clearly

$$\int h_n \, d\lambda = 0 \qquad (n = 1, 2, \dots),$$

so that

$$\| Q_n - Q \| = \int_{\{h_n > 0\}} h_n \, d\lambda - \int_{\{h_n \leqslant 0\}} h_n \, d\lambda = 2 \int_{\{h_n > 0\}} h_n \, d\lambda = 2 \int h_n \cdot I_{\{h_n > 0\}} \, d\lambda. \quad (2.3)$$

The last integrand in (2.3) is nonnegative and bounded above by q. Since it converges to zero almost everywhere, its integral converges to zero. Q.E.D.

LEMMA 2.2. *Let* S *be a separable metric space, and let* Q *be a probability measure on* S. *For every positive* ϵ *there exists a countable family* $\{A_k : k = 1, 2, \dots\}$ *of pairwise disjoint Borel subsets of* S *such that* (*i*) $\cup \{A_k : k = 1, 2, \dots\} = S$, (*ii*) *the diameter of* A_k *is less than* ϵ *for every* k, *and* (*iii*) *every* A_k *is a* Q-*continuity set*.

Packing

Proof. For each x in S there are uncountably many balls $\{B(x:\delta) : 0 < \delta < \epsilon/2\}$ (perhaps not all distinct). Since

$$\partial B(x:\delta) \subset \{ y : \rho(x,y) = \delta \}, \quad (2.4)$$

the collection $\{\partial B(x:\delta) : 0 < \delta < \epsilon/2\}$ is pairwise disjoint. But given any pairwise disjoint collection of Borel sets, those with positive Q-probability form a countable subcollection. Hence there exists a ball $B(x)$ in $\{B(x:\delta) : 0 < \delta < \epsilon/2\}$ whose boundary has zero Q-probability. The collection $\{B(x) : x \in S\}$ is an open cover of S and, by separability, admits a countable subcover $\{B(x_k) : k = 1, 2, \dots\}$, say. Now define

$$A_1 = B(x_1), \qquad A_2 = B(x_2) \setminus B(x_1), \dots, \qquad A_k = B(x_k) \setminus \left(\cup_{i=1}^{k-1} B(x_i) \right), \dots. \quad (2.5)$$

Clearly each A_k has diameter less than ϵ. Since

$$\partial A = \partial (S \setminus A), \qquad \partial (A \cap B) \subset (\partial A) \cup (\partial B), \quad (2.6)$$

for arbitrary subsets A, B of S, it follows that each A_k is a Q-continuity set. Q.E.D.

To state the next lemma, we define, in addition to $\omega_f(\cdot : \epsilon)$, the *oscillation* $\omega_f(A)$ *of* a complex-valued function f *on a set* A by

$$\omega_f(A) = \sup \{ |f(x) - f(y)| : x, y \in A \} \qquad (A \subset S). \quad (2.7)$$

For $b \in S$, $A \subset S$, $\omega_f(b, A) := \sup\{|f(b) - f(a)|; a \in A\}$

For $a \in A$

$\|f\|_A - |f(a)| \leq \omega_f(a, A) \leq \omega_f(A) \leq 2\,\omega_f(a, A) \leq |f(a)| + \|f\|_A$

$\|f\|_A := \sup\{|f(a)|, a \in A\} \leq \omega_f(a, A) + |f(a)|$

Clearly,

$$\omega_f(x:\epsilon)=\omega_f(B(x:\epsilon)) \qquad (x\in S, \quad \epsilon>0). \tag{2.8}$$

LEMMA 2.3. *Let* f *be a complex-valued, bounded, Borel-measurable function on a separable metric space* S. *For each pair of positive numbers* ϵ *and* δ *there exists a countable collection of pairwise disjoint Borel sets* $\{N_k : k=1,2,\ldots\}$ *such that (i)* $\cup\{N_k : k=1,2,\ldots\}\supset\{x:\omega_f(x:\epsilon)>\delta\}$, *(ii) the diameter of* N_k *is less than* 6ϵ *for every* k, *and (iii)* $\omega_f(N_k)>\delta$ *for every* k.

Proof. Let $G=\{x:\omega_f(x:\epsilon)>\delta\}$. Let $\{y_n : n=1,2,\ldots\}$ be dense in G, so that $G\subset\bigcup_n\{B(y_n:\epsilon):n=1,2,\ldots\}$. Let $x_1=y_1$; let x_2 be the first y_n whose distance from x_1 is at least 2ϵ; let x_3 be the first y_n beyond x_2 whose distance from $\{x_1,x_2\}$ is at least 2ϵ, and continue to get a countable set $\{x_n : n=1,2,\ldots\}$ such that $\{B(x_n:\epsilon):n=1,2,\ldots\}$ is a pairwise disjoint collection. Also, since each y_n is within a distance of 2ϵ from some x_j,

$$G\subset\cup\{B(x_n:3\epsilon):n=1,2,\ldots\}.$$

Let $B=\cup\{B(x_n:\epsilon):n=1,2,\ldots\}$ and define

$$M_k=B(x_k:3\epsilon)\setminus\big(B\cup\ \cup\{B(x_{k'}:3\epsilon):1\leqslant k'<k\}\big),$$

$$N_k=B(x_k:\epsilon)\cup M_k.$$

Note that the M_k's are pairwise disjoint and $\cup\{M_k : k=1,2,\ldots\}\supset G\setminus B$; also, M_k is disjoint from B. Hence $\{N_k : k=1,2,\ldots\}$ is a pairwise disjoint collection of sets whose union contains G. Since $x_k\in G$ and $B(x_k:\epsilon)\subset N_k$, it follows that $\omega_f(N_k)\geqslant\omega_f(x_k:\epsilon)>\delta$. Finally, $N_k\subset B(x_k:3\epsilon)$, so that the diameter of N_k is not more than 6ϵ. Q.E.D.

We are now ready to prove the main theorem of this section.

THEOREM 2.4. *Let* S *be a separable metric space and* Q *a probability measure on it. A family* $\mathfrak{F}$ *of complex-valued, bounded, Borel-measurable functions on* S *is a* Q*-uniformity class if and only if*

$$\text{(i)}\quad \sup_{f\in\mathfrak{F}}\omega_f(S)<\infty,$$

$$\text{(ii)}\quad \lim_{\epsilon\downarrow 0}\ \sup_{f\in\mathfrak{F}}\{Q(\{x:\omega_f(x:\epsilon)>\delta\})\}=0 \text{ for every positive } \delta. \tag{2.9}$$

Proof. The theorem is proved, without any essential loss of generality, for a class $\mathfrak{F}$ of *real-valued*, bounded, Borel-measurable functions.

SUFFICIENCY. Assume that (2.9) holds. Let c, ϵ be two positive numbers. Let $\{A_k : k=1,2,\dots\}$ be a partition of S by Borel-measurable Q-continuity sets each of diameter less than ϵ. Lemma 2.2 makes such a choice possible. Define the class of functions $\mathcal{F}_{c,\epsilon}$ by

$$\mathcal{F}_{c,\epsilon} = \left\{ \sum_k c_k I_{A_k} : |c_k| \leqslant c \text{ for all } k \right\}. \tag{2.10}$$

Then $\mathcal{F}_{c,\epsilon}$ is a Q-uniformity class. To prove this, suppose that $\{Q_n\}$ is a sequence of probability measures converging weakly to Q. Define functions q_n $(n=1,2,\dots)$, q on the set of all positive integers by

$$q_n(k) = Q_n(A_k), \qquad q(k) = Q(A_k), \qquad (k=1,2,\dots). \tag{2.11}$$

The functions q_n, q are densities (i.e., Radon–Nikodym derivatives) of probability measures on the set of all positive integers (endowed with the sigma-field of all subsets) with respect to the counting measure that assigns mass one to each singleton. Since each A_k is a Q-continuity set, $\{q_n(k)\}$ converges to $q(k)$ for every k. Hence, by Scheffé's theorem (Lemma 2.1),

$$\lim_n \sum_k |q_n(k) - q(k)| = 0. \tag{2.12}$$

Therefore

$$\begin{aligned}
&\sup\left\{ \left| \int f\,dQ_n - \int f\,dQ \right| : f \in \mathcal{F}_{c,\epsilon} \right\} \\
&\quad = \sup\left\{ \left| \sum_k c_k q_n(k) - \sum_k c_k q(k) \right| : |c_k| \leqslant c \quad \text{for all } k \right\} \\
&\quad \leqslant c \sum_k |q_n(k) - q(k)| \to 0 \qquad \text{as } n \to \infty.
\end{aligned} \tag{2.13}$$

Thus we have shown that $\mathcal{F}_{c,\epsilon}$ is a Q-uniformity class. We now assume, without loss of generality, that every function f in $\mathcal{F}$ is centered; that is,

$$\inf_{x \in S} f(x) = -\sup_{x \in S} f(x), \tag{2.14}$$

by subtracting from each f the midpoint c_f of its range. This is permissible because

$$\sup_{f \in \mathcal{F}} \left| \int f\,d(Q_n - Q) \right| = \sup_{f \in \mathcal{F}} \left| \int (f - c_f)\,d(Q_n - Q) \right| \tag{2.15}$$

whatever the probability measure Q_n. For each $f \in \mathcal{F}$ define

$$g_f(x) = \inf_{x \in A_k} f(x) \qquad \text{for} \quad x \in A_k \qquad (k=1,2,\ldots),$$

$$h_f(x) = \sup_{x \in A_k} f(x) \qquad \text{for} \quad x \in A_k \qquad (k=1,2,\ldots). \tag{2.16}$$

Note that $g, h \in \mathcal{F}_{c,\epsilon}$,

$$g_f \leqslant f \leqslant h_f, \qquad h_f(x) - g_f(x) = \omega_f(A_k) \qquad \text{for} \quad x \in A_k \qquad (k=1,2,\ldots). \tag{2.17}$$

It follows that

$$\begin{aligned}\int g_f d(Q_n - Q) - \int (h_f - g_f)\, dQ &= \int g_f dQ_n - \int h_f dQ \\ &\leqslant \int f dQ_n - \int f dQ \leqslant \int h_f dQ_n - \int g_f dQ \\ &= \int h_f d(Q_n - Q) + \int (h_f - g_f)\, dQ,\end{aligned} \tag{2.18}$$

so that

$$\sup_{f \in \mathcal{F}} \left| \int f d(Q_n - Q) \right| \leqslant \sup_{f' \in \mathcal{F}_{c,\epsilon}} \left| \int f' d(Q_n - Q) \right| + \sup_{f \in \mathcal{F}} \int (h_f - g_f)\, dQ. \tag{2.19}$$

Since $\mathcal{F}_{c,\epsilon}$ is a Q-uniformity class,

$$\begin{aligned}\overline{\lim_n} \sup_{f \in \mathcal{F}} \left| \int f d(Q_n - Q) \right| &\leqslant \sup_{f \in \mathcal{F}} \int (h_f - g_f)\, dQ \\ &= \sup_{f \in \mathcal{F}} \sum_k \omega_f(A_k) Q(A_k) \\ &\leqslant \sup_{f \in \mathcal{F}} \left[c \sum_{\{\omega_f(A_k) > \delta\}} Q(A_k) + \delta \right] \\ &\leqslant c \sup_{f \in \mathcal{F}} Q(\{x : \omega_f(x:\epsilon) > \delta\}) + \delta,\end{aligned} \tag{2.20}$$

since the diameter of each A_k is less than ϵ. First let $\epsilon \downarrow 0$ and then let $\delta \downarrow 0$. This proves sufficiency.

NECESSITY. Assume (2.9i) does not hold. Then there exists a sequence $\{f_n\} \subset \mathfrak{F}$ such that

$$\omega_{f_n}(S) > n \qquad (n=1,2,\ldots). \tag{2.21}$$

Thus if we write

$$\alpha_n = \inf\{f_n(x) : x \in S\}, \qquad \beta_n = \sup\{f_n(x) : x \in S\}, \tag{2.22}$$

then $\beta_n - \alpha_n > n$ for all n. Divide the closed interval $[\alpha_n, \beta_n]$ into n disjoint subintervals of equal length. There exists a subinterval I_n such that

$$Q(f_n^{-1}(I_n)) \geqslant \frac{1}{n} \qquad (n=1,2,\ldots). \tag{2.23}$$

Also, since $\beta_n - \alpha_n > n$, there exists a point x_n in S for which

$$|f_n(x_n) - t| \geqslant \frac{n-1}{2} \qquad \text{for all} \quad t \in \mathrm{Cl}(I_n) \qquad (n=1,2,\ldots). \tag{2.24}$$

Now define a probability measure Q_n by adding a point mass $1/n$ at x_n and subtracting this mass proportionately from subsets of $f_n^{-1}(I_n)$; that is,

$$\begin{aligned} Q_n(A) &= Q(A \setminus f_n^{-1}(I_n)) + \frac{1}{n}\delta_{x_n}(A) \\ &\quad + \left[1 - \frac{1}{nQ(f_n^{-1}(I_n))}\right] Q(A \cap f_n^{-1}(I_n)), \qquad (A \in \mathfrak{B}), \end{aligned} \tag{2.25}$$

where δ_{x_n} is the *probability measure degenerate at* x_n (i.e., $\delta_{x_n}(\{x_n\}) = 1$). Note that

$$\|Q_n - Q\| = \frac{2}{n}, \tag{2.26}$$

so that Q_n converges in variation norm and, therefore, weakly to Q. But

$$\begin{aligned} \left|\int f_n \, dQ_n - \int f_n \, dQ\right| &= \frac{1}{n}\left| f_n(x_n) - \frac{1}{Q(f_n^{-1}(I_n))} \int_{f_n^{-1}(I_n)} f_n \, dQ \right| \\ &= \frac{1}{n}|f_n(x_n) - t| \end{aligned}$$

for some t in $\mathrm{Cl}(I_n)$. Thus, by (2.26),

$$\left|\int f_n\,dQ_n-\int f_n\,dQ\right|\geqslant\frac{n-1}{2n}\tag{2.28}$$

for all n, implying that $\mathcal{F}$ is not a Q-uniformity class.

Next assume (2.9ii) does not hold. This means that there exist positive numbers δ and η, a sequence $\{\epsilon_n\}$ of positive reals converging to zero, and a sequence $\{f_n\}\subset\mathcal{F}$ such that

$$Q(\{x:\omega_{f_n}(x:\epsilon_n)>\delta\})>\eta>0\qquad(n=1,2,\ldots).\tag{2.29}$$

Let $\{N_{k,n}:k=1,2,\ldots\}$ be a countable collection of pairwise disjoint Borel sets satisfying

(i) $\cup\{N_{k,n}:k=1,2,\ldots\}\supset\{x:\omega_{f_n}(x:\epsilon_n)>\delta\}$,
(ii) diameter of $N_{k,n}\leqslant 6\epsilon_n$ for each k,
(iii) $\omega_{f_n}(N_{k,n})>\delta$ for each k $(n=1,2,\ldots)$.

Such a collection exists (for each n) by Lemma 2.3. Given n, for each k choose two points $x_{k,n}$, $y_{k,n}$ in $N_{k,n}$ such that

$$f_n(y_{k,n})-f_n(x_{k,n})>\delta\qquad(k=1,2,\ldots).\tag{2.30}$$

Thus

$$\sum_k Q(N_{k,n})f_n(y_{k,n})-\sum_k Q(N_{k,n})f_n(x_{k,n})>\delta\eta,$$

which implies that either

$$\sum_k\int_{N_{k,n}}f_n\,dQ-\sum_k Q(N_{k,n})f_n(x_{k,n})>\frac{\delta\eta}{2}\qquad(n=1,2,\ldots),\tag{2.31}$$

or

$$\sum_k Q(N_{k,n})f_n(y_{k,n})-\sum_k\int_{N_{k,n}}f_n\,dQ>\frac{\delta\eta}{2}\qquad(n=1,2,\ldots).\tag{2.32}$$

If (2.31) holds, then define Q_n by

$$Q_n(A)=Q(A\setminus\cup\{N_{k,n}:k=1,2,\ldots\})+\sum_k Q(N_{k,n})\,\delta_{x_{k,n}}(A)\qquad(A\in\mathcal{B});\tag{2.33}$$

if (2.32) holds, then define Q_n by

$$Q_n(A) = Q(A \setminus \cup \{N_{k,n} : k = 1, 2, \ldots\}) + \sum_k Q(N_{k,n})\, \delta_{y_{k,n}}(A) \qquad (A \in \mathfrak{B}). \tag{2.34}$$

Suppose (2.31) holds. Let f be a uniformly continuous complex-valued function on S. Then

$$\begin{aligned}\left| \int f\,dQn - \int f\,dQ \right| &= \left| \sum_k \int_{N_{k,n}} (f(x_{k,n}) - f)\,dQ \right| \\ &\leqslant \sum_k \omega_f(N_{k,n}) Q(N_{k,n}) \\ &\leqslant \sup\{\omega_f(N_{k,n}) : k = 1, 2, \ldots\}. \end{aligned} \tag{2.35}$$

The right side of (2.35) goes to zero as n goes to infinity, since the diameter of $N_{k,n}$ is, for all k, not more than $6\epsilon_n$ (which goes to zero). Hence $\{Q_n\}$ converges weakly to Q by Theorem 1.1. On the other hand

$$\int f_n\,dQ - \int f_n\,dQ_n = \sum_k \left(\int_{N_{k,n}} f_n\,dQ - Q(N_{k,n}) f_n(x_{k,n}) \right) > \frac{\delta\eta}{2} \qquad (n = 1, 2, \ldots),$$

by (2.31), which shows that $\mathfrak{F}$ is not a Q-uniformity class. A similar argument applies if (2.32) holds. Q.E.D.

Remark. Let S be a separable metric space, with Q_1 and Q_2 two probability measures on it such that Q_2 is absolutely continuous with respect to Q_1. It follows from Theorem A.3.1 (see Appendix), which characterizes absolute continuity, that every Q_1-uniformity class of functions is also a Q_2-uniformity class.

The following variant of Theorem 2.4 is also useful.

THEOREM 2.5. *Let* S *be a separable metric space and* Q *a probability measure on it. A family* $\mathfrak{F}$ *of complex-valued, bounded, Borel-measurable functions on* S *is a* Q*-uniformity class if and only if*

$$\text{(i)}\quad \sup_{f \in \mathfrak{F}} \omega_f(S) < \infty,$$
$$\text{(ii)}\quad \lim_{\epsilon \downarrow 0} \sup_{f \in \mathfrak{F}} \int \omega_f(x : \epsilon) Q(dx) = 0. \tag{2.36}$$

Proof. Suppose $\mathfrak{F}$ is a Q-uniformity class. By Theorem 2.4, (2.9) holds. Let $c = \sup\{\omega_f(S) : f \in \mathfrak{F}\}$. Given a positive δ there exists a positive

(d) $S=\mathbb{R}$, $A\subset B^\epsilon \not\Rightarrow A^{-\epsilon}\subset B$: $A=(0,2)\setminus\{1\}$, $B=(\frac{1}{2},\frac{3}{2})$ $\epsilon=1/2$

$B^\epsilon=(0,2)$, $A^{-\epsilon}=\{\frac{1}{2},\frac{3}{2}\}\notin$

But $A^{-\epsilon}\subset clB$. See ✱

number $\epsilon_0(\delta)$ such that

(e) But $A^\epsilon\subset B\Rightarrow$
$A\subset B^{-\epsilon}$:

Pf: $a\in A$
$\Rightarrow B(a,\epsilon)\subset A^\epsilon\subset B$ ✱

$$\sup_{f\in\mathcal{F}} Q\left(\left\{x:\omega_f(x:\epsilon)>\frac{\delta}{2}\right\}\right)<\frac{\delta}{2c} \qquad (2.37)$$

for every ϵ less than $\epsilon_0(\delta)$. Hence for all f in $\mathcal{F}$

$A^{-\epsilon}$ cvx if A is, in $\mathbb{R}$

$$\int \omega_f(x:\epsilon)Q(dx)\leq \int_{\{x:\omega_f(x:\epsilon)<\frac{\delta}{2}\}}\omega_f(x:\epsilon)Q(dx)+c\left(\frac{\delta}{2c}\right)\quad (p\,24)$$

✱ (d)
$A\subset B^\epsilon$
B cld cvx
$\Rightarrow$
$A^{-\epsilon}\subset B$:

Pf: $B(x,\epsilon)\subset B^\epsilon\Rightarrow x\in B$. Pf:
$\exists 1$ $b\in B \ni \|x-b\|=d(x,B)=:\delta$
$<\epsilon$ (x∈B̄.)
$\forall \delta>0$
set $u=(x-b)/\delta$
$y=x+(\epsilon-\delta)u\in B(x,\epsilon)\subset B^\epsilon$.
But Then $\epsilon>$
$d(y,B)$
$=\|y-b\|_2$
$=\|u(\delta+\epsilon-\delta)\|$
$=\epsilon+\delta/2>\epsilon$
✱ $\therefore \delta=0$
$\therefore x\in clB=B$

$$\leq \delta$$

whenever ϵ is less than $\epsilon_0(\delta)$. This proves necessity of (2.36).

Conversely, suppose (2.36) holds. Choose and fix a positive number δ. Given a positive η there exists a positive number $\epsilon_1(\eta)$ such that

✱ $\epsilon=1$, on $\mathbb{R}$
$A=\{|x|\leq 2\}$
$B=\{0\}\cup\{|x|=$
$B^\epsilon=\{|x|<2.$
$A^{-\epsilon}=\{|x|\leq1\}\notin$
$=clB.$
✱ OK for B

$$\sup_{f\in\mathcal{F}}\int \omega_f(x:\epsilon)Q(dx)<\delta\eta$$

for all ϵ less than $\epsilon_1(\eta)$. Hence for every f in $\mathcal{F}$,

$$Q(\{x:\omega_f(x:\epsilon)>\delta\})\leq\frac{1}{\delta}\int\omega_f(x:\epsilon)Q(dx)\leq\eta$$

for all ϵ less than $\epsilon_1(\eta)$. Thus (2.9) holds.   Q.E.D.

In order to specialize the above theorem to Q-uniformity classes of sets, we define

$$A^\epsilon=\{x:\rho(x,A)<\epsilon\}=\cup\{B(x:\epsilon):x\in A\},$$
$$A^{-\epsilon}=\{x:B(x:\epsilon)\subset A\}=S\setminus(S\setminus A)^\epsilon,\qquad (A\subset S,\ \epsilon>0). \quad (2.38)$$

open ball $=\{x:\rho(x,S\setminus A)\geq\epsilon\}$

The set A^ϵ was also defined earlier in (1.16). Note that A^ϵ is open and $A^{-\epsilon}$ is closed.

COROLLARY 2.6. *A class $\mathcal{A}$ of Borel subsets of a separable metric space S is a Q-uniformity class if and only if*

$(A^\epsilon)^\delta\subsetneqq A^{\epsilon+\delta}$
Pf: $\subset$ by
$\triangle\neq$.
$\neq$: $S=\{0,1\}$
$\epsilon=\delta=\frac{3}{4}$
$A=\{0\}=A^\epsilon$
$=(A^\epsilon)^\delta\neq A^{\epsilon+\delta}$
$=S.$

$$\lim_{\epsilon\downarrow0}\sup_{A\in\mathcal{A}}Q(A^\epsilon\setminus A^{-\epsilon})=0. \qquad (2.39)$$

Given A, $\epsilon>0$, $B(x,\epsilon)$ is open $\forall x\in A^\epsilon\setminus$

If every open ball of S is connected, then   Given A, $\epsilon>0$ $B(a,\epsilon)$ is int $\forall x\in A^\epsilon\setminus$

$$A^\epsilon\setminus A^{-\epsilon}=(\partial A)^\epsilon\qquad (A\subset S,\ \epsilon>0), \qquad (2.40)$$

then $\supset$ always
$\subset$ under ✱ Eg:

(a) $S=(0,1)\cup(1,2)\Rightarrow \partial A=\phi=(\partial A)^\epsilon$ (need this: in genl, $S^\epsilon=S=S^{-\epsilon}$, $\partial S=$
But $A^{-\epsilon}=(0,1)\cup(1,1+\epsilon)$, $A^{-\epsilon}=(0,1-\epsilon]$ $A^\epsilon\setminus A^{-\epsilon}=(1-\epsilon,1+\epsilon)\setminus\{1\}\neq\phi$

(b) $S=\mathbb{R}\setminus\{1\}$, $A=(0,1)\Rightarrow \partial A=\{0\}$, $A^\epsilon=(-\epsilon,1+\epsilon)\setminus\{1\}$ $A^{-\epsilon}=[\epsilon,1-\epsilon]$.

so that in this case $\mathscr{Q}$ *is a Q-uniformity class if and only if*

$$\lim_{\epsilon\downarrow 0} \sup_{A\in\mathscr{Q}} Q\left((\partial A)^{\epsilon}\right)=0. \tag{2.41}$$

Proof. For any arbitrary set A, $\omega_{I_A}(x:\epsilon)$ is one if and only if $B(x:\epsilon)$ intersects both A and $S\backslash A$; otherwise it has the value zero. Therefore

$$\omega_{I_A}(x:\epsilon)=I_{A^{\epsilon}\backslash A^{-\epsilon}}(x) \qquad (x\in S,\quad \epsilon>0). \tag{2.42}$$

Hence

$$\int \omega_{I_A}(x:\epsilon)Q(dx)=Q(A^{\epsilon}\backslash A^{-\epsilon}) \qquad (A\subset S,\quad \epsilon>0). \tag{2.43}$$

Since $\omega_{I_A}(S)\leqslant 1$ for all sets A, it follows from Theorem 2.5 that (2.39) is a necessary and sufficient condition for $\mathscr{Q}$ to be a Q-uniformity class.

We now prove (2.40) under the hypothesis that every open ball of the metric space S [separability is not needed for the validity of (2.40)] is connected. First we show that the relation

$$(\partial A)^{\epsilon}\subset A^{\epsilon}\backslash A^{-\epsilon} \qquad (A\subset S,\quad \epsilon>0) \tag{2.44}$$

is valid in every metric space S. For suppose $x\in(\partial A)^{\epsilon}$. Then there exist a positive ϵ' smaller than ϵ and a point y in ∂A such that $\rho(x,y)<\epsilon'$. Since y is a boundary point of A, there exist two points z_1 and z_2, with z_1 in A and z_2 in $S\backslash A$, such that $\rho(y,z_i)<\epsilon-\epsilon'$ for $i=1,2$. Thus $\rho(x,z_i)\leqslant\rho(x,y)+\rho(y,z_i)<\epsilon$ for $i=1,2$. This means that $x\in A^{\epsilon}$ and $x\not\in A^{-\epsilon}$, which proves (2.44). Next we assume that every open ball of S is connected. If $x\in A^{\epsilon}\backslash A^{-\epsilon}$, then

$$A\cap B(x:\epsilon)\neq\phi, \qquad (S\backslash A)\cap B(x:\epsilon)\neq\phi. \tag{2.45}$$

We now suppose that $x\not\in(\partial A)^{\epsilon}$ (and derive a contradition). This means

$$B(x:\epsilon)\cap\partial A=\phi, \tag{2.46}$$

so that

$$B(x:\epsilon)=((S\backslash \mathrm{Cl}(A))\cap B(x:\epsilon))\cup(\mathrm{Int}(A)\cap B(x:\epsilon)), \tag{2.47}$$

since $S=(S\backslash\mathrm{Cl}(A))\cup\mathrm{Int}(A)\cup\partial A$. The right side of (2.47) is the disjoint union of two open sets. These two sets are nonempty because of (2.45), (2.46), and the relations (which hold in every topological space)

$$(S\backslash\mathrm{Cl}(A))\cup\partial A\supset S\backslash A, \qquad \mathrm{Int}(A)\cup\partial A\supset A.$$

However this would imply that $B(x:\epsilon)$ is not connected. We have reached a contradiction. Hence $x \in (\partial A)^{\epsilon}$, and (2.40) is proved. The relation (2.41) is therefore equivalent to (2.39). Q.E.D.

COROLLARY 2.7. *Let* S *be a separable metric space. A class* $\mathcal{F}$ *of bounded functions is a* Q-*uniformity class for* every *probability measure* Q *on* S *if and only if* (*i*) $\sup\{\omega_f(S): f \in \mathcal{F}\} < \infty$, *and* (*ii*) $\mathcal{F}$ *is equicontinuous at every point of* S; *that is,*

$$\lim_{\epsilon \downarrow 0} \sup_{f \in \mathcal{F}} \omega_f(x:\epsilon) = 0 \qquad \text{for all } x \in S. \tag{2.48}$$

Proof. We assume, without loss of generality, that the functions in $\mathcal{F}$ are real-valued. Suppose that (i) and (ii) hold. Let Q be a probability measure on S. Whatever the positive numbers δ and ϵ are,

$$\sup_{f \in \mathcal{F}} Q(\{x:\omega_f(x:\epsilon) > \delta\}) \leq Q\left(\left\{x: \sup_{f \in \mathcal{F}} \omega_f(x:\epsilon) > \delta\right\}\right).^{\dagger} \tag{2.49}$$

For every positive δ the right side goes to zero as $\epsilon \downarrow 0$. Therefore, by Theorem 2.4, $\mathcal{F}$ is a Q-uniformity class. Necessity of (i) also follows immediately from Theorem 2.4. To prove the necessity of (ii), assume that there exist a positive number δ and a point x_0 in S such that

$$\sup_{f \in \mathcal{F}} \omega_f(x_0:\epsilon) > \delta \qquad \text{for all } \epsilon > 0.$$

This implies the existence of a sequence $\{x_n\}$ of points in S converging to x_0 and a sequence of functions $\{f_n\} \subset \mathcal{F}$ such that

$$|f_n(x_n) - f_n(x_0)| > \frac{\delta}{2} \qquad (n = 1, 2, \ldots).$$

Let $Q = \delta_{x_0}$ and $Q_n = \delta_{x_n}$ $(n = 1, 2, \ldots)$. Clearly, $\{Q_n\}$ converges weakly to Q, but

$$\left|\int f_n \, dQ_n - f_n \, dQ\right| = |f_n(x_n) - f_n(x_0)| > \frac{\delta}{2} \qquad (n = 1, 2, \ldots).$$

Hence $\mathcal{F}$ is not a Q-uniformity class. Q.E.D.

We have previously remarked that the weak topology on the set $\mathcal{P}$ of all probability measures on a separable metric space is metrizable, and the

---

†The set $\{x:\sup\{\omega_f(x:\epsilon): f \in \mathcal{F}\} > \delta\} = \cup\{\{x:\omega_f(x:\epsilon) > \delta\}: f \in \mathcal{F}\}$ is open (see Section 11) and, therefore, measurable.

Prokhorov distance d_p metrizes it. Another interesting metrization is provided by the next corollary. For every pair of positive numbers c, d, define a class $L(c,d)$ of Lipschitzian functions on S by

$$L(c,d)=\{f:\omega_f(S)\leqslant c, |f(x)-f(y)|\leqslant d\rho(x,y) \text{ for all } x,y\in S\}. \quad (2.50)$$

Now define the *bounded Lipschitzian distance* d_{BL} by

$$d_{BL}(Q_1,Q_2)=\sup_{f\in L(1,1)}\left|\int f\,dQ_1-\int f\,dQ_2\right| \qquad (Q_1,Q_2\in\mathcal{P}). \quad (2.51)$$

COROLLARY 2.8. *If* S *is a separable metric space, then* d_{BL} *metrizes the weak topology on* $\mathcal{P}$.

Proof. By Corollary 2.7, $L(1,1)$ is a Q-uniformity class for every probability measure Q on S. Hence if $\{Q_n\}$ is a sequence of probability measures converging weakly to Q, then

$$\lim_n d_{BL}(Q_n,Q)=0. \quad (2.52)$$

Conversely, suppose (2.52) holds. We shall show that $\{Q_n\}$ converges weakly to Q. It follows from (2.52) that

$$\lim_n\left|\int f\,dQ_n-\int f\,dQ\right|=0 \qquad \text{for every bounded Lipschitzian function } f. \quad (2.53)$$

For, if $|f(x)-f(y)|\leqslant d\rho(x,y)$ for all $x,y\in S$,

$$\int f\,dQ_n-\int f\,dQ=c\left(\int f'\,dQ_n-\int f'\,dQ\right),$$

where $c=\max\{\omega_f(S),d\}$ and $f'=f/c\in L(1,1)$. Let F be any nonempty closed subset of S. We now prove

$$\overline{\lim_n}\; Q_n(F)\leqslant Q(F). \quad (2.54)$$

For $\epsilon>0$ define the real-valued function f_ϵ on S by

$$f_\epsilon(x)=\psi(\epsilon^{-1}\rho(x,F)) \qquad (x\in S), \quad (2.55)$$

where ψ is defined on $[0,\infty)$ by

$$\psi(t)=\begin{cases} 1-t & \text{if } \quad 0\leqslant t\leqslant 1, \\ 0 & \text{if } \quad t>1. \end{cases} \tag{2.56}$$

Note that f_ϵ is, for every positive ϵ, a bounded Lipschitzian function satisfying

$$\omega_{f_\epsilon}(S)\leqslant 1, \qquad |f_\epsilon(x)-f_\epsilon(y)|\leqslant\left|\frac{1}{\epsilon}\rho(x,F)-\frac{1}{\epsilon}\rho(y,F)\right|$$
$$\leqslant\frac{1}{\epsilon}\rho(x,y) \qquad (x,y\in S),$$

so that, by (2.53),

$$\lim_n\int f_\epsilon\,dQ_n=\int f_\epsilon\,dQ \qquad (\epsilon>0). \tag{2.57}$$

Since $I_F\leqslant f_\epsilon$ for every positive ϵ,

$$\overline{\lim_n}\; Q_n(F)\leqslant\overline{\lim_n}\int f_\epsilon\,dQ_n=\int f_\epsilon\,dQ \qquad (\epsilon>0). \tag{2.58}$$

Also, $\lim_{\epsilon\downarrow 0} f_\epsilon(x)=I_F(x)$ for all x in S. Hence

$$\lim_{\epsilon\downarrow 0}\int f_\epsilon\,dQ=Q(F). \tag{2.59}$$

By Theorem 1.1 $\{Q_n\}$ converges weakly to Q. Finally, it is easy to check that d_{BL} is a distance function on $\mathcal{P}$. Q.E.D.

Remark. The distance d_{BL} may be defined on the class $\mathcal{M}$ of all finite signed measures on S by letting Q_1, Q_2 be finite signed measures in (2.51). The function $\mu\to d_{BL}(\mu,0)$ is a norm on the vector space $\mathcal{M}$. The topology induced by this norm is, in general, weaker than the one induced by the variation norm (1.5). It should also be pointed out that the proof of Corollary 2.8 presupposes metrizability of the weak topology on $\mathcal{P}$ and merely provides a suitable metric [as an alternative to the Prokhorov distance d_p defined by (1.16)]. This justifies the use of sequences (rather than nets) in the proof.

One can construct many interesting examples of uniformity classes beyond those provided by Corollaries 2.7 and 2.8. We give one example now. The rest of this section will be devoted to another example (Theorem 2.11) of considerable interest from our point of view.

Example Let $S=R^2$. Let $\mathcal{D}(l)$ be the class of all Borel-measurable subsets of R^2 each having a boundary contained in some rectifiable curve[†] of length not exceeding a given positive number l. We now show that $\mathcal{D}(l)$ is a Q-uniformity class for every probability measure Q that is absolutely continuous with respect to the Lebesgue measure λ_2 on R^2. Let $A \in \mathcal{D}(l)$ and let $\partial A \subset J$, where J is a rectifiable curve of length l. There exist k points $z_0, z_1, \ldots, z_{k-1}$ on J such that (i) z_0 and z_{k-1} are the end points of J (may coincide), (ii) $k \leqslant l/\epsilon + 2$, and (iii) $\|z_i - z_{i-1}\| \leqslant \epsilon$ for $i = 1, 2, \ldots, k-1$ ($\|\cdot\|$ denotes euclidean norm). The k open balls having z_i's as centers and radii 2ϵ cover J^ϵ. Hence

$$\begin{aligned}\lambda_2((\partial A)^\epsilon) \leqslant \lambda_2(J^\epsilon) &\leqslant \left(\frac{l}{\epsilon} + 2\right)\pi(2\epsilon)^2 \\ &= 4\pi l\epsilon + 8\pi\epsilon^2 \qquad [A \in \mathcal{D}(l)]. \end{aligned} \tag{2.60}$$

Let Q be a probability measure that is absolutely continuous with respect to λ_2. Then (2.60) implies (in view of Theorem A.3.1)

$$\lim_{\epsilon \downarrow 0} \left(\sup\{Q((\partial A)^\epsilon) : A \in \mathcal{D}(l)\}\right) = 0. \tag{2.61}$$

By Corollary 2.6, $\mathcal{D}(l)$ is a Q-uniformity class.

We need some preparation before proving the next theorem. Let S be a metric space. Define the *Hausdorff distance* Δ between two closed bounded subsets A, B of S by

$$\Delta(A, B) = \inf\{\epsilon : \epsilon > 0,\ A \subset B^\epsilon,\ B \subset A^\epsilon\}. \tag{2.62}$$

The class $\mathcal{C}$ of all closed bounded subsets of S is a metric space with metric Δ.

LEMMA 2.9. *Let $\mathcal{M}$ be a compact subset of $\mathcal{C}$. Then for every probability measure* Q *on* S *one has*

$$\lim_{\epsilon \downarrow 0} \sup\{Q(M^\epsilon) : M \in \mathcal{M}\} = \sup\{Q(M) : M \in \mathcal{M}\}. \tag{2.63}$$

Proof. Let η be a given positive number. For every $M \in \mathcal{M}$ there exists a positive number ϵ_M such that

$$Q(M^{\epsilon_M}) < Q(M) + \eta,$$

[†]A *rectifiable curve* in R^2 is a subset of R^2 of the form $\{z(t) : 0 \leqslant t \leqslant 1\}$, where $t \to z(t) = (x(t), y(t))$ is a continuous function of bounded variation on $[0,1]$ into R^2; $z(0)$, $z(1)$ are called the *end points* of the curve.

since $M^{\epsilon}\downarrow M$ as $\epsilon\downarrow 0$. By compactness of $\mathfrak{M}$ there exists a finite collection of sets $\{M_1,\ldots,M_r\}$ in $\mathfrak{M}$ such that every set in $\mathfrak{M}$ is within a Δ-distance $\epsilon_{M_i}/2$ from M_i for some i, $1\leqslant i\leqslant r$. Let

$$\epsilon_0=\min\left\{\frac{\epsilon_{M_i}}{2}:1\leqslant i\leqslant r\right\}.$$

Then

$$\begin{aligned}\sup\{Q(M^{\epsilon_0}):M\in\mathfrak{M}\}&\leqslant\sup\{Q(M_i^{\epsilon_0+\epsilon_{M_i}/2}):1\leqslant i\leqslant r\}\\&\leqslant\sup\{Q(M_i^{\epsilon_{M_i}}):1\leqslant i\leqslant r\}\\&\leqslant\sup\{Q(M_i)+\eta:1\leqslant i\leqslant r\}\\&\leqslant\sup\{Q(M):M\in\mathfrak{M}\}+\eta.\end{aligned}$$

Q.E.D.

A subset C of R^k is *convex* if $x_1+(1-\alpha)x_2\in C$ for all $x_1,x_2\in C$ and all $\alpha\in[0,1]$. A *hyperplane* H in R^k is a set of the form

$$H=\{x:\langle u,x\rangle=c\} = \{\vec{u}\}^{\perp} + c\vec{u} \tag{2.64}$$

where c is a real number and u is a *unit vector* of R^k; that is, $\|u\|=1$, and $\langle,\rangle$ denotes euclidean inner product

$$\langle u,x\rangle=\sum_{i=1}^{k}u_ix_i, \tag{2.65}$$

$$\|u\|=\left(\sum_{i=1}^{k}u_i^2\right)^{1/2}\qquad\left[u=(u_1,\ldots,u_k),\quad x=(x_1,\ldots,x_k)\in R^k\right].$$

A *closed half space* E is a set of the form

$$E=\{x:x\in R^k,\langle u,x\rangle\leqslant c\}\qquad\left[c\in R^1,\quad u\in R^k,\quad\|u\|=1\right]. \tag{2.66}$$

A hyperplane $H=\{y:\langle u,y\rangle=c\}$ is said to be a *supporting hyperplane* for a set A at $x\in A$ if

$$\langle u,x\rangle=c,\qquad A\subset\{z:\langle u,z\rangle\leqslant c\}, \tag{2.67}$$

that is, if the hyperplane H passes through the point x of A and has A on one side. Note that the sign $\leqslant$ in (2.66) and (2.67) may be replaced by $\geqslant$

(by merely changing u to $-u$). It is a well-known fact that if A is a compact convex set, then there exists a supporting hyperplane for A at each $x \in \partial A$.†

LEMMA 2.10. *Let* A, B *be two closed, bounded, convex subsets of* R^k. *Then*

$$\Delta(A,B) = \Delta(\partial A, \partial B). \tag{2.68}$$

Proof. Suppose $\Delta(\partial A, \partial B) = \epsilon$. Let ϵ' be any number larger than ϵ. Let $x \in A$. There exist x_1, $x_2 \in \partial A$ and $\alpha \in [0,1]$ such that $x = \alpha x_1 + (1-\alpha)x_2$, since the intersection of A with any line through x is a closed line segment whose end points (x_1, x_2, for example) lie on ∂A. Let y_1, $y_2 \in \partial B$ be such that $x_i - y_i < \epsilon'$ for $i = 1,2$. Then letting $y = \alpha y_1 + (1-\alpha) y_2$ yields $y \in B$ and

$$\|x-y\| \leqslant \alpha\|x_1 - y_1\| + (1-\alpha)\|x_2 - y_2\| < \epsilon'.$$

Thus $A \subset B^{\epsilon'}$. Similarly $B \subset A^{\epsilon'}$. Since this is true for every $\epsilon' > \epsilon$,

$$\Delta(A,B) \leqslant \Delta(\partial A, \partial B). \tag{2.69}$$

To prove the opposite inequality, suppose that $\epsilon > \Delta(A,B)$ and let η be any positive number. Let $x \in \partial A$. Let $\{z : \langle l, z\rangle = c\}$ be a supporting hyperplane for A at x. Then the half space $H = \{z : \langle l, z\rangle \leqslant c + \epsilon\}$ contains A^{ϵ}. If $z \in A$ and $\|z' - z\| < \epsilon$, then

$$\langle l, z'\rangle = \langle l, z\rangle + \langle l, z' - z\rangle \leqslant \langle l, z\rangle + \|z' - z\| \leqslant c + \epsilon.$$

Hence $H \supset A^{\epsilon} \supset B$. The ball $B(x : \epsilon + \eta)$ intersects $R^k \backslash H$. This is because the point $x + (\epsilon + \eta/2) l$ of this ball satisfies

$$\left\langle l, x + \left(\epsilon + \frac{\eta}{2}\right) l \right\rangle = \langle l, x\rangle + \left(\epsilon + \frac{\eta}{2}\right) = c + \epsilon + \frac{\eta}{2},$$

and therefore lies in the complement of H. It follows that $B(x : \epsilon + \eta)$ intersects $R^k \backslash B$. But, since $A \subset B^{\epsilon}$ and $x \in A$, $B(x : \epsilon + \eta)$ certainly intersects B. It follows that $B(x : \epsilon + \eta)$ intersects ∂B, so that $(\partial B)^{\epsilon + \eta} \supset \partial A$. Similarly $(\partial A)^{\epsilon + \eta} \supset \partial B$. Therefore

$$\Delta(\partial A, \partial B) \leqslant \epsilon + \eta$$

---

†Eggleston [1], p. 20.

for every $\epsilon > \Delta(A,B)$ and every positive η. Hence

$$\Delta(\partial A, \partial B) \leqslant \Delta(A,B). \tag{2.70}$$

The inequalities (2.69) and (2.70) together yield (2.68). Q.E.D.

THEOREM 1.11. *Let $\mathcal{C}$ denote the class of all Borel-measurable convex subsets of* R^k. *Let* Q *be a probability measure on* R^k. *The class $\mathcal{C}$ is a* Q-*uniformity class if and only if it is a* Q-*continuity class, that is, if and only if*

$$Q(\partial C)=0 \qquad \text{for all } C \in \mathcal{C}. \tag{2.71}$$

Proof. If $\mathcal{C}$ is a Q-uniformity class, then, by relation (2.41) in Corollary 2.6, $\mathcal{C}$ is a Q-continuity class. Suppose, conversely, that (2.71) holds. Since the characterization (2.41) of Q-uniformity is in terms of boundaries, and since $\partial C = \partial(\mathrm{Cl}(C))$ for every convex set [note that $\mathrm{Cl}(C) = \mathrm{Int}(C) \cup \partial C$ and that ∂C has empty interior for convex C], it follows that $\mathcal{C}$ is a Q-uniformity class if and only if $\bar{\mathcal{C}} = \{\mathrm{Cl}(C) : C \in \mathcal{C}\}$ is. Given $\eta > 0$, let r be so chosen that

$$Q(\{x : \|x\| > r\}) < \frac{\eta}{2}. \tag{2.72}$$

Write

$$\bar{\mathcal{C}}_r = \left\{ C : C \in \bar{\mathcal{C}}, \, C \subset \mathrm{Cl}(B(0:r)) \right\}.$$

By a well-known theorem of Blaschke† $\bar{\mathcal{C}}_r$ is compact in the Hausdorff metric Δ defined by (2.62). Lemma 2.10 shows that the compactness of $\bar{\mathcal{C}}_r$ is equivalent to the compactness of $\{\partial C : C \in \bar{\mathcal{C}}_r\}$. Lemma 2.9 and the hypothesis (2.71) now yield

$$\lim_{\epsilon \downarrow 0} \sup \left\{ Q((\partial C)^{\epsilon}) : C \in \bar{\mathcal{C}}_r \right\} = 0. \tag{2.73}$$

By Corollary 2.6, $\bar{\mathcal{C}}_r$ is a Q-uniformity class. Let $\{Q_n\}$ be a sequence of probability measures weakly converging to Q. Then

$$\begin{aligned}
&\overline{\lim_n} \left(\sup \left\{ |Q_n(C) - Q(C)| : C \in \bar{\mathcal{C}} \right\} \right) \\
&\qquad \leqslant \overline{\lim_n} \left(\sup \left\{ |Q_n(C) - Q(C)| : C \in \bar{\mathcal{C}}_r \right\} \right) \\
&\qquad\quad + \overline{\lim_n} \; Q_n(\{x : \|x\| > r\}) + Q(\{x : \|x\| > r\}) \\
&\qquad = 2Q(\{x : \|x\| > r\}) < \eta.
\end{aligned} \tag{2.74}$$

---

†See Eggleston [1], pp. 34–67.

Note that the last equality follows from the fact that $\{x: \|x\| > r\}$ is a Q-continuity set, since its complement is [although, given any probability measure Q and a positive η, one can always find r such that $\{x: \|x\| > r\}$ is a Q-continuity set and (2.72) holds]. Since η is an arbitrary positive number, it follows from (2.74) that $\bar{\mathcal{C}}$ is a Q-uniformity class. Consequently, $\mathcal{C}$ is a Q-uniformity class. Q.E.D.

Remark. It follows from the above theorem that if Q is absolutely continuous with respect to Lebesgue measure on R^k, then $\mathcal{C}$ is a Q-uniformity class. In particular, $\mathcal{C}$ is a Φ-uniformity class, Φ being the standard normal distribution in R^k. We shall obtain a refinement of this last statement in the next section.

It is easy to see from the proof of Theorem 2.11 that the class $\mathcal{C}$ in its statement may be replaced by any class $\mathcal{Q}$ of Borel-measurable convex sets with the property that

$$\bar{\mathcal{Q}}_r = \{\mathrm{Cl}(A) \cap \mathrm{Cl}(B(0:r)) : A \in \mathcal{Q}\}$$

is compact in the Hausdorff metric for all positive r. In particular, by letting

$$\mathcal{Q} = \{(-\infty, x_1] \times (-\infty, x_2] \times \cdots \times (-\infty, x_k] : x = (x_1, \ldots, x_k) \in R^k\},$$

we get Polya's result: *Let* P *be a probability measure on* R^k *whose distribution function* F, *defined by*

$$F(x) = P((-\infty, x_1] \times \cdots \times (-\infty, x_k]) \qquad [x = (x_1, \ldots, x_k) \in R^k], \tag{2.75}$$

is continuous on R^k. *If a sequence of probability measures* $\{\mathrm{P_n}\}$ *with distribution functions* $\{\mathrm{F_n}\}$ *converges weakly to* P, *then*

$$\sup\{|F_n(x) - F(x)| : x \in R^k\} \to 0 \tag{2.76}$$

as $\mathrm{n} \to \infty$. The left side of (2.76) is sometimes called the *Kolmogorov distance between* $\mathrm{P_n}$ *and* P. The converse of this result is also true: if (2.76) holds, then $\{P_n\}$ converges weakly to P. In fact, if $\{F_n\}$ converges to F at all points of continuity of F, then $\{P_n\}$ converges weakly to P.†

3. INEQUALITIES FOR INTEGRALS OVER CONVEX SHELLS

It is not difficult to check that if C is convex then so are $\mathrm{Int}(C), \mathrm{Cl}(C)$. The *convex hull* $c(B)$ of a subset B of R^k is the intersection of all convex sets containing B. Clearly $c(B)$ is convex. If C is a closed and bounded convex subset of R^k, then it is the convex hull of its boundary; that is,

$$c(\partial C) = C.$$

Clearly $c(\partial C) \subset C$. On the other hand, if $x \in C$, $x \not\in \partial C$, then every line through x intersects ∂C at two points and x is a convex combination of these two points. Thus $c(\partial C) = C$. If C is convex and $\epsilon > 0$, then C^{ϵ} is

†See Billingsley [1], pp. 17–18.

convex and open, and $C^{-\epsilon}$ is convex and closed. The main theorem of this section is the following:

THEOREM 3.1. *Let g be a nonnegative differentiable function on* $[0,\infty)$ *such that*

(i) $b \equiv \int_0^\infty |g'(t)| t^{k-1}\, dt < \infty$,

(ii) $\lim_{t\to\infty} g(t) = 0$. $\therefore\ g(t) = -\int_t^\infty g'(s)\,ds$

Then for every convex subset C of R^k *and every pair of positive numbers* ϵ, ρ,

$$\int_{C^{\epsilon}\setminus C^{-\rho}} g(\|x\|)\, dx \leqslant b\alpha_k(\epsilon+\rho) \tag{3.1}$$

where

$$\alpha_k = \frac{k\pi^{k/2}}{\Gamma((k+2)/2)} = \frac{2\pi^{k/2}}{\Gamma(k/2)}, \tag{3.2}$$

is the surface area of the unit sphere in R^k.

COROLLARY 3.2. *Let* $s \geqslant 0$, $k > 1$, *and*

$$f(x) = (2\pi)^{-k/2} \|x\|^s \exp\left\{-\frac{\|x\|^2}{2}\right\} \qquad (x \in R^k).$$

Then for all convex subsets C of R^k *and every pair of positive numbers* ϵ, ρ,

$$\int_{C^{\epsilon}\setminus C^{-\rho}} f(x)\, dx \leqslant 2^{(s-1)/2}(2s+k-1)\frac{\Gamma((k+s-1)/2)}{\Gamma(k/2)}(\epsilon+\rho). \tag{3.3}$$

Proof. Here one takes (in Theorem 3.1)

$$g(t) = (2\pi)^{-k/2}\, t^s \exp\{-t^2/2\} \qquad t \in [0,\infty).$$

Then

$$g'(t) = (2\pi)^{-k/2}(st^{s-1} - t^{s+1}) \exp\left\{-\frac{t^2}{2}\right\},$$

$$b \leqslant (2\pi)^{-k/2} \int_0^\infty (st^{k+s-2} + t^{k+s}) \exp\left\{-\frac{t^2}{2}\right\} dt$$

[A := C cvx]
* For $x, y \in A^{-\varepsilon}$ & $z := px + qy$, some $0 < p = 1-q < 1$, spse $z \notin A^{-\varepsilon}$.
Then $\exists\, b \notin A \ni d(z,b) < \varepsilon$. But $v := x + (b-z) \notin A$
$\Rightarrow d(x, R^k\setminus A) \leq d(x,v) = |b-z| < \varepsilon$ ✗ $\therefore v \in A$. Also $w := y + b - z \in A$. ∴
$b = pv + qw \in A$, ✗ $\therefore z \in A^{-\varepsilon}$.

$$=(2\pi)^{-k/2}\left(s\cdot 2^{(k+s-3)/2}\Gamma\left(\frac{k+s-1}{2}\right)+2^{(k+s-1)/2}\Gamma\left(\frac{k+s+1}{2}\right)\right)$$

$$=(2\pi)^{-k/2}2^{(k+s-3)/2}(2s+k-1)\,\Gamma\left(\frac{k+s-1}{2}\right)$$

$$=2^{(s-3)/2}(2s+k-1)\,\pi^{-k/2}\,\Gamma\left(\frac{k+s-1}{2}\right), \tag{3.4}$$

which gives (3.3) on substitution in (3.1). Q.E.D.

The rest of this section is devoted to the development of the material needed for the proof of Theorem 3.1.

For $k=1$, $C^{\epsilon}\setminus C^{-\rho}$ is contained in the union of two disjoint intervals each of length $\epsilon+\rho$. Hence

$$\int_{C^{\epsilon}\setminus C^{-\rho}} g(|x|)\,dx \leqslant 2(\epsilon+\rho)\Big(\sup_{x\geqslant 0} g(x)\Big) \leqslant 2(\epsilon+\rho)\int_0^{\infty}|g'(t)|\,dt = 2(\epsilon+\rho)b$$
$$= b\alpha_1(\epsilon+\rho). \tag{3.5}$$

For $k>1$ a more intricate argument is needed. For the rest of the section *we assume* $k>1$. A *polyhedron* is a closed, bounded, convex set with nonempty interior that is the intersection of a finite number of closed half spaces. If P is a polyhedron, a *face* of P is a set of the form $H\cap\partial P$, where H is a hyperplane such that $H\cap\partial P$ has nonempty interior in H.

LEMMA 3.3. *Let a polyhedron* P *be given by*

$$P=\{x:\langle u_j,x\rangle\leqslant d_j,\ 1\leqslant j\leqslant m\}, \tag{3.6}$$

where u_j*'s are distinct unit vectors. Let*

$$L_j=\{x:x\in P,\ \langle u_j,x\rangle=d_j\}\qquad (1\leqslant j<m). \tag{3.7}$$

Then $\partial P=\cup L_j$. *If* F *is a face of* P, *then* $F=L_j$ *for some* j. *Moreover,*

$$P=\{x:\langle u_j,x\rangle\leqslant d_j \text{ for all } j \text{ for which } L_j \text{ is a face of } P\}. \tag{3.8}$$

Proof. The first assertion is obvious. If $F=H\cap\partial P=\cup(H\cap L_j)$ is a face of P, then $H=\{x:\langle u_j,x\rangle=d_j\}=H_j$, say, for exactly one j, since the intersection of H with any hyperplane distinct from H has empty interior in H. This proves the second assertion. For the third, note that

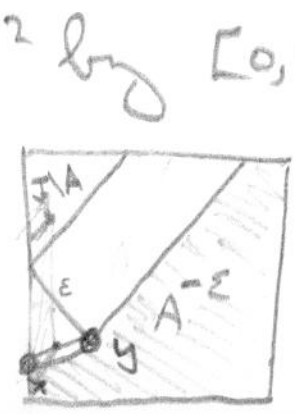

$L_j = H_j \cap \partial P$. It is clear that the interior of L_j in H_j is

$$\{x : \langle u_r, x\rangle < d_r \text{ for } r \neq j,\ \langle u_j, x\rangle = d_j\},$$

so that if L_j is not a face of P, then $L_j \subset \bigcup_{r\neq j} L_r$. This implies $\partial P \subset \partial Q$, where Q is defined by

$$Q = \{x : \langle u_j, x\rangle \leqslant d_j \text{ for all } j \text{ for which } L_j \text{ is a face of } P\}.$$

Clearly $P \subset Q$. It is sufficient to show that $\partial P \subset \partial Q$ implies $P = Q$. Let $x_1 \in \mathrm{Int}(P)$. Assume there exists $x_2 \in \partial Q \setminus \partial P$. Consider the line segment $[x_1, x_2]$ joining x_1 and x_2. This line segment meets ∂P at x_3, say. Clearly, $[x_1, x_2) \subset \mathrm{Int}(Q)$ and $x_3 \neq x_2$, so that $x_3 \in \mathrm{Int}(Q) \cap \partial P$, which contradicts the fact $\partial P \subset \partial Q$. Q.E.D.

LEMMA 3.4. *A polyhedron* P *has a finite number of faces. If* $\mathrm{F}_1, \ldots, \mathrm{F}_m$ *are the faces of* P, *then* $\partial \mathrm{P} = \cup \mathrm{F}_j$. *Moreover, there exist unique unit vectors* u_j *and constants* d_j *such that* $\mathrm{F}_j = \{x : x \in \mathrm{P},\ \langle u_j, x\rangle = d_j\}$, *and* $\mathrm{P} \subset \{x : \langle u_j, x\rangle \leqslant d_j\}$. *Also,* P *then has the representation*

$$P = \{x : \langle u_j, x\rangle \leqslant d_j,\ 1 \leqslant j \leqslant m\}. \tag{3.9}$$

Proof. This lemma follows easily from Lemma 3.3. Note that the representation (3.9) is unique up to a permutation of unit vectors u_j's and the corresponding constants d_j's. Q.E.D.

Remark. A polyhedron is the convex hull of a finite number of points. In fact, since each face is a polyhedron in a lower dimensional affine space, this follows by induction on k. Conversely, it is known that the *convex hull of a finite set of points can be expressed as the intersection of a finite number of closed half spaces.*†

There are two main steps in the proof of Theorem 3.1. The first one (Lemma 3.9) is to express the surface integral as a derivative of volume in volume integrals. The second step (Lemma 3.10) is to get a uniform bound for the surface integral of a fixed function over the boundary of a polyhedron. The ideas involved here belong naturally to the domain of surface area and surface integrals. In the following paragraphs we develop the material needed for the proofs. The development here is self-contained except for the use of Cauchy's formula, which is stated but not proved.

We begin with some notation. Let λ_k denote the Lebesgue measure on R^k normalized by the euclidean distance on R^k; that is, if $y_1, \ldots, y_k$ is an

†See Eggleston [1], pp. 29–30.

For L3.9 use Federer

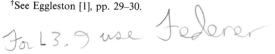

orthonormal basis and A is a cube with respect to them or $A=\{\sum t_i y_i : a_i \leqslant t_i \leqslant b_i \text{ for all } i\}$, then $\lambda_k(A)=(b_1-a_1)\cdots(b_k-a_k)$. We also refer to λ_k as the k-dimensional Lebesgue measure on R^k. On any hyperplane of R^k there is a $(k-1)$-dimensional Lebesgue measure normalized the same way. We denote this measure as λ_{k-1} and call it the $(k-1)$-dimensional Lebesgue measure. For example, if H is a hyperplane, it can be written in the form $H=x_0+Ry_1+\cdots+Ry_{k-1}$, where $x_0,y_1,\dots,y_{k-1}\in H$ and $y_1,\dots,y_{k-1}$ are $(k-1)$ orthonormal vectors in R^k. If f is a function with compact support in R^k, then

$$\int_H f\,d\lambda_{k-1}=\int f(x_0+t_1y_1+\cdots+t_{k-1}y_{k-1})\,dt_1\cdots dt_{k-1}. \tag{3.10}$$

Next we denote σ_{k-1} as the surface area measure on the unit sphere S_{k-1}. One can write an explicit formula for σ_{k-1} by using Eulerian angles to parametrize points of S_{k-1}.

It then follows that $\sigma_{k-1}(S_{k-1})=\alpha_k$, where α_k is given by (3.2).

Let P be a polyhedron with faces F_i $(1\leqslant i\leqslant m)$. Then the *surface area of* P is defined as

$$\lambda_{k-1}(\partial P)=\sum_{i=1}^{m}\lambda_{k-1}(F_i). \tag{3.11}$$

For a bounded Borel-measurable function f on R^k we define the *surface integral of* f *on* ∂P by

$$\int_{\partial P} f\,d\lambda_{k-1}=\sum_{i=1}^{m}\int_{F_i} f\,d\lambda_{k-1} \tag{3.12}$$

and if A is a Borel subset of ∂P, then the *surface area of A* is defined by

$$\lambda_{k-1}(A)=\sum_i\lambda_{k-1}(A\cap F_i). \tag{3.13}$$

Remark. Let P be given as $P=\{x:\langle u_j,x\rangle\leqslant d_j,\quad 1\leqslant j\leqslant m\}$. Then

$$\int_{\partial P} f\,d\lambda_{k-1}=\sum_{1\leqslant j\leqslant m}\int_{L_j} f\,d\lambda_{k-1}, \tag{3.14}$$

where $L_j=\{x:x\in P,\langle u_j,x\rangle=d_j\}$. This follows from the fact that $\lambda_{k-1}(L_j)=0$ if L_j is not a face of P.

CAUCHY'S FORMULA *Let* C *be a polyhedron. For each unit vector* $u \in R^k$, *let* $\lambda_{k-1}(C:u)$ *denote the* λ_{k-1}*-measure of the orthogonal projection of* C *on the hyperplane* $\{x:\langle u,x\rangle=0\}$. *Then*

$$\lambda_{k-1}(\partial C)=\beta_{k-1}^{-1}\int_{S_{k-1}}\lambda_{k-1}(C:u)\sigma_{k-1}(du), \tag{3.15}$$

where $\beta_{k-1}=\alpha_{k-1}/(k-1)$ *is the volume (Lebesgue measure in* R^{k-1}) *of the unit ball in* R^{k-1}.

COROLLARY 1. *Let* P *and* Q *be polyhedra such that* $P\subset Q$. *Then*

$$\lambda_{k-1}(\partial P)\leqslant\lambda_{k-1}(\partial Q). \tag{3.16}$$

Proof. This follows from Cauchy's formula if it is noted that $\lambda_{k-1}(P:u)\leqslant\lambda_{k-1}(Q:u)$ for any u. Q.E.D.

COROLLARY 2. *If* P *is a polyhedron and* $P\subset B(0:t)$, *then*

$$\lambda_{k-1}(\partial P)\leqslant\alpha_k t^{k-1} \tag{3.17}$$

Proof. The projection of $B(0:t)$ on the hyperplane $\{y:\langle u,y\rangle=0\}$ is again a ball of radius t in that hyperplane, so that $\lambda_{k-1}(B(0:t):u)=t^{k-1}\beta_{k-1}$. The estimate now follows from Cauchy's formula. Q.E.D.

These two corollaries of Cauchy's formula are actually used in the proof of Lemma 3.10.

Let P be a polyhedron given by

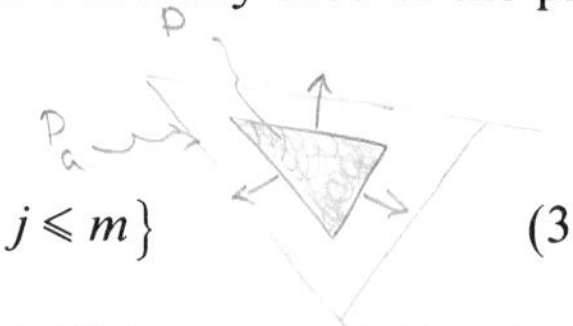

$$P=\{x:\langle u_j,x\rangle\leqslant d_j,\ 1\leqslant j\leqslant m\} \tag{3.18}$$

where the unit vectors u_j are assumed distinct. This representation is *kept fixed throughout*. For each real a define the convex set P_a by

$$P_a=\{x:\langle u_j,x\rangle\leqslant d_j+a,\ 1\leqslant j\leqslant m\}. \tag{3.19}$$

Then P_a has nonempty interior for $a\geqslant 0$, since $P\subset P_a$ for $a\geqslant 0$. In fact, there exists an $a_0<0$ such that P_a has nonempty interior if and only if $a>a_0$. Thus P_a is a polyhedron for $a>a_0$, if we show that it is bounded. A more precise result is given by the following:

LEMMA 3.5. *Let* $x_0\in \mathrm{Int}(P)$ *and suppose that* $P\subset B(x_0:\rho_0)$. *Then* $P_a\subset B(x_0:c\rho_0)$ *for* $a\geqslant 0$, *where* $c=1+a/d$, $d=\min\{d_j-\langle u_j,x_0\rangle:1\leqslant j\leqslant m\}$.

Proof. The proof uses the notion of a gauge, or support, function of a convex set. Let C be a closed convex set with nonempty interior and

$x_0 \in \mathrm{Int}(C)$. Then

$$F_C(x) = \inf\left\{\alpha : \alpha > 0,\ x_0 + \frac{x - x_0}{\alpha} \in C\right\} \tag{3.20}$$

is called the *gauge function of* C *with respect to* x_0. It is easy to see that if x_0 is an interior point of each of two closed convex sets C_1, C_2, then $C_1 \subset C_2$ if and only if $F_{C_2}(x) \leqslant F_{C_1}(x)$ for all x, where both gauge functions are with respect to x_0.

Returning to P and defining F_P with respect to x_0, an easy calculation shows that [since $x_0 + (x - x_0)/\alpha \in \partial P$ if $\alpha = F_P(x)$]

$$F_P(x) = \max_{1 \leqslant j \leqslant m} \frac{\langle u_j, x - x_0\rangle}{d_j - \langle u_j, x_0\rangle},$$

$$F_{P_a}(x) = \max_{1 \leqslant j \leqslant m} \frac{\langle u_j, x - x_0\rangle}{d_j + a - \langle u_j, x_0\rangle} \qquad (a \geqslant 0).$$

Clearly, $F_{P_a}(x) \geqslant c_1 F_P(x)$ for all x, where

$$c_1 = \min_{1 \leqslant j \leqslant m} \frac{d_j - \langle u_j, x_0\rangle}{d_j + a - \langle u_j, x_0\rangle}.$$

The gauge function of $\mathrm{Cl}(B(x_0 : \rho_0))$ with respect to x_0 is $\|x - x_0\|/\rho_0$, so that $P \subset B(x_0 : \rho_0)$ implies

$$F_P(x) \geqslant \frac{1}{\rho_0}\|x - x_0\|.$$

Thus

$$F_{P_a}(x) \geqslant \frac{c_1}{\rho_0}\|x - x_0\|,$$

which implies that $P_a \subset B(x_0 : \rho_0/c_1)$. Now

$$c_1^{-1} = \max_{1 \leqslant j \leqslant m} \frac{d_j + a - \langle u_j, x_0\rangle}{d_j - \langle u_j, x_0\rangle}$$

$$= \max_{1 \leqslant j \leqslant m}\left\{1 + \frac{a}{d_j - \langle u_j, x_0\rangle}\right\} = 1 + \frac{a}{d}.$$

Q.E.D.

LEMMA 3.6. *Let* P *be a polyhedron defined by* (3.18) *and let* P_a *be defined by* (3.19) *for all* $a \in R^1$. *Then*

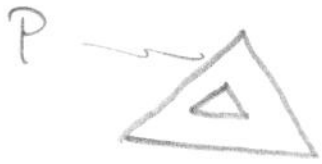

$$P_a \supset P^a \qquad \text{if } a>0,$$

$$P_a \subseteq P^a \qquad \text{if } a<0.$$

* ⊆ for L_2 metric

Proof. Suppose $a>0$ and $x \in P^a$. Then there exists $y \in P$ such that $\|x-y\|<a$. Hence

$$\begin{aligned}\langle u_j,x\rangle &= \langle u_j,y\rangle + \langle u_j,x-y\rangle \leqslant d_j + \langle u_j,x-y\rangle \\ &\leqslant d_j + \|u_j\|\cdot\|x-y\| \leqslant d_j + a \qquad (1 \leqslant j \leqslant m),\end{aligned}$$

so that $x \in P_a$. If $a=-b<0$, $x \in P_a$, and $\|y-x\|<b$, then

$$\begin{aligned}\langle u_j,y\rangle &= \langle u_j,x\rangle + \langle u_j,y-x\rangle \leqslant d_j + a + \|y-x\| \\ &\leqslant d_j \qquad (1 \leqslant j \leqslant m),\end{aligned}$$

implying $y \in P$. Hence $x \in P^a = P^{-b}$. Q.E.D.

LEMMA 3.7. *Let* v *be a unit vector in* R^k, *and* $A = B(x_0:\rho) \cap \{x : \alpha \leqslant \langle v,x\rangle \leqslant \alpha + h\}$ *for some* $\rho>0$, $h \geqslant 0$, $\alpha \in R^1$. *Then*

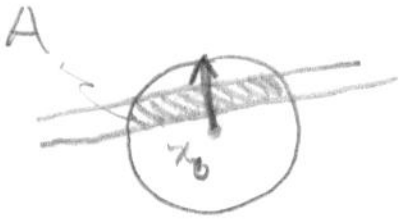

$$\lambda_k(A) \leqslant h \frac{\pi^{(k-1)/2}}{\Gamma\left(\frac{k+1}{2}\right)} \rho^{k-1}. \tag{3.21}$$

Proof. Choose an orthonormal basis $v_1,\ldots,v_k$ of R^k with $v_1 = v$. Let U denote the orthogonal transformation taking v_i to e_i, $1 \leqslant i \leqslant k$, where $(e_1,\ldots,e_k)$ is the standard euclidean basis. Then

$$UA = B(x_0':\rho) \cap \{x=(x_1,\ldots,x_k) : \alpha \leqslant x_1 \leqslant \alpha + h\} \qquad (x_0' = Ux_0),$$

$$\lambda_k(A) = \lambda_k(UA) \leqslant h\lambda_{k-1}(\{u \in R^{k-1} : \|u\| \leqslant \rho\})$$

$$= h \frac{\pi^{(k-1)/2}}{\Gamma\left(\frac{k+1}{2}\right)} \rho^{k-1}$$

this fails for L_∞ metric

Q.E.D.

$a=-b<0$

(*) $x \in P^{-b} \Rightarrow B(x,b) \subset P \Rightarrow x + \beta u_j \in P \ \forall j, \forall \beta <$
$\Rightarrow \langle x+\beta u_j, u_j\rangle \leq d_j \Rightarrow \langle x,u_j\rangle \leq d_j - \beta \ \forall \beta < b \ \forall j$
$\Rightarrow \langle x,u_j\rangle \leq d_j - b = d_j + a \ \forall j \Rightarrow x \in P_{-b} = P_a.$

LEMMA 3.8. *Let* v_1, v_2 *be two linearly independent unit vectors*, $h \geqslant 0$, *and* $\alpha_1, \alpha_2 \in R^1$. *Let*

$$A = B(y_0 : \rho) \cap \{x : \alpha_i \leqslant \langle v_i, x\rangle \leqslant \alpha_i + h \text{ for } i = 1,2\}$$

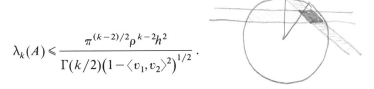

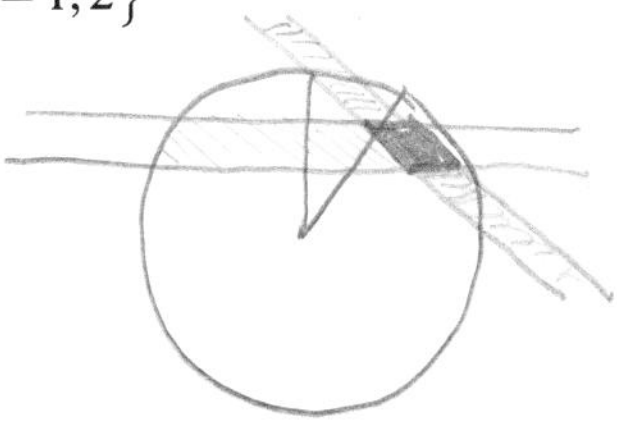

Then

$$\lambda_k(A) \leqslant \frac{\pi^{(k-2)/2}\rho^{k-2}h^2}{\Gamma(k/2)\left(1-\langle v_1, v_2\rangle^2\right)^{1/2}}.$$

Proof. Choose orthonormal vectors $p_1, p_2, \ldots, p_k$ such that $p_1 = v_1$,

$$p_2 = \left(1 - \langle v_1, v_2\rangle^2\right)^{-1/2}\left(-\langle v_2, v_1\rangle v_1 + v_2\right).$$

Writing $y_i = \langle p_i, x\rangle$, $1 \leqslant i \leqslant k$,

$$\begin{aligned}\lambda_k(A) = \lambda_k\big(B(y_0' : \rho) \cap \{ y = (y_1, \ldots, y_k) : \alpha_1 \leqslant y_1 \leqslant \alpha_1 + h,\\ \alpha_2 \leqslant \gamma y_1 + \delta y_2 \leqslant \alpha_2 + h\}\big),\end{aligned} \tag{3.22}$$

where $y_0' = (y_{01}', \ldots, y_{0k}')$ is the image of y_0 under the map $x = (x_1, \ldots, x_k) \to (\langle p_1, x\rangle, \ldots, \langle p_k, x\rangle) = (y_1, \ldots, y_k) = y$, and $\gamma = \langle v_2, v_1\rangle$, $\delta = (1 - \langle v_1, v_2\rangle^2)^{1/2}$. Thus

$$\begin{aligned}\lambda_k(A) &\leqslant \lambda_k\big(\{ y = (y_1, \ldots, y_k) :\\ &\qquad (y_3 - y_{03}')^2 + \cdots + (y_k - y_{0k}')^2 \leqslant \rho^2,\ \alpha_1 \leqslant y_1 \leqslant \alpha_1 + h,\\ &\qquad \delta^{-1}(\alpha_2 - \gamma y_1) \leqslant y_2 \leqslant \delta^{-1}(\alpha_2 + h - \gamma y_1)\}\big)\\ &= \delta^{-1} h^2 \lambda_{k-2}\big(\{u \in R^{k-2} : \|u\| \leqslant \rho\}\big)\\ &= \delta^{-1} h^2 \frac{\pi^{(k-2)/2}}{\Gamma(k/2)} \rho^{k-2}.\end{aligned}$$

Q.E.D.

LEMMA 3.9. *Let* f *be a continuous function on* R^k *and let*

$$M(a) = \int_{P_a} f(x)\,dx \qquad (a \in R^1), \tag{3.23}$$

where P *is a polyhedron defined by* (3.18), *and* P_a *is defined by* (3.19). *Then the function* M *is differentiable on* R^1 *with a continuous derivative given by*

$$M'(a)=\int_{\partial P_a} f\,d\lambda_{k-1} \qquad (a\in R^1). \tag{3.24}$$

Proof. Since P_a is a bounded set for each a, we assume (without loss of generality) that f is bounded on R^k. We assume throughout that

$$0\leqslant h\leqslant 1. \tag{3.25}$$

If there exist u_{j_1}, u_{j_2}, where $1\leqslant j_1,j_2\leqslant m$, such that $u_{j_1}+u_{j_2}=0$, then

$$d_{j_1}+d_{j_2}>0, \tag{3.26}$$

in order that P may not be empty or have an empty interior (as is the case if $d_{j_1}+d_{j_2}=0$). Similarly, if $d_{j_1}+a+d_{j_2}+a$ is negative, P_{a+h} is empty for sufficiently small h, and M identically vanishes in $(-\infty, a+h_0)$, say. Thus (3.24) is trivially true, both sides being identically equal to zero. We therefore assume that

$$a\geqslant -\tfrac{1}{2}\beta, \tag{3.27}$$

where

$$\beta=\min\{d_{j_1}+d_{j_2}: u_{j_1}+u_{j_2}=0,\ 1\leqslant j_1,j_2\leqslant m\} \tag{3.28}$$

if there exist j_1,j_2 such that $u_{j_1}+u_{j_2}=0$, and $\beta=\infty$ otherwise.

Let b be a positive number. By Lemma 3.5 there exists a positive number c_1 such that

$$P_a\subset B(0:c_1) \qquad \text{for all } a\leqslant b+1. \tag{3.29}$$

The definition of P_a gives

$$P_{a+h}\backslash P_a=\cup\, Q_{j_1,\ldots,j_r}, \tag{3.30}$$

where the union is over all r, $1\leqslant r\leqslant m$, and for fixed r over all r-tuples of indices $(j_1,\ldots,j_r)$ satisfying $1\leqslant j_1<j_2<\ldots<j_r\leqslant m$, and

$$\begin{aligned}Q_{j_1,\ldots,j_r}=\{x:\langle u_j,x\rangle\leqslant d_j+a \text{ for all } j\not\in\{j_1,\ldots,j_r\}\\ \text{and } d_j+a<\langle u_j,x\rangle\leqslant d_j+a+h \quad \text{for } j\in\{j_1,\ldots,j_r\}\}.\end{aligned} \tag{3.31}$$

Then

$$M(a+h)-M(a)=\int_{P_{a+h}\backslash P_a} f\,dx=\sum\int_{Q_{j_1,\ldots,j_r}} f\,dx, \tag{3.32}$$

where the summation is over all r-tuples $(j_1,\ldots,j_r)$, $1 \leqslant r \leqslant m$, as specified above. Let

$$L_i(a)=\{x : x\in P_a, \langle u_i,x\rangle = d_i + a\}. \tag{3.33}$$

We shall prove the following inequalities:

$$\lambda_k(Q_{j_1,\ldots,j_r}) \leqslant c_2h^2 \qquad \text{if } r\geqslant 2, \tag{3.34}$$

and

$$\left|\int_{Q_i} f(x)\,dx - h\int_{L_i(a)} f\,d\lambda_{k-1}\right| \leqslant c_3h^2 + c_4V(h)h \qquad (1\leqslant i\leqslant m), \tag{3.35}$$

where c_2, c_3 are constants depending only on k, P, and b, and

$$V(h)=\sup\{|f(x)-f(x')| : x,x'\in B(0:c_1), \|x-x'\|\leqslant h\}. \tag{3.36}$$

To prove (3.34), note that if any two of the vectors $u_{j_1},\ldots,u_{j_r}$ are linearly independent, then (3.34) follows from Lemma 3.8 and (3.29). The only remaining case occurs when $r=2$ and $u_{j_1}+u_{j_2}=0$. In this case

$$\begin{aligned} Q_{j_1,j_2}\subset\{x : \max[-d_{j_1}-a-h, d_{j_2}+a] &\leqslant \langle u_{j_2},x\rangle \\ &\leqslant \min[-d_{j_1}-a, d_{j_2}+a+h]\}. \end{aligned}$$

Thus $\lambda_k(Q_{j_1,j_2})=0$ unless

$$d_{j_2}+a < -d_{j_1}-a, \qquad \text{or,} \qquad a<-\tfrac{1}{2}(d_{j_1}+d_{j_2}). \tag{3.37}$$

But (3.37) is ruled out by (3.27). Hence (3.34) is proved. Now define

$$\begin{aligned} \alpha_{ij}&=\langle u_i,u_j\rangle, \qquad \gamma_{ij}=\left(1-\alpha_{ij}^2\right)^{1/2}, \\ \gamma&=\min\{\gamma_{ij} : \gamma_{ij}\neq 0,\ 1\leqslant i,j\leqslant m\}, \\ \alpha&=\max\{|\alpha_{ij}| : 1\leqslant i,j\leqslant m\}. \end{aligned} \tag{3.38}$$

If $u_i+u_j=0$, that is, if $\gamma_{ij}=0$, then, by (3.27),

$$\begin{aligned} \{x : \langle u_i,x\rangle > d_i+a\} &= \{x : \langle u_j,x\rangle < -d_i-a\} \\ &\subset \{x : \langle u_j,x\rangle \leqslant d_j+a\}, \end{aligned}$$

so that one may write

$$\begin{aligned} Q_i = \{ x : \langle u_j, x \rangle \leqslant d_j + a \text{ for all } j \text{ with } \gamma_{ij} \neq 0, \\ d_i + a < \langle u_i, x \rangle \leqslant d_i + a + h \}, \end{aligned} \tag{3.39}$$

$$\begin{aligned} L_i(a) = \{ x : \langle u_j, x \rangle \leqslant d_j + a \text{ for all } j \text{ with } \gamma_{ij} \neq 0, \\ \langle u_i, x \rangle = d_i + a \} \qquad (1 \leqslant i \leqslant m). \end{aligned}$$

In the following discussion we fix the integer i. Let $v_1, \ldots, v_k$ be an orthonormal basis such that $v_1 = u_i$. The transformation $x = (x_1, \ldots, x_k) \to y = (y_1, \ldots, y_k)$ given by

$$y_j = \langle v_j, x \rangle \qquad (1 \leqslant j \leqslant k),$$

enables one to write $x = y_1 v_1 + y'$, where $y' = \sum_{j=2}^{k} y_j v_j$. For $j \neq i$ let u_j' be the unit vector defined by

$$u_j = \langle u_j, u_i \rangle u_i + \left(1 - \langle u_j, u_i \rangle^2\right)^{1/2} u_j' = \alpha_{ji} v_1 + \gamma_{ji} u_j'.$$

Then, since u_j' is orthogonal to u_i,

$$\langle u_j, x \rangle = \alpha_{ji} y_1 + \gamma_{ji} \langle u_j', y' \rangle.$$

Thus in terms of the new coordinates Q_i, $L_i(a)$ become $\tilde{Q}_i$, $\tilde{L}_i(a)$, respectively, where

$$\begin{aligned} \tilde{Q}_i = \Bigg\{ y = (y_1, \ldots, y_k) : \langle u_j', y' \rangle \leqslant c_j - \frac{\alpha_{ji}(y_1 - d_i - a)}{\gamma_{ji}} \\ \text{for all } j \text{ with } \gamma_{ji} \neq 0, \text{ and } d_i + a < y_1 \leqslant d_i + a + h \Bigg\}, \end{aligned} \tag{3.40}$$

$$\begin{aligned} \tilde{L}_i(a) = \{ (d_i + a, y_2, \ldots, y_k) : \langle u_j', y' \rangle \leqslant c_j \text{ for all } j \\ \text{with } \gamma_{ji} \neq 0 \}, \end{aligned}$$

where

$$c_j = \left(\frac{d_j + a - \alpha_{ji}(d_i + a)}{\gamma_{ji}} \right).$$

Let $\tilde{Q}_i(y_1)$ denote the y_1-section of $\tilde{Q}_i$. Note that

$$\int_{\tilde{Q}_i} f(x)\,dx = \int_{d_i+a}^{d_i+a+h} dy_1 \int_{\tilde{Q}_i(y_1)} f(y_1 v_1 + y')\,dy_2 \ldots dy_k,$$

$$\int_{\tilde{L}_i(a)} f\,d\lambda_{k-1} = \int_{\tilde{Q}_i(d_i+a)} f((d_i+a)v_1 + y')\,dy_2 \ldots dy_k. \tag{3.41}$$

Now since $\tilde{Q}_i(y_1)$ is a section of $\tilde{Q}_i$ that is the image (under an orthogonal transformation) of $Q_i \subset P_{a+h} \subset B(0:c_1)$, it follows that

$$\int_{\tilde{Q}_i(y_1)} |f(y_1 v_1 + y') - f((d_i+a)v_1 + y')|\,dy_2 \ldots dy_k$$
$$\leqslant V(h)\lambda_{k-1}(\tilde{Q}_i(y_1)) \leqslant c_5 V(h), \tag{3.42}$$

say, for all y_1 satisfying $d_i + a \leqslant y_1 \leqslant d_i + a + h$. Next the symmetric difference $\tilde{Q}_i(y_1) \Delta \tilde{Q}_i(d_i + a)$ is contained in

$$\bigcup_{\{j:\gamma_{ji}\neq 0\}} \{(y_2,\ldots,y_k): \|(y_2,\ldots,y_k)\| \leqslant c_1\} \cap \left\{ (y_2,\ldots,y_k): |\langle u_j', y'\rangle - c_j| \leqslant \frac{|\alpha_{ji}|}{\gamma_{ji}} h \right\}$$

It follows from Lemma 3.7 that

$$\lambda_{k-1}(\tilde{Q}_i(y_1)\Delta \tilde{Q}_i(d_i+a)) \leqslant c_6 h$$

for a suitable constant c_6, and so

$$\left| \int_{\tilde{Q}_i(y_1)} f(v_1 y_1 + y')\,dy_2 \ldots dy_k - \int_{\tilde{Q}_i(d_i+a)} f(v_1 y_1 + y')\,dy_2 \ldots dy_k \right|$$
$$\leqslant c_6 \|f\| h. \tag{3.43}$$

The inequality (3.35) now follows from (3.41), (3.42), and (3.43). We thus have

$$\left| M(a+h) - M(a) - h\int_{\partial P_a} f\,d\lambda_{k-1} \right| \leqslant c_7(\|f\|+1)h^2 + c_8 V(h)h.$$

Since $V(h) \to 0$ as $h \to 0$, (3.24) is proved. Continuity of this derivative follows from that of the function $a \to \int_{\tilde{L}_i(a)} f\,d\lambda_{k-1}$, which is an immediate consequence of (3.42) and (3.43). Q.E.D.

LEMMA 3.10. *Let* g *be a differentiable function on* $[0,\infty)$ *such that*

$$\lim_{t\to\infty} g(t)=0$$

$$b \equiv \int_0^\infty |g'(t)|t^{k-1}\,dt < \infty. \tag{3.44}$$

Then for every polyhedron P

$$\int_{\partial P} g(\|x\|)\,d\lambda_{k-1} \leqslant \alpha_k \cdot b. \tag{3.45}$$

Proof. Define the function F on $[0,\infty)$ by

$$F(t)=\lambda_{k-1}(\partial P \cap \mathrm{Cl}(B(0:t))). \tag{3.46}$$

Then F is the distribution function of the measure $\lambda_{k-1}\circ h^{-1}$ induced on $[0,\infty)$ by the map $x \xrightarrow{h} \|x\|$ on ∂P (endowed with the surface area measure λ_{k-1}). Note that

$$\begin{aligned} &F(0)=0,\\ &F(t)=\lambda_{k-1}(\partial P) \qquad \text{for all sufficiently large } t. \end{aligned} \tag{3.47}$$

By changing variables and integrating by parts,[†]

$$\begin{aligned} \int_{\partial P} g(\|x\|)\lambda_{k-1}(dx) &= \int_0^\infty g(t)\lambda_{k-1}\circ h^{-1}(dt)\\ &= \int_0^\infty g(t)\,dF(t)\\ &= \lim_{t\to\infty}(F(t)g(t)) - F(0)g(0) - \int_0^\infty g'(t)F(t)\,dt\\ &= -\int_0^\infty g'(t)F(t)\,dt, \end{aligned} \tag{3.48}$$

from (3.47). Fix $t>0$. Let Q be a polyhedron such that $Q \subset \mathrm{Cl}\,B(0:t)$.

[†]See Dieudonné [1], Vol. II, p. 218.

Then $Q\cap\partial P\subset\partial(P\cap Q)$. Since $P\cap Q$ is a polyhedron and λ_{k-1} is a measure on the boundary of a polyhedron, it follows that $\lambda_{k-1}(Q\cap\partial P)\leqslant\lambda_{k-1}(\partial(P\cap Q))$. On the other hand, (3.16) implies that $\lambda_{k-1}(\partial(P\cap Q))\leqslant\lambda_{k-1}(\partial Q)$. Thus by (3.17) Cauchy

$$\lambda_{k-1}(Q\cap\partial P)\leqslant\lambda_{k-1}(\partial Q)<\alpha_k t^{k-1}. \tag{3.49}$$

Choose a countable dense subset $\{x_1,x_2,\ldots\}$ of $\partial B(0:t)$ and let Q_n be the convex hull of $x_1,\ldots,x_n$. Then $Q_n\uparrow \mathrm{Cl}(B(0:t))$, and (3.49) yields

$$F(t)=\lim\lambda_{k-1}(\partial P\cap Q_n)\leqslant\alpha_k t^{k-1}. \tag{3.50}$$

Substituting in (3.48) we get the lemma. Q.E.D.

We are now ready to prove Theorem 3.1.

Proof (of Theorem 3.1). First suppose that C is bounded, $\mathrm{Int}(C)\neq\phi$. Given $\delta>0$, choose $x_1,\ldots,x_n\in\partial C$ such that $\partial C\subset\{x_1,\ldots,x_n\}^\delta$. Let P be the convex hull of $\{x_1,\ldots,x_n\}$. By taking δ smaller than the radius of a ball contained in C, we ensure that P has a nonempty interior. Also, clearly,

$$P\subset C\subset P^\delta,$$

so that

$$C^\epsilon\subset P^{\epsilon+\delta}.$$

By Lemma 3.6

$$(P_a)^\rho\subset P_{a+\rho}$$

for all $\rho>0$, all $a\in R^1$. Thus $(P_{-\rho})^\rho\subset P\subset C$, or $P_{-\rho}\subset C^{-\rho}$, so that

$$C^\epsilon\backslash C^{-\rho}\subset(P^\delta)^\epsilon\backslash P_{-\rho}\subset P_{\epsilon+\delta}\backslash P_{-\rho}\qquad(\rho>0),$$

and

$$\int_{C^\epsilon\backslash C^{-\rho}}g(\|x\|)\,dx\leqslant\int_{P_{\epsilon+\delta}\backslash P_{-\rho}}g(\|x\|)\,dx$$

$$=\int_{-\rho}^{\epsilon+\delta}da\int_{\partial P_a}g(\|x\|)\lambda_{k-1}(dx)\qquad(\rho>0)$$

see Federer proof of 3.2.34: $\int_{C^\varepsilon\backslash C} g(x)\,d\mathcal{L}^m x = \int_0^\varepsilon \int_{\{x\mid \mathrm{dist}(x,C)=y\}} g(x)\,d\mathcal{H}^{m-1}x\,d\mathcal{L}^1 y$

by Lemma 3.9. From Lemma 3.10 we have

$$\int_{C^{\epsilon}\setminus C^{-\rho}} g(\|x\|)\,dx \leqslant (\epsilon+\delta+\rho)b\,\alpha_k. \tag{3.51}$$

Since δ is arbitrary, we are done. If C is bounded, $\mathrm{Int}(C)=\phi$, then $C^{-\rho}=\phi$ for $\rho>0$, and for all $\delta>0$

$$\int_{C^{\epsilon}\setminus C^{-\rho}} g(\|x\|)\,dx = \int_{C^{\epsilon}} g(\|x\|)\,dx \leqslant \int_{(C^{\delta})^{\epsilon}} g(\|x\|)\,dx$$

$$= \int_{(C^{\delta})^{\epsilon}\setminus(C^{\delta})^{-\delta}} g(\|x\|)\,dx \leqslant (\epsilon+\delta)\,b\alpha_k, \tag{3.52}$$

since C^{δ} has a nonempty interior. Since δ is arbitrary, (3.1) is true for all bounded convex sets C. If C is unbounded, look at $C_r = C\cap B(0:r)$ and let r increase to infinity. Since $C_r^{\epsilon}\uparrow C^{\epsilon}$ and $C_r^{-\rho}\uparrow C^{-\rho}$ as $r\uparrow\infty$, and the right side of (3.1) does not depend on the particular convex set, the proof is complete. Q.E.D.

NOTES

Section 1. Detailed accounts of the theory of weak convergence of probability measures may be found in Billingsley [1] and Parathasarathy [1].

Section 2. A systematic study of uniformity classes was initiated by Rao [3], who proved, in particular, Corollary 2.7 and Theorem 2.11. The theory was advanced and completed by Billingsley and Topsøe [1], who proved the main theorem Theorem 2.4, Corollary 2.6, Lemmas 2.2, 2.3, 2.9, 2.10, and gave the example following Corollary 2.8. These two articles also contain many results and applications not included here. One significant application of the theory is in generalizing and strengthening the classical Gilvenko–Cantelli Theorem. For this, see Rao [3] and Topsøe [1]. The useful Lemma 2.1 was proved by Scheffé [1]. The bounded Lipschitzian distance has been studied by Dudley [1], and Corollary 2.8 is due to him. Theorem 2.5, which is a convenient variant of Theorem 2.4, is due to Bhattacharya [3].

Section 3. Rao [1] was the first to prove the existence of a constant $c(k)$, depending only on k, such that $\Phi((\partial C)^{\epsilon})\leqslant c(k)\epsilon$ for all convex subsets C of R^k. The particular estimation given by Theorem 3.1 is essentially due to von Bahr [3]. It is easy to check that for $k=1$ and a nonincreasing g [on $[0,\infty)$], this estimate cannot be improved upon. The proof presented here provides perhaps the first detailed formal derivation of the estimate. The development (in particular, Lemma 3.9) may be of some independent interest.

CHAPTER 2

Fourier Transforms and Expansions of Characteristics Functions

The main mathematical tool used in this monograph is the *Fourier transform* and its extension, the *Fourier–Stieltjes transform*. The Fourier–Stieltjes transform of a probability measure on R^k is better known in probability literature as its *characteristic function*. In Sections 4 and 5 we present a summary of some basic facts about these transforms. Section 6 introduces *moments* and *cumulants* of probability measures on R^k and presents some inequalities concerning them. In Section 7 the Cramér–Edgeworth polynomials associated with a set of cumulants are studied. The principal results of Chapter 2 are the asymptotic expansions of (derivatives of) characteristic functions of normalized sums of independent random vectors. These expansions, in terms of the Cramér–Edgeworth polynomials, are developed in detail in Sections 8 and 9. In Section 10 some classes of probability measures, which are used as smoothing kernels in Chapters 3 and 4, are introduced.

4. THE FOURIER TRANSFORM

In this section we collect some standard results on Fourier transforms without proofs.

Let $L^p(R^k)$, $1 \leqslant p < \infty$, denote the Banach space of equivalence classes of complex-valued, Lebesgue-measurable functions f on R^k satisfying

$$\int |f|^p \, dx < \infty$$

with norm

$$\|f\|_p=\left(\int|f|^p\,dx\right)^{1/p}, \tag{4.1}$$

where $\int g\,dx$ denotes the integral of g on R^k with respect to Lebesgue measure on R^k. As usual, two functions f and g on R^k are said to be equivalent if they are equal almost everywhere with respect to Lebesgue measure. The space $L^2(R^k)$ is a Hilbert space endowed with the inner product $\langle\cdot,\cdot\rangle_2$

$$\langle f,g\rangle_2=\int f\bar{g}\,dx. \tag{4.2}$$

Here $\bar{g}$ denotes the complex conjugate of g. For nonnegative integral vectors $\alpha=(\alpha_1,\dots,\alpha_k)$ we write

$$x^\alpha=x_1^{\alpha_1}\cdots x_k^{\alpha_k} \qquad \left[x=(x_1,\dots,x_k)\in R^k\right],$$

$$D_j=\frac{\partial}{\partial x_j},\quad D^\alpha=D_1^{\alpha_1}\cdots D_k^{\alpha_k} \qquad \left[1\leqslant j\leqslant k,\quad \alpha=(\alpha_1,\dots,\alpha_k)\in(\mathbf{Z}^+)^k\right], \tag{4.3}$$

and say that $D^\alpha f$ is the *αth derivative of the function f.* Also, for each complex-valued function f on R^k we define the function $\tilde{f}$ by

$$\tilde{f}(x)=f(-x) \qquad (x\in R^k). \tag{4.4}$$

The function is said to be *symmetric* if $f=\tilde{f}$.

For $f\in L^1(R^k)$ the *Fourier transform* $\hat{f}$ of f is a complex-valued function on R^k defined by

$$\hat{f}(t)=\int e^{i\langle t,x\rangle}f(x)\,dx \qquad (t\in R^k), \tag{4.5}$$

where $\langle\cdot,\cdot\rangle$ is the usual inner product in $\mathbf{C}^k$:

$$\langle t,x\rangle=\sum_{j=1}^{k}t_j\bar{x}_j \qquad \left[t=(t_1,\dots,t_k),\ x=(x_1,\dots,x_k)\in\mathbf{C}^k\right], \tag{4.6}$$

with $\bar{c}$ the complex conjugate of the complex number c, giving rise to the norm $\|\cdot\|$,

$$\|t\|=\left(\sum_{j=1}^{k}|t_j|^2\right)^{1/2} \qquad (t\in\mathbf{C}^k). \tag{4.7}$$

Here $|c|^2 = c\bar{c}$ for complex numbers c. We shall sometimes use a different norm $|\cdot|$ in $\mathbf{C}^k$:

$$|x| = \sum_{j=1}^{k} |x_j| \qquad (x \in \mathbf{C}^k). \tag{4.8}$$

THEOREM 4.1. *Suppose* $f \in L^1(R^k)$.

(i) $\hat{f}$ *is uniformly continuous on* R^k.

(ii) $|\hat{f}(t)| \leqslant \|f\|_1$ *for all* $t \in R^k$.

(iii) (Riemann–Lebesgue Lemma) $|\hat{f}(t)| \to 0$ *as* $\|t\| \to \infty$.

(iv) (Fourier Inversion Theorem) *If* (*the equivalence class of*) $\hat{f} \in L^1(R^k)$, *then one can recover* (*the continuous version of*) f *from* $\hat{f}$ *by the formula*

$$f(x) = (2\pi)^{-k} \int e^{-i\langle t,x\rangle} \hat{f}(t)\, dt \qquad (x \in R^k).$$

Hence

$$\tilde{f} = (2\pi)^{-k} \hat{\hat{f}}, \qquad f = (2\pi)^{-k} \tilde{\hat{\hat{f}}}.$$

(v) *Let* $\alpha = (\alpha_1, \ldots, \alpha_k)$ *be a nonnegative integral vector. Define* g *by*

$$g(x) = x^\alpha f(x) \qquad (x \in R^k).$$

If $g \in L^1(R^k)$, *then* $D^\alpha \hat{f}$ *exists and*

$$\hat{g} = (-i)^{|\alpha|} D^\alpha \hat{f}.$$

(vi) *If* $f \in L^1(R^k) \cap L^2(R^k)$, *then* $\hat{f} \in L^2(R^k)$ *and*

$$\|f\|_2 = (2\pi)^{-k/2} \|\hat{f}\|_2.$$

The last result (vi) says that the map defined by $f \to (2\pi)^{-k/2}\hat{f}$, regarded as a map on the subset $L^1(R^k) \cap L^2(R^k)$ of $L^2(R^k)$ into $L^2(R^k)$, is an isometry. Since $L^1(R^k) \cap L^2(R^k)$ is a dense subset of $L^2(R^k)$, one can extend the above isometry on all of $L^2(R^k)$, and $\hat{f}$ so extended is still called the *Fourier transform* of f $[\in L^2(R^k)]$. The following theorem is then immediate.

THEOREM 4.2 (Plancherel Theorem). *The map* $f \to (2\pi)^{-k/2}\hat{f}$ *is a linear isometry on* $L^2(R^k)$, *and*

$$\langle f, g\rangle_2 = (2\pi)^{-k} \langle \hat{f}, \hat{g}\rangle_2 \qquad \left[f, g \in L^2(R^k)\right].$$

For $f,\ g \in L^1(R^k)$ one defines the (equivalence class of) function(s) $f*g$, called the *convolution* of f and g by

$$(f*g)(x) = \int f(x-y)\,g(y)\,dy \qquad (x \in R^k). \tag{4.9}$$

It follows from Fubini's theorem that $f*g \in L^1(R^k)$ and that

$$\|f*g\|_1 \leqslant \|f\|_1 \cdot \|g\|_1 \qquad \left[f, g \in L^1(R^k)\right]. \tag{4.10}$$

The convolution operation is clearly commutative and associative. The n-*fold convolution* f^{*n} is defined recursively by

$$f^{*1} = f, \qquad f^{*n} = f^{*(n-1)} * f \qquad \left[n > 1,\ f \in L^1(R^k)\right]. \tag{4.11}$$

5. THE FOURIER–STIELTJES TRANSFORM

We shall now extend the concept of the Fourier transform to the set $\mathfrak{M}$ of all finite signed measures on the Borel sigma-field $\mathfrak{B}^k$ of R^k. Let μ be a finite signed measure. We define the finite signed measure $\tilde{\mu}$ by

$$\tilde{\mu}(B) = \mu(-B) \qquad (B \in \mathfrak{B}^k). \tag{5.1}$$

A finite signed measure μ is called *symmetric* if $\mu = \tilde{\mu}$.

The Fourier–Stieltjes transform $\hat{\mu}$ of a finite signed measure μ is a complex-valued function on R^k defined by

$$\hat{\mu}(t) = \int e^{i\langle t, x\rangle}\mu(dx) \qquad (t \in R^k,\quad \mu \in \mathfrak{M}), \tag{5.2}$$

where, as usual, the *integral is over the whole space* R^k. If μ is a probability measure, $\hat{\mu}$ is also called the *characteristic function* of μ. Note that if μ is absolutely continuous (finite signed measure) with respect to Lebesgue measure, having *density* (i.e., the Radon–Nikodym derivative) f, then

$$\hat{\mu} = \hat{f}. \tag{5.3}$$

The *convolution* $\mu * \nu$ of two finite signed measures is a finite signed measure defined by

$$(\mu * \nu)(B) = \int \mu(B - x)\nu(dx) \qquad (B \in \mathfrak{B}^k), \tag{5.4}$$

where for $A \subset R^k$, $y \in R^k$, the *translate* $A+y$ is defined by

$$A+y=\{z=u+y : u \in A\}. \tag{5.5}$$

It is clear that convolution is commutative and associative. One defines the *n-fold convolution* μ^{*n} by

$$\mu^{*1}=\mu, \quad \mu^{*n}=\mu^{*(n-1)}*\mu \qquad (n>1, \quad \mu \in \mathfrak{M}). \tag{5.6}$$

Let μ be a signed measure on R^k. For any measurable map T on the space $(R^k, \mathfrak{B}^k, \mu)$ into $(R^s, \mathfrak{B}^s)$ one defines the *induced signed measure* $\mu \circ T^{-1}$ by

$$(\mu \circ T^{-1})(B)=\mu(T^{-1}(B)) \qquad (B \in \mathfrak{B}^s). \tag{5.7}$$

THEOREM 5.1

(i) (Uniqueness Theorem). *The map* $\mu \to \hat{\mu}$ *is one-to-one on* $\mathfrak{M}$.

(ii) *For* $\mu \in \mathfrak{M}$, $\hat{\mu}$ *is uniformly continuous and*

$$\hat{\mu}(0)=\mu(R^k), \qquad |\hat{\mu}(t)| \leqslant \|\mu\| \qquad (t \in R^k).$$

(iii) *For* μ, $\nu \in \mathfrak{M}$

$$\widehat{\mu * \nu}=\hat{\mu}\cdot\hat{\nu}.$$

(iv) *If* $\mu \in \mathfrak{M}$, *then*

$$\hat{\tilde{\mu}}=\bar{\hat{\mu}},$$

so that μ *is symmetric if and only if* $\hat{\mu}$ *is real-valued. The symmetrization* $\mu * \tilde{\mu}$ *of* μ *is always symmetric since*

$$\widehat{\mu * \tilde{\mu}}=|\hat{\mu}|^2.$$

(v) *If* T *is an affine transformation on* R^k *into* R^s: $\mathrm{Tx}=\mathrm{a}+\mathrm{Bx}$, $\mathrm{x} \in \mathrm{R}^k$, *then*

$$\widehat{\mu \circ T^{-1}}(t)=e^{i\langle t,a\rangle}\hat{\mu}(B't) \qquad (t \in R^s, \quad \mu \in \mathfrak{M}),$$

where B′ *is the adjoint (or transpose) of* B.

(vi) *If* $\mu \in \mathfrak{M}$ *and*

$$\int |x^{\alpha}|\,|\mu|(dx) < \infty$$

for a nonnegative integral vector $\alpha = (\alpha_1, \ldots, \alpha_k)$, *then* $D^{\alpha}\hat{\mu}$ *exists and is given by*

$$(D^{\alpha}\hat{\mu})(t) = (i)^{|\alpha|} \int x^{\alpha} e^{i\langle t, x\rangle} \mu(dx) \qquad (t \in R^k).$$

(vii) *If* $\hat{\mu} \in L^1(R^k)$, *then* μ *is absolutely continuous with respect to Lebesgue measure with a uniformly continuous and bounded density.*

(viii) (Parseval's Relation). *Let* μ, ν *be two finite signed measures on* R^k. *Then*

$$\int \hat{\mu}\, d\nu = \int \hat{\nu}\, d\mu.$$

The following theorem (due to Cramér and Lévy) gives a useful characterization of weak convergence of probability measures on R^k.

THEOREM 5.2 *Let* $\{G_n : n \geq 1\}$ *be a sequence of probability measures on* R^k. *If the sequence converges weakly to a probability measure* G, *then* $\{\hat{G}_n\}$ *converges pointwise to* $\hat{G}$. *Conversely, if* $\{\hat{G}_n\}$ *converges pointwise to a continuous limit* h, *then there exists a probability measure* H *such that* $\{\hat{G}_n\}$ *converges weakly to* H *and* $\hat{H} = h$.

It should be noted that if $\{G_n\}$ converges weakly to G, then $\{\hat{G}_n\}$ converges to $\hat{G}$ not merely pointwise, but uniformly on compact subsets of R^k. This is because the class of functions

$$\{e^{i\langle t, \cdot\rangle} : t \in K\},$$

where K is a compact subset of R^k, is equicontinuous and uniformly bounded; hence, by Corollary 2.7, it is a uniformity class for every probability measure G.

6. MOMENTS, CUMULANTS, AND NORMAL DISTRIBUTION

Let G be a probability measure on R^k. If $\nu = (\nu_1, \nu_2, \ldots, \nu_k)$ is a nonnegative integral vector such that $|x^{\nu}|$ is integrable with respect to G, one defines the ν-th *moment* μ_{ν} *of* G by

$E X_1^{\nu_1} \cdots X_k^{\nu_k} =$ $$\mu_{\nu} = \int x^{\nu} G(dx). \tag{6.1}$$

$x^{\nu} := x_1^{\nu_1} \cdots x_k^{\nu_k}$ (see 7.1)

$\| \quad E \vec{X}^{\nu}$

For every nonnegative real s one defines the s-th *absolute moment* ρ_s *of* G by

$$\rho_s = \int \|x\|^s G(dx) \qquad (s \geqslant 0). \tag{6.2}$$

Note that, by Theorem 5.1 (vi), if μ_ν is finite for some nonnegative integral vector ν, then

$$i^{|\nu|}\mu_\nu = (D^\nu \hat{G})(0) \tag{6.3}$$

and, therefore, one has the Taylor expansion† p 57–8 ∴ Cor 8.3.

$$\hat{G}(t) = 1 + \sum_{|\nu| \leqslant s} \mu_\nu \frac{(it)^\nu}{\nu!} + o(\|t\|^s) \qquad (t \to 0), \tag{6.4}$$

if ρ_s is finite for some positive integer s. Here

$$\nu! = \nu_1!\nu_2!\cdots\nu_k! \qquad [\nu = (\nu_1, \ldots, \nu_k)]. \tag{6.5}$$

If $k = 1$,

$$\mu_j = \int x^j G(dx) \qquad (j = 1, 2, \ldots). \tag{6.6}$$

For a positive real number x, $\log x$ denotes the natural logarithm. In other words, the function $x \to \log x$ on $(0, \infty)$ is defined as the inverse of the function $x \to \exp\{x\} = 1 + x + x^2/2! + x^3/3! + \cdots$ on R^1. For a nonzero complex number $z = r\exp\{i\theta\}$ $(r > 0, \ -\pi < \theta \leqslant \pi)$, we define

$$\log z = \log r + i\theta \qquad (r > 0, \ -\pi < \theta \leqslant \pi). \tag{6.7}$$

Thus we always take the so-called *principal branch* of the logarithm. The characteristic function $\hat{G}$ of a probability measure G on R^k is continuous and has a value of one at 0. Hence in a neighborhood of zero one has the Taylor expansion

$$\log \hat{G}(t) = \sum_{|\nu| \leqslant s} \chi_\nu \frac{(it)^\nu}{\nu!} + o(\|t\|^s) \qquad (t \to 0), \tag{6.8}$$

if ρ_s is finite for some positive integer s. Here the summation is over

†Note that $|x^\nu| \leqslant \|x\|^{|\nu|}$, so that $|\mu_\nu| \leqslant \rho_{|\nu|}$. Also, see Corollary 8.3.

nonnegative integral vectors ν, and (see Corollary 8.3)

$$i^{|\nu|}\chi_\nu = (D^\nu \log \hat{G})(0). \tag{6.9}$$

The coefficient χ_ν is called the ν-th *cumulant* of G. A simple calculation gives

$$\chi_0 = 0, \qquad \chi_\nu = \mu_\nu \quad \text{if } |\nu| = 1. \tag{6.10}$$

The following formal identity enables one to express a cumulant χ_ν in terms of moments by equating coefficients of t^ν on both sides:

Key

$$\sum_{|\nu| \geq 1} \chi_\nu \frac{(it)^\nu}{\nu!} = \sum_{s=1}^{\infty} (-1)^{s+1} \frac{1}{s} \left(\sum_{|\nu| \geq 1} \mu_\nu \frac{(it)^\nu}{\nu!} \right)^s. \tag{6.11}$$

This is obtained by noting

$$\log(1+z) = \sum_{s=1}^{\infty} (-1)^{s+1} \frac{1}{s} z^s \qquad (z \in \mathbf{C}, \ |z| < 1). \tag{6.12}$$

For example, in one dimension $\quad \chi_1 = \mu_1, \ \chi_0 = 0$

$$\sigma^2 \approx \chi_2 = \mu_2 - \mu_1^2, \qquad \chi_3 = \mu_3 - 3\mu_2\mu_1 + 2\mu_1^3 = E(Y-\mu)^3$$

$$\chi_4 = \mu_4 - 4\mu_3\mu_1 - 3\mu_2^2 + 12\mu_2\mu_1^2 - 6\mu_1^4, = E(Y-\mu)^4 - 3(\text{var} Y)^2 \tag{6.13}$$

$$\chi_5 = \mu_5 - 5\mu_4\mu_1 - 10\mu_3\mu_2 + 20\mu_3\mu_1^2 + 30\mu_2^2\mu_1 - 60\mu_2\mu_1^3 + 24\mu_1^5.$$

In general, in R^k one may write

$$\chi_\nu = \sum c(\nu_1, \ldots, \nu_s; j_1, \ldots, j_s) \mu_{\nu_1}^{j_1} \cdots \mu_{\nu_s}^{j_s}, \tag{6.14}$$

where the summation is over $s = 1, 2, \ldots, |\nu|$ and for each s over all s-tuples of nonnegative integral vectors $(\nu_1, \ldots, \nu_s)$ and s-tuples of nonnegative integers $(j_1, \ldots, j_s)$ satisfying $\quad \nu_j = (\nu_{j1}, \ldots, \nu_{jk}), \text{ vector}$

$$\sum_{i=1}^{s} j_i \nu_i = \nu, \tag{6.15}$$

and $c(\nu_1, \ldots, \nu_s; j_1, \ldots, j_s)$ is a constant depending only on $(\nu_1, \ldots, \nu_s)$ and $(j_1, \ldots, j_s)$.

For a given probability measure G on R^k with cumulants $\{\chi_\nu : |\nu| \leq s\}$,

$* \{$ Skewness: $\quad \chi_3/\sigma^3 = E(Y^*)^3$

$\quad$ Kurtosis: $\quad \chi_4/\sigma^4 = E(Y^*)^4 - 3$

we define the *polynomial* $\chi_s(z)$ in $z=(z_1,\dots,z_k)\in\mathbf{C}^k$ by

$$\chi_s(z)=\sum_{|\nu|=s}\frac{s!}{\nu!}\chi_\nu z^\nu \qquad (z\in\mathbf{C}^k;\quad s=1,2,\dots). \tag{6.16}$$

We point out that for $t\in R^k$, $\chi_s(t)$ is the sth cumulant of the probability measure on R^1 induced by the map $x\to\langle t,x\rangle$ defined on the probability space $(R^k,\mathfrak{B}^k,G)$ into $(R^1,\mathfrak{B}^1)$. This follows because the function $u\to\log\hat G(ut)$ [$u\in(-a,a)$ for some $a>0$] is the logarithm of the characteristic function of this induced probability measure on R^1 and because [by (6.8)]

$$\left.\frac{d^s}{du^s}\log\hat G(ut)\right|_{u=0}=s!\sum_{|\nu|=s}\chi_\nu\frac{(it)^\nu}{\nu!}.$$

Often we find it convenient to state results in terms of random variables and random vectors (although for purposes of this monograph it would be possible to avoid such notions). The νth *moment* $\mu_\nu(\mathbf{X})$, the νth *cumulant* $\chi_\nu(\mathbf{X})$, and the sth *absolute moment* $\rho_s(\mathbf{X})$ of a random vector $\mathbf{X}$ in R^k are defined to be the corresponding characteristics of the distribution $P_{\mathbf{X}}$ of $\mathbf{X}$. Note that for a random vector $\mathbf{X}$ with a finite covariance matrix,

$$\operatorname{Cov}(T\mathbf{X}+a)=\operatorname{Cov}(T\mathbf{X})=T\operatorname{Cov}(\mathbf{X})T', \tag{6.17}$$

where T is an $m\times k$ matrix $(1\leqslant m\leqslant k)$, T' is its transpose, and $a\in R^m$. Also, if $\{\mathbf{X}_1,\dots,\mathbf{X}_n\}$ are n independent random vectors in R^k and $\rho_2(\mathbf{X}_i)<\infty$ for $1\leqslant i\leqslant n$, then

$$\operatorname{Cov}(\mathbf{X}_1+\cdots+\mathbf{X}_n)=\operatorname{Cov}(\mathbf{X}_1)+\cdots+\operatorname{Cov}(\mathbf{X}_n). \tag{6.18}$$

If $\mathbf{X}$ is a random vector in R^k and has a finite νth moment (and cumulant), then for every $c\in R^1$,

$$\begin{aligned}\mu_\nu(c\mathbf{X})&=c^{|\nu|}\mu_\nu(\mathbf{X})\\ \chi_\nu(c\mathbf{X})&=c^{|\nu|}\chi_\nu(\mathbf{X})\end{aligned}\qquad(|\nu|\geqslant 1), \tag{6.19}$$

and for all $c\in R^1$, $b\in R^k$,

$$\begin{cases}\chi_\nu(c\mathbf{X}+b)=\chi_\nu(c\mathbf{X})=c^{|\nu|}\chi_\nu(\mathbf{X}) & \text{if } |\nu|\geqslant 2,\\ \chi_\nu(c\mathbf{X}+b)=\mu_\nu(c\mathbf{X}+b)=c\mu_\nu(\mathbf{X})+\langle b,\nu\rangle & \\ \qquad\qquad =c\chi_\nu(\mathbf{X})+\langle b,\nu\rangle & \text{if } |\nu|=1.\end{cases} \tag{6.20}$$

$Y=X+b\sim H$. $\hat H(t)=e^{i t\cdot b}\hat G(t)$. So

$i^{|\nu|}\chi_\nu(X+b)=D_\nu(\log\hat H)(0)=D^\nu\, i\,t\cdot b\big|_{t=0}+i^{|\nu|}\chi_\nu(X)$

$\Rightarrow \chi_\nu(X+b)=\chi_\nu(X)+\nu\cdot b\,[\,|\nu|=1\,]$.

It may also be shown, either by direct computation, as in (6.13), or by looking at (6.14) and (6.15), that

$$\chi_\nu(\mathbf{X}) = \mu_\nu(\mathbf{X}) \quad \text{for } |\nu| = 2, 3, \qquad \text{if } E(\mathbf{X}) = 0. \tag{6.21}$$

The relations (6.19) and (6.20) follow easily from definitions; note that in a neighborhood of $t=0$

$$\log \hat{P}_{c\mathbf{X}+b}(t) = i\langle t, b\rangle + \log \hat{P}_{\mathbf{X}}(ct). \tag{6.22}$$

Since the cumulants χ_ν are invariant under changes of origin for $|\nu| \geqslant 2$, they are sometimes called *semi-invariants*. The effect on χ_ν (and μ_ν) of a change of scale, as given by (6.19), is also quite simple. However the most important property of cumulants from our point of view is the following: *If* $\{\mathbf{X}_1, \ldots, \mathbf{X}_n\}$ *are independent random vectors in* R^k *each having a finite* νth *cumulant, then*

$$\chi_\nu(\mathbf{X}_1 + \cdots + \mathbf{X}_n) = \chi_\nu(\mathbf{X}_1) + \cdots + \chi_\nu(\mathbf{X}_n). \tag{6.23}$$

This follows from

$$\log \hat{P}_{\mathbf{X}_1 + \cdots + \mathbf{X}_n}(t) = \sum_{j=1}^{n} \log \hat{P}_{\mathbf{X}_j}(t), \tag{6.24}$$

which holds in a neighborhood of $t=0$.

We now derive some inequalities for moments and cumulants of a random vector $\mathbf{X}$. Henceforth, if there is no possibility of confusion, *we simply write* μ_ν, ρ_s, χ_ν *for* $\mu_\nu(\mathbf{X})$, $\rho_s(\mathbf{X})$, $\chi_\nu(\mathbf{X})$, *and so forth, respectively*.

The following basic inequality is stated without proof.† It is used fairly often in this book.

LEMMA 6.1 (Generalized Hölder Inequality). *Let* f_i, $1 \leqslant i \leqslant m$, *be measurable functions on a measure space* $(\Omega, \mathcal{A}, \mu)$ *into* $(R^1, \mathcal{B}^1)$ *such that* $|f_i|^{p_i}$, $1 \leqslant i \leqslant m$, *are integrable for some* m-*tuple of positive reals* $(p_1, \ldots, p_m)$ *satisfying* $p_1^{-1} + \cdots + p_m^{-1} = 1$. *Then*

$$\int |f_1 \cdots f_m| \, d\mu \leqslant \|f_1\|_{p_1} \cdots \|f_m\|_{p_m}.$$

LEMMA 6.2. *Let* $\mathbf{X}$ *be a random vector in* R^k *having a finite* s*th absolute moment* ρ_s *for some positive* s. *If* $\mathbf{X}$ *is not degenerate at* 0,‡

†See Hardy, Littlewood, and Polya [1] for several different proofs of this important inequality.
‡That is, $\mathrm{Prob}(X=0) \neq 1$.

(i) $r \to \log \rho_r$ *is a convex function on* $[0, s]$.
(ii) $r \to \rho_r^{1/r}$ *is nondecreasing on* $(0, s]$.
(iii) $r \to (\rho_r / \rho_2^{r/2})^{1/(r-2)}$ *is nondecreasing on* $(2, s]$ *if* $s > 2$.

Proof. We may assume that $\mathbf{X}$ is not degenerate at 0, so that $\log \rho_r$ is defined for $0 \leqslant r \leqslant s$.

(i) Let $0 < \alpha < 1$, $0 \leqslant r_1, r_2 \leqslant s$. Then

$$\rho_{\alpha r_1 + (1-\alpha) r_2} = E\left(\|\mathbf{X}\|^{\alpha r_1 + (1-\alpha) r_2}\right) = E(\mathbf{YZ}),$$

where $\mathbf{Y} = \|\mathbf{X}\|^{\alpha r_1}$, $\mathbf{Z} = \|\mathbf{X}\|^{(1-\alpha) r_2}$. Since $E(\mathbf{Y}^{1/\alpha})$ and $E(\mathbf{Z}^{1/(1-\alpha)})$ are both finite, by Lemma 6.1 (with $m = 2$),

$$E(\mathbf{YZ}) \leqslant \left(E(\mathbf{Y}^{1/\alpha})\right)^{\alpha} \left(E(\mathbf{Z}^{1/(1-\alpha)})\right)^{1-\alpha}$$
$$= \rho_{r_1}^{\alpha} \rho_{r_2}^{1-\alpha},$$

so that

$$\log \rho_{\alpha r_1 + (1-\alpha) r_2} \leqslant \alpha \log \rho_{r_1} + (1-\alpha) \log \rho_{r_2}.$$

(ii) Since $\log \rho_0 = 0$, and $(1/r) \log \rho_r$ is the slope of the line segment joining $(0, \log \rho_0)$ and $(r, \log \rho_r)$, it follows by the convexity of $r \to \log \rho_r$ that the function

$$r \to \log \rho_r^{1/r} = \frac{1}{r} \log \rho_r$$

is nondecreasing.

(iii) If $\rho_2 = 1$, then the function

$$r \to \log \rho_r^{1/(r-2)} = (r-2)^{-1} \log \rho_r = (r-2)^{-1} (\log \rho_r - \log \rho_2) \qquad (6.25)$$

is increasing in $(2, s]$, since the expression on the extreme right of (6.25) is the slope of the line segment joining $(2, \log \rho_2)$ and $(r, \log \rho_r)$, and $r \to \log \rho_r$ is convex in $[0, s]$ by (i). In the general case, apply this argument to $\mathbf{Y} = \mathbf{X} / \rho_2^{1/2}$. Q.E.D.

A simple application of Lemma 6.2(ii) yields

$$\left| \sum_{i=1}^{m} a_i \right|^r \leqslant m^{r-1} \sum_{i=1}^{m} |a_i|^r \qquad (a_i \in R^1, \quad 1 \leqslant i \leqslant m), \qquad (6.26)$$

for every $r \geqslant 1$ and every positive integer m. It also holds, trivially, for $r = 0$.

LEMMA 6.3. *Let* **X** *be a random vector in* R^k *having a finite* s*th absolute moment* ρ_s *for some positive integer* s. *Then for nonnegative integral vectors* ν *satisfying* $|\nu| \leqslant s$,

$$|\mu_\nu| \leqslant E|\mathbf{X}^\nu| \leqslant \rho_{|\nu|}, \tag{6.27}$$

and there exists a constant $c_1(\nu)$ *depending only on* ν *such that*

$$|\chi_\nu| \leqslant c_1(\nu)\rho_{|\nu|}. \tag{6.28}$$

Proof. The inequalities (6.27) follow from the simple inequality

$$|x^\nu| \leqslant \|x\|^{|\nu|}. \tag{6.29}$$

The inequality (6.28) follows from (6.14) and (6.15) by noting that

$$|\mu_{\nu_1}^{j_1} \cdots \mu_{\nu_s}^{j_s}| \leqslant \rho_{|\nu_1|}^{j_1} \cdots \rho_{|\nu_s|}^{j_s} \leqslant \rho_{|\nu|}^{\sum_{i=1}^{s} j_i|\nu_i|/|\nu|} = \rho_{|\nu|}. \tag{6.30}$$

The first inequality in (6.30) follows from (6.27), the second from Lemma 6.2(ii). Q.E.D.

Before concluding this section, we mention a few properties of the single most important probability measure on R^k—the *normal distribution* $\Phi_{m,V}$, whose density $\phi_{m,V}$ (with respect to Lebesgue measure) is given by

$$\phi_{m,V}(x) = (2\pi)^{-k/2}(\operatorname{Det} V)^{-1/2}\exp\{-\tfrac{1}{2}\langle x, V^{-1}x\rangle\} \qquad (x \in R^k). \tag{6.31}$$

Of the two parameters, $m \in R^k$ and V is a symmetric positive-definite $k \times k$ matrix. The notation $\operatorname{Det} V$ stands for *determinant of* V, and V^{-1} is, of course, the inverse of V. It is well known that†

$$\hat{\Phi}_{m,V}(t) = \exp\{i\langle t, m\rangle - \tfrac{1}{2}\langle t, Vt\rangle\} \qquad (t \in R^k). \tag{6.32}$$

From this it follows that m is the mean and V is the covariance matrix of $\Phi_{m,V}$. For the computation of cumulants χ_ν for $|\nu| \geqslant 2$, it is convenient to take $m = 0$ (for $|\nu| \geqslant 2$ a change of origin does not affect the cumulants χ_ν). This yields

$$\log \hat{\Phi}_{0,V}(t) = -\tfrac{1}{2}\langle t, Vt\rangle \tag{6.33}$$

†See Cramer [4], pp. 118, 119.

which shows that

$$\chi_\nu = (i,j) \text{ element of } V \text{ if } \nu = e_i + e_j, \tag{6.34}$$

where e_i is the vector with 1 for the ith coordinate and 0 for others, $1 \leqslant i \leqslant k$. Also,

$$\chi_\nu = 0 \qquad \text{if } |\nu| > 2. \tag{6.35}$$

Another important property of the normal distribution is

$$\Phi_{m_1,V_1} * \Phi_{m_2,V_2} * \cdots * \Phi_{m_n,V_n} = \Phi_{m,V}, \tag{6.36}$$

where

$$m = m_1 + m_2 + \cdots + m_n,$$

$$V = V_1 + V_2 + \cdots + V_n. \tag{6.37}$$

This follows from (6.32) and Theorem 5.1(i), (iii). The normal distribution $\Phi_{0,I}$, where I is the $k \times k$ identity matrix, is called the *standard normal distribution on* R^k and is denoted by Φ; the density of Φ is denoted by ϕ. Lastly, if $\mathbf{X} = (\mathbf{X}_1, \ldots, \mathbf{X}_k)$ is a random vector with distribution $\Phi_{m,V}$, then, for every $a \in R^k$, $a \neq 0$, the random variable $\langle a, \mathbf{X} \rangle$ has the one-dimensional normal distribution with mean $\langle a, m \rangle$ and variance $\langle a, Va \rangle = \sum_{i,j=1}^{k} a_i a_j \sigma_{ij}$, where

$$\sigma_{ij} = (i,j) \text{ element of } V = \operatorname{cov}(\mathbf{X}_i, \mathbf{X}_j) \qquad (i,j = 1, \ldots, k).$$

7. THE POLYNOMIALS $\tilde{P}_s$ AND THE SIGNED MEASURES P_s

Throughout this section $\nu = (\nu_1, \ldots, \nu_k)$ is a nonnegative integral vector in R^k. Consider the polynomials

$$\chi_s(z) = s! \sum_{|\nu| = s} \frac{\chi_\nu}{\nu!} z^\nu \qquad (z^\nu = z_1^{\nu_1} \cdots z_k^{\nu_k}) \tag{7.1}$$

in k *variables* $z_1, z_2, \ldots, z_k$ (real or complex) for a given set of real *constants* χ_ν. We define the formal polynomials $\tilde{P}_s(z : \{\chi_\nu\})$ in $z_1, \ldots, z_k$ by means of the following identity between two formal power series (in the real *variable* u).

$$1 + \sum_{s=1}^{\infty} \tilde{P}_s(z : \{\chi_\nu\}) u^s = \exp\left\{ \sum_{s=1}^{\infty} \frac{\chi_{s+2}(z)}{(s+2)!} u^s \right\}$$

$$= 1 + \sum_{m=1}^{\infty} \frac{1}{m!} \left[\sum_{s=1}^{\infty} \frac{\chi_{s+2}(z)}{(s+2)!} u^s \right]^m. \tag{7.2}$$

In other words, $\tilde{P}_s(z:\{\chi_\nu\})$ is the coefficient of u^s in the series on the extreme right. Thus

$$\tilde{P}_s(z:\{\chi_\nu\}) = \sum_{m=1}^{s} \frac{1}{m!} \left\{ \sum_{j_1,\ldots,j_m}^{*} \frac{\chi_{j_1+2}(z)}{(j_1+2)!} \frac{\chi_{j_2+2}(z)}{(j_2+2)!} \cdots \frac{\chi_{j_m+2}(z)}{(j_m+2)!} \right\}$$

$$= \sum_{m=1}^{s} \frac{1}{m!} \left\{ \sum_{j_1,\ldots,j_m}^{*} \left(\sum^{**} \frac{\chi_{\nu_1} \cdots \chi_{\nu_m}}{\nu_1! \cdots \nu_m!} z^{\nu_1 + \cdots + \nu_m} \right) \right\}$$

$$\tilde{P}_0 \equiv 0 \quad (s = 1, 2, \ldots), \tag{7.3}$$

where the summation Σ^* is over all m-tuples of positive integers $(j_1, \ldots, j_m)$ satisfying

$$\sum_{i=1}^{m} j_i = s, \qquad j_i = 1, 2, \ldots, s \qquad (1 \leqslant i \leqslant m), \tag{7.4}$$

and Σ^{**} denotes summation over all m-tuples of nonnegative integral vectors $(\nu_1, \ldots, \nu_m)$ satisfying

$$|\nu_i| = j_i + 2 \qquad (1 \leqslant i \leqslant m). \tag{7.5}$$

In particular,

$$\begin{cases} \tilde{P}_1(z:\{\chi_\nu\}) = \dfrac{\chi_3(z)}{3!} = \displaystyle\sum_{|\nu|=3} \frac{\chi_\nu}{\nu!} z^\nu, \\ \tilde{P}_2(z:\{\chi_\nu\}) = \dfrac{\chi_4(z)}{4!} + \dfrac{\chi_3^2(z)}{2!(3!)^2}, \\ \tilde{P}_3(z:\{\chi_\nu\}) = \dfrac{\chi_5(z)}{5!} + \dfrac{\chi_4(z)\chi_3(z)}{3!4!} + \dfrac{\chi_3^3(z)}{(3!)^4}. \end{cases} \tag{7.6}$$

LEMMA 7.1. *The degree of the polynomial* $\tilde{P}_s(z:\{\chi_\nu\})$ *is* $3s$, *and the smallest order of the terms in the polynomial is* $s+2$. *The coefficients of* $\tilde{P}_s(z:\{\chi_\nu\})$ *only involve* χ_ν*'s with* $|\nu| \leqslant s+2$.

Proof. This follows immediately from (7.3). Q.E.D.

The notation above has been chosen to suggest that eventually we use *cumulants* of some probability measure for χ_ν's in the expression for $\tilde{P}_s(z:\{\chi_\nu\})$. The polynomial $\tilde{P}_s$ can be defined in this sense only if the

$(s+2)$th absolute moment ρ_{s+2} is finite. The role of the polynomials $\tilde{P}_s$ in the theory of normal approximation is briefly indicated now. More precise results are given in Section 9.

Let G be a probability measure on R^k with zero mean, positive-definite covariance matrix V, and finite sth absolute moment ρ_s for some integer $s \geqslant 3$. Then [see (6.8), (6.16), and (6.23)]

$$\log \hat{G}^n\left(\frac{t}{n^{1/2}}\right) = n \log \hat{G}\left(\frac{t}{n^{1/2}}\right)$$

$$= -\tfrac{1}{2}\langle t, Vt\rangle + \sum_{r=1}^{s-2} \frac{\chi_{r+2}(it)}{(r+2)!} n^{-r/2} + n \cdot o\left(\left\|\frac{t}{n^{1/2}}\right\|^s\right)$$

$$\left(\frac{t}{n^{1/2}} \to 0\right). \qquad (7.7)$$

Thus for any fixed $t \in R^k$,

$$\hat{G}^n\left(\frac{t}{n^{1/2}}\right) = \exp\{-\tfrac{1}{2}\langle t, Vt\rangle\} \cdot \exp\left\{\sum_{r=1}^{s-2} \frac{\chi_{r+2}(it)}{(r+2)!} n^{-r/2} + o(n^{-(s-2)/2})\right\}$$

$$= \exp\{-\tfrac{1}{2}\langle t, Vt\rangle\} \cdot \left[1 + \sum_{r=1}^{s-2} n^{-r/2} \tilde{P}_r(it : \{\chi_\nu\})\right]$$

$$\times \left(1 + o(n^{-(s-2)/2})\right) \qquad (n \to \infty), \qquad (7.8)$$

where, in the evaluation of $\tilde{P}_r(it : \{\chi_\nu\})$, one uses the cumulants χ_ν of G. Thus, for each $t \in R^k$, one has an *asymptotic expansion* of $\hat{G}^n(t/n^{1/2})$ in powers of $n^{-1/2}$, in the sense that the remainder is of smaller order of magnitude than the last term in the expansion. The first term in the asymptotic expansion is the characteristic function of the normal distribution $\Phi_{0,V}$. The function (of t) that appears as the coefficient of $n^{-r/2}$ in the asymptotic expansion is the Fourier transform of a function that we denote by $P_r(-\phi_{0,V} : \{\chi_\nu\})$. The reason for such a notation is supplied by the following lemma.

LEMMA 7.2. *The function*

$$t \to \tilde{P}_r(it : \{\chi_\nu\}) \exp\{-\tfrac{1}{2}\langle t, Vt\rangle\} \qquad (t \in R^k), \qquad (7.9)$$

is the Fourier transform of the function $\mathrm{P}_r(-\phi_{0,\mathrm{V}} : \{\chi_\nu\})$ *obtained by formally substituting*

$$(-1)^{|\nu|} D^{\nu} \phi_{0,V} \qquad \text{for } (it)^{\nu} \tag{7.10}$$

for each ν in the polynomial $\tilde{P}_r(\text{it} : \{\chi_\nu\})$. Here $\phi_{0,V}$ is the normal density in R^k with mean zero and covariance matrix V. Thus one has the formal identity

$$P_r(-\phi_{0,V} : \{\chi_\nu\}) = \tilde{P}_r(-D : \{\chi_\nu\}) \phi_{0,V}, \tag{7.11}$$

where $-D = (-D_1, \ldots, -D_k)$.

Proof. The Fourier transform of $\phi_{0,V}$ is given by [see (6.32)]

$$\hat{\phi}_{0,V}(t) = \exp\left\{ -\tfrac{1}{2} \langle t, Vt \rangle \right\} \qquad (t \in R^k). \tag{7.12}$$

Also

$$\widehat{D^{\nu} \phi_{0,V}}(t) = (-it)^{\nu} \hat{\phi}_{0,V}(t) \qquad (t \in R^k), \tag{7.13}$$

which is obtained by taking the νth derivatives with respect to x on both sides of (the Fourier inversion formula)

$$\phi_{0,V}(x) = (2\pi)^{-k} \int \exp\left\{ -i\langle t, x \rangle \right\} \hat{\phi}_{0,V}(t) \, dt \qquad (x \in R^k) \tag{7.14}$$

[or, by Theorem 4.1(iv), (v)].   Q.E.D.

We define $P_r(-\Phi_{0,V} : \{\chi_\nu\})$ as the finite signed measure on R^k whose density is $P_r(-\phi_{0,V} : \{\chi_\nu\})$. For any given finite signed measure μ on R^k, we define $\mu(\cdot)$, *the distribution function of μ,* by

$$\mu(x) = \mu((-\infty, x]) \qquad (x \in R^k), \tag{7.15}$$

where

$$(-\infty, x] = (-\infty, x_1] \times \cdots \times (-\infty, x_k] \qquad \left[x = (x_1, \ldots, x_k) \in R^k \right]. \tag{7.16}$$

Note that

$$
\begin{aligned}
D_1 \cdots D_k P_r(-\Phi_{0,V} : \{\chi_\nu\})(x) &= P_r(-\phi_{0,V} : \{\chi_\nu\})(x) \\
&= \tilde{P}_r(-D : \{\chi_\nu\}) \phi_{0,V}(x) \\
&= \tilde{P}_r(-D : \{\chi_\nu\})(D_1 \cdots D_k \Phi_{0,V})(x) \\
&= D_1 \cdots D_k (\tilde{P}_r(-D : \{\chi_\nu\}) \Phi_{0,V})(x). \tag{7.17}
\end{aligned}
$$

So $P_0(-\phi_{0,V} ; \chi) = \phi_{0,V}$.

Thus the distribution function of $P_r(-\Phi_{0,V}:\{\chi_\nu\})$ is obtained by using the operator $\tilde{P}_r(-D:\{\chi_\nu\})$ on the normal distribution function $\Phi_{0,V}(\cdot)$. The last equality in (7.17) follows from the fact that the differential operators $\tilde{P}_r(-D:\{\chi_\nu\})$ and $D_1D_2\cdots D_k$ commute.

Let us write down $P_1(-\phi_{0,V}:\{\chi_\nu\})$ explicitly. By (7.6),

$$\tilde{P}_1(it:\{\chi_\nu\})=\sum_{|\nu|=3}\frac{\chi_\nu}{\nu!}(it)^\nu \qquad (t\in R^k), \tag{7.18}$$

so that (by Lemma 7.2)

$$\begin{aligned}
&P_1(-\phi_{0,V}:\{\chi_\nu\})(x)\\
&=-\sum_{|\nu|=3}\frac{\chi_\nu}{\nu!}D^\nu\phi_{0,V}(x)\\
&=\Bigg\{-\tfrac{1}{6}\Bigg[\chi_{(3,0,\ldots,0)}\Bigg[-\Bigg(\sum_{j=1}^k v^{1j}x_j\Bigg)^3+3v^{11}\sum_{j=1}^k v^{1j}x_j\Bigg]\\
&\quad+\cdots+\chi_{(0,\ldots,0,3)}\Bigg[-\Bigg(\sum_{j=1}^k v^{kj}x_j\Bigg)^3+3v^{kk}\sum_{j=1}^k v^{kj}x_j\Bigg]\Bigg]\\
&\quad-\tfrac{1}{2}\Bigg[\chi_{(2,1,0,\ldots,0)}\Bigg[-\Bigg(\sum_{j=1}^k v^{1j}x_j\Bigg)^2\Bigg(\sum_{j=1}^k v^{2j}x_j\Bigg)\\
&\qquad+2v^{12}\sum_{j=1}^k v^{1j}x_j+v^{11}\sum_{j=1}^k v^{2j}x_j\Bigg]\\
&\quad+\cdots+\chi_{(0,\ldots,0,1,2)}\Bigg[-\Bigg(\sum_{j=1}^k v^{kj}x_j\Bigg)^2\Bigg(\sum_{j=1}^k v^{k-1,j}x_j\Bigg)\\
&\quad+2v^{k,k-1}\sum_{j=1}^k v^{kj}x_j+v^{kk}\sum_{j=1}^k v^{k-1,j}x_j\Bigg]\Bigg]
\end{aligned}$$

$$-\left[\chi_{(1,1,1,0,\ldots,0)}\left[-\left(\sum_{j=1}^{k} v^{1j}x_j\right)\left(\sum_{j=1}^{k} v^{2j}x_j\right)\left(\sum_{j=1}^{k} v^{3j}x_j\right)\right.\right.$$

$$\left.+v^{12}\sum_{j=1}^{k} v^{3j}x_j+v^{13}\sum_{j=1}^{k} v^{2j}x_j+v^{23}\sum_{j=1}^{k} v^{1j}x_j\right]+\cdots$$

$$+\chi_{(0,\ldots,0,1,1,1)}\left[-\left(\sum_{j=1}^{k} v^{k-2,j}x_j\right)\left(\sum_{j=1}^{k} v^{k-1,j}x_j\right)\left(\sum_{j=1}^{k} v^{kj}x_j\right)\right.$$

$$+v^{k-2,k-1}\sum_{j=1}^{k} v^{kj}x_j+v^{k-2,k}\sum_{j=1}^{k} v^{k-1,j}x_j$$

$$\left.\left.\left.+v^{k-1,k}\sum_{j=1}^{k} v^{k-2,j}k_j\right]\right]\right\}\phi_{0,V}(x)$$

$$\left[V^{-1}=((v^{ij})),\quad x=(x_1,\ldots,x_k)\in R^k\right]. \tag{7.19}$$

If one takes $V=I$, then (7.19) reduces to

$$P_1(-\phi:\{\chi_\nu\})(x)$$

$$=\left\{-\tfrac{1}{6}\left[\chi_{(3,0,\ldots,0)}(-x_1^3+3x_1)+\cdots+\chi_{(0,\ldots,0,3)}(-x_k^3+3x_k)\right]\right.$$

$$-\tfrac{1}{2}\left[\chi_{(2,1,0,\ldots,0)}(-x_1^2x_2+x_2)+\cdots+\chi_{(0,\ldots,0,1,2)}(-x_k^2x_{k-1}+x_{k-1})\right]$$

$$\left.-\left[\chi_{(1,1,1,0,\ldots,0)}(-x_1x_2x_3)+\cdots+\chi_{(0,\ldots,0,1,1,1)}(-x_kx_{k-1}x_{k-2})\right]\right\}\phi(x)$$

$$(x\in R^k), \tag{7.20}$$

where $\phi=\phi_{0,I}$ is the standard normal density in R^k. If $k=1$, by letting χ_j be the jth cumulant of a probability measure G on $R^1 (j=1,2,3)$ having zero mean, one gets [using (6.13)]

$$P_1(-\phi:\{\chi_\nu\})(x)=\tfrac{1}{6}\,\mu_3(x^3-3x)\phi(x)\qquad(x\in R^1), \tag{7.21}$$

where μ_3 is the third moment of G.

Finally, note that whatever the numbers $\{\chi_\nu\}$ and the positive definite symmetric matrix V are,

$$\int P_s(-\phi_{0,V}:\{\chi_\nu\})(x)\,dx=P_s(-\Phi_{0,V}:\{\chi_\nu\})(R^k)$$

$$=0, \tag{7.22}$$

for all s $\geqslant 1$. This follows from

$$\int (D^{\beta}\phi_{0,V})(x)\,dx = 0 \qquad \text{for } |\beta| \geqslant 1. \tag{7.23}$$

The relation (7.23) is a consequence of the fact that $\phi_{0,V}$ and all its derivatives are integrable and vanish at infinity.

8. APPROXIMATION OF CHARACTERISTIC FUNCTIONS OF NORMALIZED SUMS OF INDEPENDENT RANDOM VECTORS

Let $\mathbf{X}_1,\ldots,\mathbf{X}_n$ be n independent random vectors in R^k each with zero mean and a finite third (or fourth) absolute moment. In this section we investigate the rate of convergence of $\hat{P}_{(\mathbf{X}_1+\cdots+\mathbf{X}_n)/n^{1/2}}$ to $\hat{\Phi}_{0,V}$, where V is the average of the covariance matrices of $\mathbf{X}_1,\ldots,\mathbf{X}_n$. The following form of Taylor's expansion will be useful to us.

LEMMA 8.1.[†] *Let* f *be a complex-valued function defined on an open interval* J *of the real line, having continuous derivatives* $f^{(r)}$ *of orders* $r=1,\ldots,s$. *If* x, $x+h \in J$, *then*

$$f(x+h) = f(x) + \sum_{r=1}^{s-1} \frac{h^r}{r!} f^{(r)}(x) + \frac{h^s}{(s-1)!}\int_0^1 (1-v)^{s-1} f^{(s)}(x+vh)\,dv. \tag{8.1}$$

COROLLARY 8.2. *For all real numbers* u *and positive integers* s

$$\left|\exp\{iu\} - 1 - iu - \cdots - \frac{(iu)^{s-1}}{(s-1)!}\right| \leqslant \frac{|u|^s}{s!}. \tag{8.2}$$

Consequently, if G *is a probability measure on* R^k *having a finite* s*th absolute moment* ρ_s *for some positive integer* s, *then*

$$\left|\hat{G}(t) - 1 - i\mu_1(t) - \cdots - \frac{i^{s-1}}{(s-1)!}\mu_{s-1}(t)\right| \leqslant \frac{\beta_s(t)}{s!} \leqslant \frac{\rho_s\|t\|^s}{s!}, \qquad (t\in R^k), \tag{8.3}$$

[†]Hardy [1], p. 327.

where for r = 1,...,s,

$$\mu_r(t) = \int \langle t,x\rangle^r G(dx) = \sum_{|\nu|=r} \frac{r!}{\nu!} \mu_\nu t^\nu, \; \beta_s(t) = \int |\langle t,x\rangle|^s \, G(dx). \quad (8.4)$$

Proof. The inequality (8.2) follows immediately from Lemma 8.1 on taking $f(u) = \exp\{iu\}$ $(u \in R^1)$ and $x = 0$, $h = u$. Inequality (8.3) is obtained on replacing u by $\langle t,x\rangle$ in (8.2) and integrating with respect to $G(dx)$. Note that

$$\beta_s(t) = \int |\langle t,x\rangle|^s G(dx) \leqslant \|t\|^s \rho_s.$$

Q.E.D.

COROLLARY 8.3. *Let* f *be a complex-valued function defined on an open subset* Ω *of* R^k, *having continuous derivatives* $\mathrm{D}^\nu \mathrm{f}$ *for* $|\nu| \leqslant \mathrm{s}$ *(on* Ω*). If the closed line segment joining* x, x + h $(\in \mathrm{R}^k)$ *lies in* Ω, *then*

$$f(x+h) = f(x) + \sum_{1 \leqslant |\nu| \leqslant s-1} \frac{h^\nu}{\nu!} (D^\nu f)(x)$$

$$+ s \sum_{|\nu|=s} \frac{h^\nu}{\nu!} \int_0^1 (1-u)^{s-1} (D^\nu f)(x+uh)\, du, \quad (8.5)$$

and therefore

$$\left| f(x+h) - f(x) - \sum_{1 \leqslant |\nu| \leqslant s-1} \frac{h^\nu}{\nu!} (D^\nu f)(x) \right|$$

$$\leqslant \sum_{|\nu|=s} \frac{|h^\nu|}{\nu!} \max\{ |(D^\nu f)(x+uh)| : 0 \leqslant u \leqslant 1 \}. \quad (8.6)$$

Proof. Define the function g on $(-\epsilon, 1+\epsilon)$ by

$$g(u) = f(x+uh), \quad (8.7)$$

ϵ being a positive number for which the line segment joining $x\text{-}\epsilon h$, $x + (1+\epsilon)h$ lies in Ω. Using the formula for differentiation of composite functions, one obtains, by induction on r,

$$g^{(r)}(u) = \sum_{|\nu|=r} \frac{r!}{\nu!} h^\nu (D^\nu f)(x+uh) \qquad (r = 1,\ldots,s). \quad (8.8)$$

Hence, by Lemma 8.1,

$$g(1)=g(0)+\sum_{r=1}^{s-1}\frac{g^{(r)}(0)}{r!}+\frac{1}{(s-1)!}\int_0^1(1-u)^{s-1}g^{(s)}(u)\,du,$$

which, on substitution from (8.7) and (8.8), yields (8.5). The inequality (8.6) follows immediately from (8.5). Q.E.D.

Let $\mathbf{X}_1,\ldots,\mathbf{X}_n$ be n independent random vectors in R^k each having a zero mean and a finite sth absolute moment for some $s \geqslant 2$. Assume that the *average covariance matrix* V, defined by

$$V=\frac{1}{n}\sum_{j=1}^{n}\operatorname{Cov}(\mathbf{X}_j)=\frac{1}{n}\sum_{j=1}^{n}V_j, \tag{8.9}$$

is nonsingular. Then we define the *Liapounov coefficient* $l_{s,n}$ by

$$l_{s,n}=\sup_{\|t\|=1}\frac{n^{-1}\sum_{j=1}^{n}E\left(|\langle t,\mathbf{X}_j\rangle|^s\right)}{\left[n^{-1}\sum_{j=1}^{n}E\left(\langle t,\mathbf{X}_j\rangle^2\right)\right]^{s/2}}n^{-(s-2)/2}\qquad(s\geqslant 2). \tag{8.10}$$

It is simple to check that $l_{s,n}$ is *independent of scale*. If B is a nonsingular $k\times k$ matrix, then $B\mathbf{X}_1,\ldots,B\mathbf{X}_n$ have the same Liapounov coefficients as $\mathbf{X}_1,\ldots,\mathbf{X}_n$ have. If one writes

$$\rho_{r,j}=E\left(\|\mathbf{X}_j\|^r\right)\qquad(1\leqslant j\leqslant n),$$

$$\rho_r=n^{-1}\sum_{j=1}^{n}\rho_{r,j}\qquad(r\geqslant 0), \tag{8.11}$$

then it is clear from (8.10) that

$$l_{s,n}\leqslant n^{-(s-2)/2}\sup_{\|t\|=1}\frac{\rho_s\|t\|^s}{\langle t,Vt\rangle^{s/2}}=\frac{\rho_s}{\lambda^{s/2}}n^{-(s-2)/2}, \tag{8.12}$$

where λ is the *smallest eigenvalue of* V. In one dimension (i.e., $k=1$)

$$l_{s,n}=\frac{\rho_s}{\rho_2^{s/2}}n^{-(s-2)/2}\qquad(s\geqslant 2). \tag{8.13}$$

We also use the following simple inequalities in the proofs of some of the theorems below. If $V=I$, then

$$E|\langle t,\mathbf{X}_j\rangle|^s \leqslant \sum_{j=1}^{n} E|\langle t,\mathbf{X}_j\rangle|^s \leqslant n^{s/2} l_{s,n}\|t\|^s. \tag{8.14}$$

In the rest of this section we use the notation (8.9)–(8.11), often without further mention.

THEOREM 8.4. *Let* $\mathbf{X}_1,\ldots,\mathbf{X}_n$ *be* n *independent random vectors (with values) in* R^k *having distributions* $\mathrm{G}_1,\ldots,\mathrm{G}_n$*, respectively. Assume that each* $\mathbf{X}_j$ *has zero mean and a finite third absolute moment. Assume also that the average covariance matrix* V *is nonsingular. Let* B *be the symmetric positive-definite matrix satisfying*

$$B^2 = V^{-1}. \tag{8.15}$$

Define

$$a(d) = \tfrac{1}{4}\sum_{r=2}^{\infty} r^{-1}\left(\frac{d^2}{2}\right)^{r-2},$$

$$b_n(d) = \tfrac{1}{2} - d\left(da(d) + \tfrac{1}{6}\right)(l_{3,n})^{2/3}. \tag{8.16}$$

Then for every $\mathrm{d}\in(0,2^{1/2})$ *and for all* t *satisfying*

$$\|t\| \leqslant d l_{3,n}^{-1/3}, \tag{8.17}$$

one has the inequality

$$\left|\prod_{j=1}^{n}\hat{G}_j\left(\frac{Bt}{n^{1/2}}\right) - \exp\left\{-\tfrac{1}{2}\|t\|^2\right\}\right|$$

$$\leqslant \left(da(d) + \tfrac{1}{6}\right) l_{3,n}\|t\|^3 \exp\left\{-b_n(d)\|t\|^2\right\}. \tag{8.18}$$

Proof. Assume first that $V=I$ and, consequently, $B=I$, where I is the identity matrix. In the given range (8.17) one has [see (8.14)]

$$\begin{aligned}
\left|\hat{G}_j\left(\frac{t}{n^{1/2}}\right) - 1\right| &\leqslant (2n)^{-1} E\langle t, \mathbf{X}_j\rangle^2 \\
&\leqslant (2n)^{-1}\left(E|\langle t,\mathbf{X}_j\rangle|^3\right)^{2/3} \\
&\leqslant (2n)^{-1}\left(\sum_{j=1}^{n} E|\langle t,\mathbf{X}_j\rangle|^3\right)^{2/3} \\
&\leqslant (2n)^{-1}\left(n^{3/2} l_{3,n}\|t\|^3\right)^{2/3} \\
&= \tfrac{1}{2} l_{3,n}^{2/3}\|t\|^2 \leqslant \tfrac{1}{2} d^2,
\end{aligned} \tag{8.19}$$

so that $\log \hat{G}_j(t/n^{1/2})$ is defined for $1 \leqslant j \leqslant n$, and [using (8.14)]

$$\begin{aligned}
&\left|\log \prod_{j=1}^{n} \hat{G}_j\left(\frac{t}{n^{1/2}}\right) + \tfrac{1}{2}\|t\|^2\right| \\
&\quad = \left|\sum_{j=1}^{n}\left[\log\left(1-\left(1-\hat{G}_j\left(\frac{t}{n^{1/2}}\right)\right)\right) + (2n)^{-1}E\langle t, \mathbf{X}_j\rangle^2\right]\right| \\
&\quad = \left|\sum_{j=1}^{n}\left[-\sum_{r=2}^{\infty} r^{-1}\left(1-\hat{G}_j\left(\frac{t}{n^{1/2}}\right)\right)^r\right.\right. \\
&\qquad\qquad \left.\left. + \hat{G}_j\left(\frac{t}{n^{1/2}}\right) - 1 + (2n)^{-1}E\langle t, \mathbf{X}_j\rangle^2\right]\right| \\
&\quad \leqslant \sum_{j=1}^{n}\left[(4n^2)^{-1}\left(E\langle t, \mathbf{X}_j\rangle^2\right)^2 \sum_{r=2}^{\infty} r^{-1}\left(\frac{d^2}{2}\right)^{r-2} + \tfrac{1}{6}n^{-3/2}E|\langle t, \mathbf{X}_j\rangle|^3\right] \\
&\quad \leqslant a(d)n^{-2}\sum_{j=1}^{n}\left(E|\langle t, \mathbf{X}_j\rangle|^3\right)^{4/3} + \tfrac{1}{6}l_{3,n}\|t\|^3 \\
&\quad \leqslant a(d)n^{-2}\left(\sum_{j=1}^{n} E|\langle t, \mathbf{X}_j\rangle|^3\right)^{4/3} + \tfrac{1}{6}l_{3,n}\|t\|^3 \\
&\quad \leqslant a(d)l_{3,n}^{4/3}\|t\|^4 + \tfrac{1}{6}l_{3,n}\|t\|^3 \leqslant \left(da(d)+\tfrac{1}{6}\right)l_{3,n}\|t\|^3 \\
&\quad \leqslant d\left(da(d)+\tfrac{1}{6}\right)l_{3,n}^{2/3}\|t\|^2. \qquad (8.20)
\end{aligned}$$

Hence, noting that $|e^x - 1| \leqslant |x|e^{|x|}$ for all complex numbers x, one has

$$\begin{aligned}
&\left|\prod_{j=1}^{n} \hat{G}_j\left(\frac{t}{n^{1/2}}\right) - \exp\left\{-\tfrac{1}{2}\|t\|^2\right\}\right| \\
&\qquad = \exp\left\{-\tfrac{1}{2}\|t\|^2\right\}\left|\exp\left\{\log\prod_{j=1}^{n}\hat{G}_j\left(\frac{t}{n^{1/2}}\right) + \tfrac{1}{2}\|t\|^2\right\} - 1\right| \\
&\qquad \leqslant \left(da(d)+\tfrac{1}{6}\right)l_{3,n}\|t\|^3 \exp\left\{-\tfrac{1}{2}\|t\|^2\right. \\
&\qquad\qquad \left. + d\left(da(d)+\tfrac{1}{6}\right)l_{3,n}^{2/3}\|t\|^2\right\} \\
&\qquad = \left(da(d)+\tfrac{1}{6}\right)l_{3,n}\|t\|^3 \exp\left\{-b_n(d)\|t\|^2\right\}. \qquad (8.21)
\end{aligned}$$

This proves the theorem when $V=I$. In the general case, look at the random vectors $B\mathbf{X}_1,\ldots,B\mathbf{X}_n$, whose average covariance matrix is I, and recall that the Liapounov coefficient is independent of scale. Q.E.D.

By going through the proof it is easy to see that if the $\mathbf{X}_j$'s in Theorem 8.4 are i.i.d. with common distribution G, then

$$\left|\hat{G}^n\left(\frac{Bt}{n^{1/2}}\right)-\exp\{-\tfrac{1}{2}\|t\|^2\}\right|\leqslant\left(da(d)+\tfrac{1}{6}\right)l_{3,n}\|t\|^3 \times\exp\left\{-\left(\tfrac{1}{2}-d^2a(d)-\frac{d}{6}\right)\|t\|^2\right\} \quad (8.22)$$

for all $d\in(0,2^{1/2})$ and for all t satisfying

$$\|t\|\leqslant dl_{3,n}^{-1}. \quad (8.23)$$

The next two theorems sharpen (8.22) and (8.18), under the additional assumption of finiteness of fourth moments.

THEOREM 8.5. *Let* $\mathbf{X}$ *be a random vector in* R^k *with destribution* G. *Suppose* $\mathbf{X}$ *has zero mean, positive-definite covariance matrix* V, *and a finite fourth absolute moment* ρ_4. *For all* t *satisfying*

$$\|t\|\leqslant\tfrac{1}{2}l_{4,n}^{-1/2} \qquad (n\geqslant 1), \quad (8.24)$$

one has the inequality

$$\left|\hat{G}^n\left(\frac{Bt}{n^{1/2}}\right)-\exp\{-\tfrac{1}{2}\|t\|^2\}\left(1+\tfrac{1}{6}i^3n^{-1/2}\mu_3(t)\right)\right|$$
$$\leqslant\left[(0.1325)n^{-1}+\tfrac{1}{24}l_{4,n}\right]\|t\|^4\exp\{-\tfrac{1}{2}\|t\|^2\}$$
$$+(0.0272)l_{3,n}^2\|t\|^6\exp\{-(0.3835)\|t\|^2\} \qquad (n\geqslant 1), \quad (8.25)$$

where $\mu_3(\mathrm{t})=\mathrm{E}\langle\mathrm{t},\mathbf{X}\rangle^3$, *and* B *is defined by* (8.15).

Proof. Assume that $V=I$. In the given range of t

$$\left|\hat{G}\left(\frac{t}{n^{1/2}}\right)-1\right|\leqslant(2n)^{-1}\|t\|^2\leqslant\tfrac{1}{8}. \quad (8.26)$$

Hence

$$\left|\log\hat{G}^n\left(\frac{t}{n^{1/2}}\right)+\tfrac{1}{2}\|t\|^2-\tfrac{1}{6}i^3n^{-1/2}\mu_3(t)\right|$$
$$=n\left|\log\left[1-\left(1-\hat{G}\left(\frac{t}{n^{1/2}}\right)\right)\right]+(2n)^{-1}\|t\|^2-\tfrac{1}{6}i^3n^{-3/2}\mu_3(t)\right|$$

$$= n\left| -\sum_{r=2}^{\infty} r^{-1}\left(1-\hat{G}\left(\frac{t}{n^{1/2}}\right)\right)^r + \hat{G}\left(\frac{t}{n^{1/2}}\right) - 1 + (2n)^{-1}\|t\|^2 \right.$$

$$\left. - \tfrac{1}{6} i^3 n^{-3/2}\mu_3(t) \right|$$

$$\leqslant (4n)^{-1}\|t\|^4 \sum_{r=2}^{\infty} r^{-1}\left(\tfrac{1}{8}\right)^{r-2} + \tfrac{1}{24} l_{4,n}\|t\|^4$$

$$\leqslant \left[(0.1325)n^{-1} + \tfrac{1}{2} l_{4,n}\right]\|t\|^4. \tag{8.27}$$

Similarly,

$$\left| n \log \hat{G}\left(\frac{t}{n^{1/2}}\right) + \tfrac{1}{2}\|t\|^2 \right| \leqslant (0.1325)n^{-1}\|t\|^4 + \tfrac{1}{6} l_{3,n}\|t\|^3$$

$$\leqslant (0.0663 + 0.1667) l_{3,n}\|t\|^3 = (0.233) l_{3,n}\|t\|^3$$

$$\leqslant (0.1165)\|t\|^2. \tag{8.28}$$

We then have

$$\left| \hat{G}^n\left(\frac{t}{n^{1/2}}\right) - \exp\left\{-\tfrac{1}{2}\|t\|^2\right\}\left(1 + \tfrac{1}{6} i^3 n^{-1/2}\mu_3(t)\right) \right|$$

$$= \exp\left\{-\tfrac{1}{2}\|t\|^2\right\} \left| \exp\left\{ \log \hat{G}^n\left(\frac{t}{n^{1/2}}\right) + \tfrac{1}{2}\|t\|^2 \right\} - \left(1 + \frac{i^3}{6} n^{-1/2}\mu_3(t)\right) \right|$$

$$\leqslant \exp\left\{-\tfrac{1}{2}\|t\|^2\right\} \left| \exp\left\{ \log \hat{G}^n\left(\frac{t}{n^{1/2}}\right) + \tfrac{1}{2}\|t\|^2 \right\} \right.$$

$$\left. - \left[1 + \left(\log \hat{G}^n\left(\frac{t}{n^{1/2}}\right) + \frac{\|t\|^2}{2}\right)\right] \right|$$

$$+ \exp\left\{-\tfrac{1}{2}\|t\|^2\right\} \left| \log \hat{G}^n\left(\frac{t}{n^{1/2}}\right) + \tfrac{1}{2}\|t\|^2 - \frac{i^3}{6} n^{-1/2}\mu_3(t) \right|$$

$$\leqslant \exp\left\{-\tfrac{1}{2}\|t\|^2\right\} \cdot \tfrac{1}{2} \left| \log \hat{G}^n\left(\frac{t}{n^{1/2}}\right) + \tfrac{1}{2}\|t\|^2 \right|^2 \exp\left\{ \left| \log \hat{G}^n\left(\frac{t}{n^{1/2}}\right) + \frac{\|t\|^2}{2} \right| \right\}$$

$$+ \exp\left\{-\tfrac{1}{2}\|t\|^2\right\}\left[(0.1325)n^{-1} + \tfrac{1}{24} l_{4,n}\right]\|t\|^4$$

$$\leqslant (0.0272) l_{3,n}^2 \|t\|^6 \exp\left\{-0.3835\|t\|^2\right\}$$

$$+ \left[(0.1325)n^{-1} + \tfrac{1}{24} l_{4,n}\right]\|t\|^4 \exp\left\{-\tfrac{1}{2}\|t\|^2\right\}. \tag{8.29}$$

We have used the inequality $|e^x - 1 - x| \leqslant \frac{1}{2}|x|^2 e^{|x|}$ in (8.29) [as well as the estimates (8.27), (8.28)]. This proves the theorem for $V = I$. In the general case look at the random vectors $B\mathbf{X}_1, \dots, B\mathbf{X}_n$. Q.E.D.

THEOREM 8.6. *Let* $\mathbf{X}_1, \dots, \mathbf{X}_n$ *be* n *independent random vectors in* R^k *having distributions* $G_1, \dots, G_n$, *respectively. Suppose that each random vector* $\mathbf{X}_j$ *has zero mean and a finite fourth absolute moment. Assume that the average covariance matrix* V *is nonsingular. Also assume*

$$l_{4,n} \leqslant 1. \tag{8.30}$$

Then for all t *satisfying*

$$\|t\| \leqslant \tfrac{1}{2} l_{4,n}^{-1/4}, \tag{8.31}$$

one has

$$\left| \prod_{j=1}^{n} \hat{G}_j\left(\frac{Bt}{n^{1/2}}\right) - \exp\{-\tfrac{1}{2}\|t\|^2\}\left(1 + \frac{i^3}{6} n^{-1/2} \mu_3(t)\right) \right|$$

$$\leqslant (0.175) l_{4,n} \|t\|^4 \exp\{-\tfrac{1}{2}\|t\|^4\}$$

$$+ \left[(0.018) l_{4,n}^2 \|t\|^8 + \tfrac{1}{36} l_{3,n}^2 \|t\|^6\right] \exp\{-(0.383)\|t\|^2\}, \tag{8.32}$$

where B *is the positive-definite symmetric matrix defined by* (8.15), *and*

$$\mu_3(t) = n^{-1} \sum_{j=1}^{n} E\langle t, \mathbf{X}_j\rangle^3.$$

Proof. We may, as before, take $V = I$. Note that [by (8.14) and (8.31)]

$$\left| \hat{G}_j\left(\frac{t}{n^{1/2}}\right) - 1 \right| \leqslant (2n)^{-1} E\langle t, \mathbf{X}_j\rangle^2 \leqslant (2n)^{-1}\left(E\langle t, \mathbf{X}_j\rangle^4\right)^{1/2}$$

$$\leqslant (2n)^{-1}\left(\sum_{j=1}^{n} E\langle t, \mathbf{X}_j\rangle^4\right)^{1/2} \leqslant \tfrac{1}{2} l_{4,n}^{1/2} \|t\|^2 \leqslant \tfrac{1}{8} \tag{8.33}$$

in the given range of t. Proceeding as in (8.28), one has

$$\left| \log \prod_{j=1}^{n} \hat{G}_j\left(\frac{t}{n^{1/2}}\right) + \tfrac{1}{2}\|t\|^2 \right| \leqslant (0.1325) n^{-2} \sum_{j=1}^{n} \left(E\langle t, \mathbf{X}_j\rangle^2\right)^2 + \tfrac{1}{6} l_{3,n} \|t\|^3$$

$$\leqslant (0.1325) l_{4,n} \|t\|^4 + (\tfrac{1}{6}) l_{3,n} \|t\|^3$$

$$\leqslant (0.117)\|t\|^2, \tag{8.34}$$

using (8.30), (8.31) in the last step. Also, as in (8.27),

$$\left|\log \prod_{j=1} \hat{G}_j\left(\frac{t}{n^{1/2}}\right) + \tfrac{1}{2}\|t\|^2 - \frac{i^3}{6} n^{-1/2}\mu_3(t)\right|$$
$$\leqslant [(0.1325) + (1/24)] l_{4,n}\|t\|^4 \leqslant (0.175) l_{4,n}\|t\|^4. \quad (8.35)$$

Hence, as in (8.30),

$$\left|\prod_{j=1}^{n} \hat{G}_j\left(\frac{t}{n^{1/2}}\right) - \exp\{-\tfrac{1}{2}\|t\|^2\}\left(1 + \frac{i^3}{6} n^{-1/2}\mu_3(t)\right)\right|$$
$$\leqslant \exp\{-\tfrac{1}{2}\|t\|^2\} \left| \exp\left\{ \sum_{j=1}^{n} \log \hat{G}_j\left(\frac{t}{n^{1/2}}\right) + \tfrac{1}{2}\|t\|^2 \right\} - 1 \right.$$
$$\left. - \left[\sum_{j=1}^{n} \log \hat{G}_j\left(\frac{t}{n^{1/2}}\right) + \tfrac{1}{2}\|t\|^2 \right] \right|$$
$$+ \exp\{-\tfrac{1}{2}\|t\|^2\} \left| \sum_{j=1}^{n} \log \hat{G}_j\left(\frac{t}{n^{1/2}}\right) + \tfrac{1}{2}\|t\|^2 \right.$$
$$\left. - \frac{i^3}{6} n^{-1/2}\mu_3(t) \right|$$
$$\leqslant (0.175) l_{4,n}\|t\|^4 \exp\{-\tfrac{1}{2}\|t\|^2\} + [(0.1325)^2 l_{4,n}^2\|t\|^8$$
$$+ \tfrac{1}{36} l_{3,n}^2 \|t\|^6] \exp\{-\tfrac{1}{2}\|t\|^2 + (0.117)\|t\|^2\}. \quad (8.36)$$

Q.E.D.

Note that [see (7.6)] if $\chi_{\nu,j}$ is the νth cumulant of $\mathbf{X}_j$ and if $\chi_\nu = n^{-1}\sum_{j=1}^{n} \chi_{\nu,j}$, then

$$\tilde{P}_1(it : \{\chi_\nu\}) = i^3 \sum_{|\nu|=3} \frac{\chi_\nu}{\nu!} t^\nu$$
$$= n^{-1} \sum_{j=1}^{n} \left(\sum_{|\nu|=3} i^3 \frac{\chi_{\nu,j}}{\nu!} t^\nu \right) = n^{-1} \sum_{j=1}^{n} \left(\sum_{|\nu|=3} \frac{i^3}{3!} \chi_{3,j}(t) \right)$$
$$= n^{-1} \sum_{j=1}^{n} \left(\sum_{|\nu|=3} \frac{i^3}{3!} \mu_{3,j}(t) \right) = \frac{i^3}{6} \mu_3(t),$$

the equality $\chi_{3,j}(t)=\mu_{3,j}(t)$ being a consequence of the fact that $\chi_{3,j}(t)$, $\mu_{3,j}(t)$ are, respectively, the third cumulant and third moment of the random variable $\langle t,\mathbf{X}_j\rangle$, which has zero mean [see (6.21)].

THEOREM 8.7. *Let* G *be a probability measure on* R^k *having zero mean, nonsingular covariance matrix* V, *and a finite third absolute moment. For all* t *satisfying*

$$\|t\|\leqslant 2^{1/2}l_{3,n}^{-1}, \tag{8.37}$$

one has

$$\left|\hat{G}^n\left(\frac{Bt}{n^{1/2}}\right)\right|\leqslant \exp\left\{-\left(\tfrac{1}{2}-\frac{2^{1/2}}{6}\right)\|t\|^2\right\}, \tag{8.38}$$

where $B=B'$, $B^2=V^{-1}$.

Proof. First take $V=I$. By Taylor expansion, if $\mathbf{X}$ has distribution G,

$$\hat{G}\left(\frac{t}{n^{1/2}}\right)=1-(2n)^{-1}\|t\|^2+\frac{\theta}{6}n^{-3/2}E|\langle t,\mathbf{X}\rangle|^3, \tag{8.39}$$

where $|\theta|\leqslant 1$. Hence, noting that $1-(2n)^{-1}\|t\|^2\geqslant 0$ in the given range of t,

$$\begin{aligned}\left|\hat{G}\left(\frac{t}{n^{1/2}}\right)\right| &\leqslant 1-(2n)^{-1}\|t\|^2+\tfrac{1}{6}n^{-1}l_{3,n}\|t\|^3\\ &\leqslant \exp\{-(2n)^{-1}\|t\|^2+\tfrac{1}{6}n^{-1}l_{3,n}\|t\|^3\}\\ &\leqslant \exp\left\{-\left(\tfrac{1}{2}-\frac{2^{1/2}}{6}\right)\frac{\|t\|^2}{n}\right\},\end{aligned} \tag{8.40}$$

which proves (8.38), when $V=I$. For the general case, look at the distribution of $B\mathbf{X}$, where $\mathbf{X}$ has distribution G. Q.E.D.

Before proving a similar result for non-i.i.d. random vectors, we need a simple lemma.

LEMMA 8.8. *Let* $\mathbf{X}$ *and* $\mathbf{Y}$ *be two independent random variables (in* R^1*) having the same distribution. If this common distribution has mean zero and a finite third absolute moment, then*

$$E|\mathbf{X}-\mathbf{Y}|^3\leqslant 4E|\mathbf{X}|^3. \tag{8.41}$$

Proof. Since

$$|\mathbf{X}-\mathbf{Y}|^3=(\mathbf{X}-\mathbf{Y})^2|\mathbf{X}-\mathbf{Y}|\leqslant(\mathbf{X}^2-2\mathbf{X}\mathbf{Y}+\mathbf{Y}^2)(|\mathbf{X}|+|\mathbf{Y}|)$$
$$=|\mathbf{X}|^3+|\mathbf{Y}|^3+\mathbf{X}^2|\mathbf{Y}|+\mathbf{Y}^2|\mathbf{X}|-2\mathbf{X}|\mathbf{X}|\mathbf{Y}-2\mathbf{Y}|\mathbf{Y}|\mathbf{X}, \tag{8.42}$$

and $E(\mathbf{X}|\mathbf{X}|\mathbf{Y})=(E(\mathbf{X}|\mathbf{X}|))E\mathbf{Y}=0=E(\mathbf{Y}|\mathbf{Y}|\mathbf{X})$, one has

$$E|\mathbf{X}-\mathbf{Y}|^3\leqslant 2E|\mathbf{X}|^3+2(E\mathbf{X}^2)E|\mathbf{Y}|=2E|\mathbf{X}|^3+2(E\mathbf{X}^2)E|\mathbf{X}|$$
$$\leqslant 4E|\mathbf{X}|^3. \tag{8.43}$$

Q.E.D.

THEOREM 8.9. *Let* $\mathbf{X}_1,\ldots,\mathbf{X}_n$ *be* n *independent random vectors in* R^k *having distributions* $G_1,\ldots,G_n$, *respectively. Suppose that each* $\mathbf{X}_j$ *has zero mean and a finite third absolute moment. Also, assume that* $V=n^{-1}\sum_{j=1}^{n}\mathrm{Cov}(\mathbf{X}_j)$ *is nonsingular. Then for every* $\delta\in(0,\frac{3}{2})$,

$$\left|\prod_{j=1}^{n}\hat{G}_j\left(\frac{Bt}{n^{1/2}}\right)\right|\leqslant\exp\left\{-\frac{\delta}{3}\|t\|^2\right\}, \tag{8.44}$$

for all t *satisfying*

$$\|t\|\leqslant(\tfrac{3}{2}-\delta)l_{3,n}^{-1}. \tag{8.45}$$

Here $B=B'$, $B^2=V^{-1}$.

Proof. We may take $V=I$. In this case, for each j, $1\leqslant j\leqslant n$,

$$\left|\hat{G}_j\left(\frac{t}{n^{1/2}}\right)\right|^2\leqslant 1-n^{-1}E\langle t,\mathbf{X}_j\rangle^2+\tfrac{2}{3}n^{-3/2}E|\langle t,\mathbf{X}_j\rangle|^3$$
$$\leqslant\exp\left\{-n^{-1}E\langle t,\mathbf{X}_j\rangle^2+\tfrac{2}{3}n^{-3/2}E|\langle t,\mathbf{X}_j\rangle|^3\right\}, \tag{8.46}$$

since $|\hat{G}_j(ut)|^2$ is the characteristic function, evaluated at u, of the random variable $\langle t,\mathbf{X}_j\rangle-\langle t,\mathbf{Y}_j\rangle$, where $\mathbf{Y}_j$ is a random vector independent of and having the same distribution as $\mathbf{X}_j$; also, by Lemma 8.8,

$$E[\langle t,\mathbf{X}_j\rangle-\langle t,\mathbf{Y}_j\rangle]^2=2E\langle t,\mathbf{X}_j\rangle^2,$$
$$E|\langle t,\mathbf{X}_j\rangle-\langle t,\mathbf{Y}_j\rangle|^3\leqslant 4E|\langle t,\mathbf{X}_j\rangle|^3. \tag{8.47}$$

Multiplying both sides of (8.46) over $j=1,\dots,n$, one gets

$$\left|\prod_{j=1}^{n}\hat{G}_j\left(\frac{t}{n^{1/2}}\right)\right|^2 \leqslant \exp\left\{-\|t\|^2+\tfrac{2}{3}l_{3,n}\|t\|^3\right\}$$

$$\leqslant \exp\left\{-\|t\|^2+\tfrac{2}{3}\left(\tfrac{3}{2}-\delta\right)\|t\|^2\right\}$$

$$=\exp\left\{-\frac{2\delta}{3}\|t\|^2\right\}, \tag{8.48}$$

in the given range (8.45). Q.E.D.

9. ASYMPTOTIC EXPANSIONS OF DERIVATIVES OF CHARACTERISTIC FUNCTIONS

We first state and prove some preliminary lemmas.

LEMMA 9.1. *Let* g *be a complex-valued function on an open subset* Ω *of* R^k *having continuous derivatives* $D^{\nu}g$ *for* $|\nu|\leqslant m$, *where* m *is a positive integer. If* g *has no zero in* Ω, *then on* Ω†

$$D^{\nu}\log g=\sum c(\{\beta^1,\dots,\beta^j\})\frac{(D^{\beta_1}g)\cdots(D^{\beta_j}g)}{g^j},$$

where the summation is over all collections of nonnegative integral vectors $\{\beta^1,\dots,\beta^j\}$ *satisfying*

$$\beta^1+\cdots+\beta^j=\nu,\qquad 1\leqslant j\leqslant|\nu|,\qquad [1\leqslant|\nu|\leqslant m],$$

and the constant $c(\{\beta^1,\dots,\beta^j)\}$ *depends only on the collection* $\{\beta^1,\dots,\beta^j\}$.

Proof. The result is obviously true for $|\nu|=1$. Assume that it holds for all ν satisfying $1\leqslant|\nu|\leqslant m$, where m is a positive integer. An immediate computation shows that the result holds for all vectors $\nu+e_i$, $1\leqslant i\leqslant k$, where e_i is the vector having one in the ith coordinate and zeros elsewhere. Thus the result holds for all ν satisfying $1\leqslant|\nu|\leqslant m+1$. Q.E.D.

LEMMA 9.2 (Cauchy's Estimate). *Let* f *be a complex-valued function on* $\bar{B}(a:R)=\{z=(z_1,\dots,z_k)\in C^k:\sum_{i=1}^{k}|z_i-a_i|^2\leqslant R^2\}$ [*where* $a=(a_1,\dots,a_k)\in C^k$

†This is a local result and is valid whatever the branch of the logarithm chosen.

and $R>0$] *given by the power series* [*absolutely convergent on* $\bar{B}(a:R)$]

$$f(z)=\sum_{\nu} c(\nu)(z-a)^{\nu} \qquad [z-a=(z_1-a_1,\dots,z_k-a_k)],$$

where the summation is over all nonnegative integral vectors $\nu\in(\mathbf{Z}^+)^k$. *Then*

$$|(D^{\nu}f)(a)|\leqslant r^{-|\nu|}\nu!\,M_r,$$

for every $r\in(0,k^{-1/2}R]$, *where* M_r *is defined by*

$$M_r=\sup\{|f(z)|:|z_i-a_i|=r,\ 1\leqslant i\leqslant k\}.$$

Proof. Since the cube $\{z:|z_i-a_i|\leqslant r,\ 1\leqslant i\leqslant k\}$ is contained in $\bar{B}(a:R)$, one can define the function g_r on $[-\pi,\pi)^k$ by

$$g_r(\theta_1,\dots,\theta_k)=f(a_1+re^{i\theta_1},\dots,a_k+re^{i\theta_k})$$
$$(-\pi\leqslant\theta_i<\pi, \qquad 1\leqslant i\leqslant k).$$

Clearly, for all ν,

$$r^{|\nu|}c(\nu)=(2\pi)^{-k}\int_{[-\pi,\pi)^k}\exp\{-i\langle\nu,\theta\rangle\}\,g_r(\theta_1,\dots,\theta_k)\,d\theta_1\cdots d\theta_k.$$

Noting that the integrand is bounded above by M_r, one has

$$|(D^{\nu}f)(a)|=\nu!|c(\nu)|\leqslant r^{-|\nu|}\nu!\,M_r.$$

Q.E.D.

For x,y in R^k we define

$$x\leqslant y \qquad \text{if } x_i\leqslant y_i \qquad \text{for } i=1,\dots,k. \tag{9.1}$$

We also write $x<y$ if $x\leqslant y$ but $y\nleqslant x$.

LEMMA 9.3. *Let* f *be a complex-valued function on an open subset* Ω *of* R^k, *having a continuous* ν*th derivative for a nonnegative integral vector* $\nu\neq 0$. *Then* $\exp\{f\}$ *also has a continuous* ν*th derivative on* Ω *given by*

$$D^{\nu}(\exp\{f\})=\exp\{f\}\sum{}^{*}c(\{j_\beta:0<\beta\leqslant\nu\})\cdot\Pi(D^{\beta}f)^{j_\beta}, \tag{9.2}$$

where the summation Σ^* *is over all collections of nonnegative integers*

$\{j_\beta : 0 < \beta \leqslant \nu\}$ *satisfying (here β's are integral vectors)*

$$\sum_{\{\beta : 0 < \beta \leqslant \nu\}} j_\beta \beta = \nu,$$

and $c(\{j_\beta : 0 < \beta \leqslant \nu\})$ *depends only on the collection* $\{j_\beta : 0 < \beta \leqslant \nu\}$. *Also,* Π *denotes the product over all* β, $0 < \beta \leqslant \nu$.

Proof. Let $e_1 = (1,0,\ldots,0),\ldots,e_k = (0,0,\ldots,0,1)$. Then

$$D^{e_i}(\exp\{f\}) = \exp\{f\}(D^{e_i}f) \qquad (1 \leqslant i \leqslant k).$$

Thus the assertion is true for all ν with $|\nu| = 1$. Suppose that it is true for all ν with $|\nu| \leqslant m$. Given ν with $|\nu| = m$, one has, for all i $(1 \leqslant i \leqslant k)$,

$$D^{\nu+e_i}(\exp\{f\}) = \exp\{f\} \sum {}^{*} c(\{j_\beta : 0 < \beta \leqslant \nu\})$$
$$\times \left[\sum_{\{\beta' : j_{\beta'} > 0\}} j_{\beta'} (D^{\beta'} f)^{j_{\beta'} - 1} \cdot (D^{\beta' + e_i} f) \Pi' (D^\beta f)^{j_\beta} \right]$$
$$+ \exp\{f\}(D^{e_i} f) \sum {}^{*} c(\{j_\beta : 0 < \beta \leqslant \nu\}) \Pi (D^\beta f)^{j_\beta},$$

where Π is as above, and Π' is the product over all β such that $0 < \beta \leqslant \nu$, $\beta \neq \beta'$. This is of the form (9.2) with ν replaced by $\nu + e_i$ $(1 \leqslant i \leqslant k)$. This proves (9.2) for all ν satisfying $|\nu| \leqslant m+1$. Q.E.D.

LEMMA 9.4. *Let* **X** *be a random vector in* R^k *with distribution* G. *Suppose* **X** *has zero mean, a covariance matrix* V, *and a finite* s*th absolute moment* ρ_s *for some positive integer* $s \geqslant 3$. *Then there exists a constant* $c_2(s)$ *depending only on* s *such that*

$$|(D^\nu \log \hat{G})(t)| \leqslant c_2(s) \rho_s \tag{9.3}$$

for all t *in* R^k *satisfying*

$$|\hat{G}(t) - 1| \leqslant \tfrac{1}{2} \tag{9.4}$$

and for all nonnegative integral vectors ν *satisfying* $|\nu| = s$.

Proof. Note that $\log \hat{G}$ is defined on an open set containing the set of t's satisfying (9.4). Also, if $\beta_1, \ldots, \beta_j$ are nonnegative integral vectors satisfying

$$\beta_1 + \cdots + \beta_j = \nu,$$

then

$$|(D^{\beta_1}\hat{G})(t)|\cdots|(D^{\beta_j}\hat{G})(t)|\leqslant(E|\mathbf{X}^{\beta_1}|)\cdots(E|\mathbf{X}^{\beta_j}|)$$

$$\leqslant\rho_{|\beta_1|}\cdots\rho_{|\beta_j|}\leqslant\rho_s^{(|\beta_1|+\cdots+|\beta_j|)/s}=\rho_s, \quad (9.5)$$

by (6.29) and Lemma 6.2(ii). The proof is now completed using Lemma 9.1. Q.E.D.

For the remainder of this section we shall consider n *independent random vectors* $\mathbf{X}_1,\ldots,\mathbf{X}_n$ *with respective distributions* $G_1,\ldots,G_n$. *We assume that* $\mathbf{X}_j$ *has mean zero, covariance matrix* V_j, *and a finite* s*th absolute moment for some integer* $s\geqslant 3$. *We write*

$$\rho_{r,j}=E\|\mathbf{X}_j\|^r, \qquad \rho_r=n^{-1}\sum_{j=1}^{n}\rho_{r,j} \qquad (r\geqslant 0),$$

$$\chi_{\nu,j}=\nu\text{th cumulant of }\mathbf{X}_j, \qquad \chi_\nu=n^{-1}\sum_{j=1}^{n}\chi_{\nu,j} \quad (9.6)$$

$$\left[\nu\in(\mathbf{Z}^+)^k,\quad 0\leqslant|\nu|\leqslant s\right]$$

$$V=n^{-1}\sum_{j=1}^{n}V_j,$$

unless otherwise specified. In case V is nonsingular, let B denote the symmetric positive-definite matrix satisfying

$$B^2=V^{-1} \quad (9.7)$$

and write

$$\eta_r=n^{-1}\sum_{j=1}^{n}E\|B\mathbf{X}_j\|^r \qquad (r>0). \quad (9.8)$$

The constants c's *that appear below depend only on their arguments.*

LEMMA 9.5. *For every nonnegative integral vector* α *satisfying* $0\leqslant|\alpha|\leqslant 3r$, *and for* $r=0,1,\ldots,s-2$,

$$|D^\alpha\tilde{P}_r(z:\{\chi_\nu\})|\leqslant c_3(\alpha,r)\left(1+\rho_2^{r(s-3)/(s-2)}\right)\left(1+\|z\|^{3r-|\alpha|}\right)\rho_s^{r/(s-2)}$$

$$(z\in\mathbf{C}^k). \quad (9.9)$$

Proof. By (7.3),

$$D^{\alpha}\tilde{P}_r(z:\{\chi_\nu\}) = \sum_{m=1}^{r} \frac{1}{m!} \sum_{j_1,\dots,j_m} \left[\sum{}^{**} \frac{\chi_{\nu_1}\cdots\chi_{\nu_m}}{\nu_1!\cdots\nu_m!} \times \frac{(\nu_1+\cdots+\nu_m)!}{(\nu_1+\cdots+\nu_m-\alpha)!} z^{\nu_1+\cdots+\nu_m-\alpha} \right], \tag{9.10}$$

where Σ^* denotes summation over all m-tuples of positive integers $(j_1,\dots,j_m)$ satisfying

$$\sum_{i=1}^{m} j_i = r, \tag{9.11}$$

and Σ^{**} denotes summation over all m-tuples of nonnegative integral vectors $(\nu_1,\dots,\nu_m)$ satisfying (7.5). By Lemma 6.3 and averaging,

$$\begin{aligned}
|\chi_{\nu_1}\cdots\chi_{\nu_m}| &\leqslant c_1(\nu_1)\cdots c_1(\nu_m)\rho_{j_1+2}\cdots\rho_{j_m+2} \\
&= c_1(\nu_1)\cdots c_1(\nu_m)\rho_2^{((j_1+2)+\cdots+(j_m+2))/2}\frac{\rho_{j_1+2}}{\rho_2^{(j_1+2)/2}}\cdots\frac{\rho_{j_m+2}}{\rho_2^{(j_m+2)/2}} \\
&\leqslant c_1(\nu_1)\cdots c_1(\nu_m)\rho_2^{\left(\frac{r}{2}+m\right)}\left(\frac{\rho_s}{\rho_2^{s/2}}\right)^{(j_1+\cdots+j_m)/(s-2)} \\
&= c_1(\nu_1)\cdots c_1(\nu_m)\rho_2^{m-r/(s-2)}\rho_s^{r/(s-2)}.
\end{aligned} \tag{9.12}$$

We have used Lemma 6.2(iii) and the fact that $\rho_{r'}$ is the r'th absolute moment of $n^{-1}(G_1+\cdots+G_n)$, where G_j is the distribution of $\mathbf{X}_j$. On using (9.12) in (9.10), the desired inequality (9.9) is obtained if one notes that

$$\|z\|^m \leqslant \|z\|^{m'} + \|z\|^{m''} \qquad (0 \leqslant m' \leqslant m \leqslant m'', \quad z\in\mathbf{C}^k). \tag{9.13}$$

Q.E.D.

LEMMA 9.6. *Let* g *be an absolutely convergent power series in* $z\in\mathbf{C}^k$ *on* $B(0:r)=\{\|z\|<r\}$ *such that*

$$|g(z)| \leqslant h(\|z\|) \qquad [z\in B(0:r)], \tag{9.14}$$

where h *is a nondecreasing function on* $[0,\infty)$. *Then for every* $c>0$,

$$|(D^{\nu}g)(z)| \leqslant \nu!\,c^{-|\nu|}\|z\|^{-|\nu|}h\big((1+k^{1/2}c)\|z\|\big) \tag{9.15}$$

for all nonnegative integral vectors ν *and for all* z *satisfying*

$$\|z\| < \frac{r}{(1+k^{1/2}c)} \qquad (z \in \mathbf{C}^k - \{0\}). \tag{9.16}$$

Proof. By Cauchy's estimate (Lemma 9.2)

$$\begin{aligned} |(D^{\nu}g)(z)| &\leq \nu!(c\|z\|)^{-|\nu|}\sup\{|g(z')| : |z'_i - z_i| = c\|z\|,\ 1 \leq i \leq k\} \\ &\leq \nu!(c\|z\|)^{-|\nu|}\sup\{|g(z')| : \|z' - z\| \leq k^{1/2}c\|z\|\} \\ &\leq \nu!(c\|z\|)^{-|\nu|}\sup\{|g(z')| : \|z'\| \leq (1+k^{1/2}c)\|z\|\} \\ &\leq \nu!(c\|z\|)^{-|\nu|}h\big((1+k^{1/2}c)\|z\|\big) \end{aligned} \tag{9.17}$$

for all $z \neq 0$ satisfying (9.16) if one takes (in Lemma 9.2) $a = z$, $r = c\|z\|$ and notes that [for z satisfying (9.16)]

$$\begin{aligned} \{z' : |z'_i - z_i| = c\|z\|, 1 \leq i \leq k\} &\subset \{z' : \|z'\| \leq (1+k^{1/2}c)\|z\|\} \\ &\subset \{\|z'\| < r\}. \end{aligned} \tag{9.18}$$

Q.E.D.

LEMMA 9.7. *Let* s *be an integer not smaller than* 3 *and let* $\{\chi_\nu : \nu \in (\mathbf{Z}^+)^k, |\nu| \leq s\}$ *be real numbers. Define*

$$\beta_s = \left(\max\{|\chi_\nu|^{1/(|\nu|-2)} : 3 \leq |\nu| \leq s\}\right)^{s-2}, \tag{9.19}$$

$$c(s,k) = \left\{\sum_{r=1}^{s-3} \sum_{|\nu|=r+2} \frac{1}{\nu!}\right\}.$$

Then for every $u \in R^1 - \{0\}$ *and all* z *satisfying*

$$\|z\| \leq \left(11c(s,k)|u|\,\beta_s^{1/(s-2)}\right)^{-1} \tag{9.20}$$

one has, in the notation of Section 7,

$$\left|D^{\alpha}\left[\exp\left\{\sum_{r=1}^{s-3} \frac{\chi_{r+2}(z)}{(r+2)!}u^r\right\} - \sum_{r=0}^{s-3} \tilde{P}_r(z : \{\chi_\nu\})u^r\right]\right|$$
$$\leq c'(s,k)|u|^{s-2}\beta_s\left(\|z\|^{s-|\alpha|} + \|z\|^{3(s-2)-|\alpha|}\right)\exp\left\{\tfrac{2}{9}\|z\|^2\right\}, \tag{9.21}$$

where $c'(s,k)$ *depends only on* s *and* k, *and* D^{α} *is the* α*th derivative with respect to* $z=(z_1,\ldots,z_k)$, α *being any nonnegative integral vector satisfying*

$$|\alpha| \leqslant s.$$

Proof. Write [see (7.1)]

$$g(u:z)=\sum_{r=1}^{s-3} \frac{u^r \chi_{r+2}(z)}{(r+2)!} = \sum_{r=1}^{s-3} u^r \sum_{|\nu|=r+2} \frac{\chi_\nu}{\nu!} z^\nu,$$

$$f(u:z)=\exp\{g(u:z)\}, \tag{9.22}$$

$$h(u:z)=f(u:z)-\sum_{r=0}^{s-3} u^r \tilde{P}_r(z:\{\chi_\nu\}) \qquad (u\in R^1, \quad z\in \mathbf{C}^k).$$

By definition of the polynomials $\tilde{P}_r$,

$$\frac{d^m}{du^m} h(u:z)\bigg|_{u=0} = 0 \qquad \text{for } m=0,1,\ldots,s-3,$$

$$\frac{d^{s-2}}{du^{s-2}} h(u:z) = \frac{d^{s-2}}{du^{s-2}} f(u:z),$$

when h and f are regarded as functions of the first argument u only. By Corollary 8.3,

$$|h(u:z)| \leqslant \frac{|u|^{s-2}}{(s-2)!} \sup\left\{\left|\frac{d^{s-2}}{da^{s-2}} f(a:z)\right| : 0 \leqslant |a| \leqslant |u|\right\}. \tag{9.23}$$

Now according to Lemma 9.3, $(d^{s-2}/du^{s-2})f(u:z)$ is a linear combination of terms like

$$f(u:z)\left(\frac{d}{du} g(u:z)\right)^{j_1} \cdots \left(\frac{d^{s-2}}{du^{s-2}} g(u:z)\right)^{j_{s-2}}, \tag{9.24}$$

where $j_1,\ldots,j_{s-2}$ are nonnegative integers satisfying

$$\sum_{m=1}^{s-2} m j_m = s-2. \tag{9.25}$$

If z satisfies

$$\|z\| \leqslant \left(8c(s,k)|u|\,\beta_s^{1/(s-2)}\right)^{-1}, \tag{9.26}$$

then, since [by (9.17)]

$$|\chi_\nu| \leqslant \beta_s^{(|\nu|-2)/(s-2)}, \tag{9.27}$$

one has

$$\begin{aligned}\left|\frac{d^m}{du^m} g(u:z)\right| &= \left|\sum_{r=m}^{s-3} r(r-1)\cdots(r-m+1)\left(\sum_{|\nu|=r+2}\frac{z^\nu}{\nu!}\chi_\nu\right)u^{r-m}\right|\\ &\leqslant \sum_{r=m}^{s-3} c_1(r,m,k)\,\beta_s^{r/(s-2)}|u|^{r-m}\cdot\|z\|^{r+2}\\ &\leqslant \sum_{r=m}^{s-3} c_1(r,m,k)\left(\beta_s^{1/(s-2)}|u|\|z\|\right)^{r-m}\beta_s^{m/(s-2)}\|z\|^{m+2}\\ &\leqslant c_2(s,m,k)\,\beta_s^{m/(s-2)}\|z\|^{m+2} \qquad (1\leqslant m\leqslant s-2),\end{aligned} \tag{9.28}$$

so that

$$\begin{aligned}&\left|\left(\frac{d}{du}g(u:z)\right)^{j_1}\cdots\left(\frac{d^{s-2}}{du^{s-2}}g(u:z)\right)^{j_{s-2}}\right|\\ &\qquad\leqslant c_3(s,k)\,\beta_s\left(\|z\|^s+\|z\|^{3(s-2)}\right).\end{aligned} \tag{9.29}$$

Also, if z satisfies (9.26), then

$$\begin{aligned}|g(u:z)| &\leqslant \sum_{r=1}^{s-3}\sum_{|\nu|=r+2}\frac{1}{\nu!}\beta_s^{r/(s-2)}\|z\|^{r+2}|u|^r\\ &\leqslant \|z\|^2\sum_{r=1}^{s-3}\sum_{|\nu|=r+2}\frac{1}{\nu!}\left(|u|\|z\|\,\beta_s^{1/(s-2)}\right)^r\\ &\leqslant \|z\|^2\sum_{r=1}^{s-3}\sum_{|\nu|=r+2}\frac{1}{\nu!}(8c(s,k))^{-r}\\ &\leqslant \frac{\|z\|^2}{(8c(s,k))}\sum_{r=1}^{s-3}\sum_{|\nu|=r+2}\frac{1}{\nu!}\\ &= \frac{\|z\|^2}{8},\end{aligned}$$

$$|f(u:z)| \leqslant \exp\left\{\tfrac{1}{8}\|z\|^2\right\}. \tag{9.30}$$

Hence, by (9.23), (9.24), (9.29), and (9.30),

$$|h(u:z)| \leqslant c_4(s,k)|u|^{s-2}\beta_s\left(\|z\|^s + \|z\|^{3(s-2)}\right)\exp\left\{\frac{\|z\|^2}{8}\right\} \qquad (9.31)$$

for all $u \neq 0$ and all z satisfying (9.26). One can now use Lemma 9.6 with r given by the right side of (9.26), $c = (3k^{1/2})^{-1}$, to obtain

$$\begin{aligned} |D^\alpha h(u:z)| &\leqslant \alpha!\left(\frac{\|z\|}{3k^{1/2}}\right)^{-|\alpha|} c_4(s,k)|u|^{s-2} \\ &\quad \times \beta_s\left[\left(\tfrac{4}{3}\|z\|\right)^s + \left(\tfrac{4}{3}\|z\|\right)^{3(s-2)}\right]\exp\left\{\tfrac{2}{9}\|z\|^2\right\} \\ &\leqslant c_5(s,k,\alpha)|u|^{s-2}\beta_s\left(\|z\|^{s-|\alpha|} + \|z\|^{3(s-2)-|\alpha|}\right)\exp\left\{\tfrac{2}{9}\|z\|^2\right\} \qquad (9.32) \end{aligned}$$

for all z satisfying (9.20), since the right side of (9.20) is smaller than $r/(1+k^{1/2}c) = \frac{3}{4}r$. Q.E.D.

If $V = I$, then by Lemmas 6.3, 6.2(iii), one has (since $\rho_2 = k$)

$$\begin{aligned} |\chi_\nu|^{1/(|\nu|-2)} &\leqslant \left(c_1(\nu)\rho_{|\nu|}\right)^{1/(|\nu|-2)} \\ &\leqslant c_1'(\nu,k)\rho_s^{1/(s-2)} \qquad (3 \leqslant |\nu| \leqslant s). \qquad (9.33) \end{aligned}$$

Taking these χ_ν's and $u = n^{-1/2}$ in Lemma 9.7, one has

LEMMA 9.8. *Let* s *be an integer not smaller than* 3. *Define*

$$c_6(s,k) = \left(11c(s,k)\max\left\{c_1'(\nu,k) : 3 \leqslant |\nu| \leqslant s\right\}\right)^{-1}. \qquad (9.34)$$

There exists a constant $c_7(s,k)$ *such that for all* t *in* R^k *satisfying*

$$\|t\| \leqslant c_6(s,k)n^{1/2}\rho_s^{-1/(s-2)}, \qquad (9.35)$$

one has, for every nonnegative integral vector α, $0 \leqslant |\alpha| \leqslant s$,

$$\begin{aligned} &\left| D^\alpha\left[\exp\left\{-\frac{\|t\|^2}{2} + \sum_{r=1}^{s-3}\frac{\chi_{r+2}(it)}{(r+2)!}n^{-r/2}\right\}\right.\right. \\ &\qquad \left.\left. - \exp\left\{-\frac{\|t\|^2}{2}\right\}\sum_{r=0}^{s-3} n^{-r/2}\tilde{P}_r(it:\{\chi_\nu\})\right]\right| \\ &\qquad\qquad \leqslant \frac{c_7(s,k)\rho_s}{n^{(s-2)/2}}\left(\|t\|^{s-|\alpha|} + \|t\|^{3(s-2)+|\alpha|}\right)\exp\left\{-\frac{\|t\|^2}{4}\right\}. \end{aligned}$$

Proof. The assertion follows from Lemma 9.7, inequality (9.34), and the following:

(i) $D^{\alpha}(\exp\{-\frac{\|t\|^2}{2}\}\cdot h(n^{-1/2}:it))$ may be expressed as a linear combination of terms like

$$\left(D^{\beta}h(n^{-1/2}:it)\right)\left(D^{\alpha-\beta}\exp\left\{-\frac{\|t\|^2}{2}\right\}\right) \qquad (0\leqslant \beta \leqslant \alpha);$$

(ii) also

$$\left|D^{\alpha-\beta}\exp\left\{-\frac{\|t\|^2}{2}\right\}\right| \leqslant c_7'(\alpha-\beta,k)(1+\|t\|^{|\alpha-\beta|})\exp\left\{-\frac{\|t\|^2}{2}\right\}.$$

Here h is the function defined in (9.22). Q.E.D.

We are now ready to prove the main theorem of this section. Before stating it, we define

$$d_n = \sup\left\{a>0:\langle t,Vt\rangle \leqslant a^2 \text{ implies } \left|\hat{G}_j\left(\frac{Bt}{n^{1/2}}\right)-1\right| \leqslant \tfrac{1}{2} \text{ for } 1\leqslant j \leqslant n\right\}. \tag{9.36}$$

THEOREM 9.9 *There exist constants* $c_8(s,k)$, $c_9(s,k)$ *such that, for all* t *in* R^k *satisfying*

$$\|t\| \leqslant d_n,\ \|t\| \leqslant c_8(s,k)n^{1/2}\eta_s^{-1/(s-2)}, \tag{9.37}$$

one has, for all nonnegative integral vectors α, $0\leqslant|\alpha|\leqslant s$,

$$\left|D^{\alpha}\left[\prod_{j=1}^{n}\hat{G}_j\left(\frac{Bt}{n^{1/2}}\right)-\exp\{-\tfrac{1}{2}\|t\|^2\}\sum_{r=0}^{s-3}n^{-r/2}\tilde{P}_r(iBt:\{\chi_\nu\})\right]\right|$$
$$\leqslant c_9(s,k)\eta_s n^{-(s-2)/2}\left[\|t\|^{s-|\alpha|}+\|t\|^{3(s-2)+|\alpha|}\right]\exp\{-\tfrac{1}{4}\|t\|^2\}. \tag{9.38}$$

Proof. First assume that $V=B=I$. In the given region logarithm of $\hat{G}_j(t/n^{1/2})$ is defined. Write

$$h_j(t)=\log\hat{G}_j\left(\frac{t}{n^{1/2}}\right)-\left\{-\frac{1}{2n}\|t\|^2+\sum_{r=1}^{s-3}\frac{\chi_{r+2,j}(it)}{(r+2)!}n^{-(r+2)/2}\right\},$$

$$h(t)=\sum_{j=1}^{n}h_j(t),$$

$$\psi(t)=-\tfrac{1}{2}\|t\|^2+\sum_{r=1}^{s-3}n^{-r/2}\chi_{r+2,j}(it)/(r+2)!$$

$$=\sum_{j=1}^{n}\log\hat{G}_j\left(\frac{t}{n^{1/2}}\right)-h(t). \tag{9.39}$$

We want to estimate

$$D^{\alpha}\left[\prod_{j=1}^{n}\hat{G}_j\left(\frac{t}{n^{1/2}}\right)-\exp\{\psi(t)\}\right]$$

$$=D^{\alpha}[(\exp\{h(t)\}-1)\exp\{\psi(t)\}]$$

$$=\sum_{0\leqslant\beta\leqslant\alpha}c_{10}(\alpha,\beta)\big(D^{\beta}\exp\{\psi(t)\}\big)\big[D^{\alpha-\beta}(\exp\{h(t)\}-1)\big]. \quad (9.40)$$

By relation (9.30),

$$|\psi(t)|\leqslant\frac{\|t\|^2}{2}+\frac{\|t\|^2}{8}=\tfrac{5}{8}\|t\|^2 \quad (9.41)$$

if t satisfies (9.37) [see (9.27)]. Also, using (9.33),

$$|(D^{\beta}\psi)(t)|=\left|D^{\beta}\left(\sum_{r=0}^{s-3}n^{-r/2}\sum_{|\nu|=r+2}\frac{\chi_\nu}{\nu!}t^{\nu}\right)\right|$$

$$=\left|\sum_{r=\max\{0,|\beta|-2\}}^{s-3}n^{-r/2}\sum_{\substack{|\nu|=r+2,\\ \nu\geqslant\beta}}\frac{\chi_\nu}{(\nu-\beta)!}t^{\nu-\beta}\right|$$

$$\leqslant\sum_{r=\max\{0,|\beta|-2\}}^{s-3}n^{-r/2}\sum_{\substack{|\nu|=r+2\\ \nu\geqslant\beta}}c_{11}(\nu,k)\rho_s^{r/(s-2)}\|t\|^{r+2-|\beta|}$$

$$\leqslant\sum_{r=\max\{0,|\beta|-2\}}^{s-3}\left(n^{-1/2}c_8(s,k)n^{1/2}\rho_s^{-1/(s-2)}\right)^{r}\rho_s^{r/(s-2)}\|t\|^{2-|\beta|}$$

$$\leqslant c_{12}(s,k,\beta)\|t\|^{2-|\beta|} \qquad \big(\beta>0,\quad t\in R^k-\{0\}\big). \quad (9.42)$$

Let $j_1,\dots,j_r$ be nonnegative integers and $\beta_1,\dots,\beta_r$ nonnegative integral vectors such that

$$\sum_{i=1}^{r}j_i\beta_i=\beta, \qquad \beta_i>0 \qquad (1\leqslant i\leqslant r).$$

By (9.42) one has [using (9.13)]

$$|(D^{\beta_1}\psi(t))^{j_1}\cdots(D^{\beta_r}\psi(t))^{j_r}| \leqslant c_{13}(s,k)\|t\|^{\sum_{i=1}^{r} j_i(2-|\beta_i|)}$$
$$\leqslant c_{13}(s,k)(\|t\|^{2-|\beta|}+\|t\|^{|\beta|}) \qquad (t\neq 0). \tag{9.43}$$

It now follows from Lemma 9.3 and inequality (9.30) that

$$|D^{\beta}\exp\{\psi(t)\}| \leqslant c_{14}(s,k)(\|t\|^{2-|\beta|}+\|t\|^{|\beta|})|\exp\{\psi(t)\}|$$
$$\leqslant c_{14}(s,k)(\|t\|^{2-|\beta|}+\|t\|^{|\beta|})\exp\{-\tfrac{1}{2}\|t\|^2+\tfrac{1}{8}\|t\|^2\}$$
$$= c_{14}(s,k)(\|t\|^{2-|\beta|}+\|t\|^{|\beta|})\exp\{-\tfrac{3}{8}\|t\|^2\} \quad (t\neq 0) \tag{9.44}$$

if t satisfies (9.37).

Next note that for any nonnegative integral vector β satisfying $0\leqslant|\beta|\leqslant s$,

$$D^{\beta'}(D^{\beta}h_j)(0)=0 \qquad [0\leqslant|\beta'|\leqslant s-|\beta|-1].$$

Hence, applying Corollary 8.3 to $g\equiv D^{\beta}h_j$, one gets

$$|D^{\beta}h_j(t)| \leqslant \sum_{|\beta'|=s-|\beta|}\frac{|t^{\beta'}|}{\beta'!}\sup\{|(D^{\beta'}g)(ut)|: 0\leqslant u\leqslant 1\}. \tag{9.45}$$

But if $|\beta'|=s-|\beta|$, then by Lemma 9.4 [note that (9.4) holds for $\|t\|\leqslant d_n$],

$$|(D^{\beta'}g)(ut)| = |(D^{\beta+\beta'}h_j)(ut)| = \left|(D^{\beta+\beta'}\log\hat{G}_j)\left(\frac{ut}{n^{1/2}}\right)\right| n^{-s/2}$$
$$\leqslant n^{-s/2}c_2(s)\rho_{s,j},$$

so that summing over $j=1,\dots,n$, in (9.45) one has

$$|D^{\beta}h(t)| \leqslant \frac{c_2(s)\rho_s}{n^{(s-2)/2}}\|t\|^{s-|\beta|}\cdot\left(\sum_{|\beta'|=s-|\beta|}\frac{1}{\beta'!}\right). \tag{9.46}$$

In particular, taking $\beta=0$ in (9.46) yields

$$|h(t)| \leqslant c_2(s)\left(\sum_{|\beta'|=s}\frac{1}{\beta'!}\right)\frac{\rho_s}{n^{(s-2)/2}}\|t\|^s \leqslant \frac{\|t\|^2}{8} \tag{9.47}$$

for an appropriate choice of $c_8(s,k)$. If $\alpha-\beta=0$, then [using both inequalities in (9.47)]

$$|D^{\alpha-\beta}(\exp\{h(t)\}-1)|=|\exp\{h(t)\}-1|\leqslant|h(t)|\exp\{|h(t)|\}$$
$$\leqslant c_{15}(s,k)\frac{\rho_s}{n^{(s-2)/2}}\|t\|^s\exp\left\{\frac{\|t\|^2}{8}\right\}. \tag{9.48}$$

If $\alpha>\beta$, then

$$D^{\alpha-\beta}(\exp\{h(t)\}-1)=D^{\alpha-\beta}(\exp\{h(t)\}), \tag{9.49}$$

which is a linear combination of terms such as

$$\left(D^{\beta_1}h(t)\right)^{j_1}\cdots\left(D^{\beta_r}h(t)\right)^{j_r}\exp\{h(t)\},$$

where $\Sigma_{i=1}^r j_i\beta_i=\alpha-\beta$. By (9.46) [and (9.13)],

$$|\left(D^{\beta_1}h(t)\right)^{j_1}\cdots\left(D^{\beta_r}h(t)\right)^{j_r}|\leqslant\frac{c_{16}(s,k)\rho_s}{n^{(s-2)/2}}\left[\frac{\|t\|}{n^{1/2}}\rho_s^{1/(s-2)}\right]^{(s-2)\left(\sum_1^r j_i-1\right)}$$
$$\cdot\|t\|^{s-2+2\sum_1^r j_i-|\alpha-\beta|}$$

$$\leqslant\frac{c_{17}(s,k)}{n^{(s-2)/2}}\rho_s\left(\|t\|^{s-|\alpha-\beta|}+\|t\|^{s+|\alpha-\beta|-2}\right). \tag{9.50}$$

Hence if $\alpha>\beta$, then

$$\left|D^{\alpha-\beta}(\exp\{h(t)\}-1)\right|\leqslant\frac{c_{18}(s,k)}{n^{(s-2)/2}}\rho_s\left(\|t\|^{s-|\alpha-\beta|}\right.$$
$$\left.+\|t\|^{s+|\alpha-\beta|+2}\right)\exp\left\{\frac{\|t\|^2}{8}\right\}. \tag{9.51}$$

Using (9.44), (9.48), and (9.51) in (9.40), one obtains

$$\left|D^{\alpha}\left[\prod_{j=1}^{n}\hat{G}_j\left(\frac{t}{n^{1/2}}\right)-\exp\left\{-\frac{\|t\|^2}{2}+\sum_{r=1}^{s-3}n^{-r/2}\frac{\chi_{r+2}(it)}{(r+2)!}\right\}\right]\right|$$
$$\leqslant\frac{c_{19}(s,k)\rho_s}{n^{(s-2)/2}}\left(\|t\|^{s-|\alpha|}+\|t\|^{s+|\alpha|+2}\right)\exp\left\{-\frac{\|t\|^2}{4}\right\}. \tag{9.52}$$

Now use Lemma 9.8 to complete the proof when $V=B=I$. If $V\neq I$, look at random vectors $B\mathbf{X}_1,\ldots,B\mathbf{X}_n$ and observe that

$$\prod_{j=1}^{n}\hat{G}_j\left(\frac{Bt}{n^{1/2}}\right)$$

is the characteristic function of $\mathbf{Z}_n=n^{-1/2}\mathbf{Y}_n$, where $\mathbf{Y}_n=B(\mathbf{X}_1+\cdots+\mathbf{X}_n)$. Also, if the $(r+2)$th cumulant of the random variable $\langle t,\mathbf{X}_j\rangle$ is denoted by $\chi_{r+2,j}(t)$, then the corresponding cumulant of $\langle t,B\mathbf{X}_j\rangle=\langle Bt,\mathbf{X}_j\rangle$ is $\chi_{r+2,j}(\mathrm{B}t)$. Q.E.D.

The following theorems are easy consequences of Theorem 9.9.

THEOREM 9.10. *Let* G *be a probability measure on* R^k *with zero mean, positive-definite covariance matrix* V, *and finite* s*th absolute moment for some integer* s *not smaller than* 3. *Then there exist two positive constants* $c_{20}(s,k)$, $c_{21}(s,k)$ *such that for all* t *in* R^k *satisfying*

$$\|t\|\leqslant c_{20}(s,k)n^{1/2}\eta_s^{-1/(s-2)}$$

one has, for all nonnegative integral vectors α, $0\leqslant|\alpha|\leqslant s$,

$$\left|D^{\alpha}\left[\hat{G}^n\left(\frac{Bt}{n^{1/2}}\right)-\exp\{-\tfrac{1}{2}\|t\|^2\}\sum_{r=0}^{s-3}n^{-r/2}\tilde{P}_r(iBt:\{\chi_\nu\})\right]\right|$$
$$\leqslant\frac{c_{21}(s,k)\eta_s}{n^{(s-2)/2}}\left[\|t\|^{s-|\alpha|}+\|t\|^{3(s-2)+|\alpha|}\right]\exp\left\{-\frac{\|t\|^2}{4}\right\}.$$

Here B *is the symmetric positive-definite matrix satisfying* (9.7),

$$\eta_s=\int\|Bx\|^sG(dx),$$

and χ_ν *is the* ν*th cumulant of* G.

Proof. Note that if t satisfies

$$\|t\|\leqslant n^{1/2}\eta_s^{-1/(s-2)},$$

then

$$\left|\hat{G}\left(\frac{Bt}{n^{1/2}}\right)-1\right|\leqslant\frac{\langle Bt,VBt\rangle}{2n}=\frac{\|t\|^2}{2n}\leqslant\tfrac{1}{2}\eta_s^{-2/(s-2)}\leqslant\tfrac{1}{2}k^{-s/(s-2)}\leqslant\tfrac{1}{2},$$

because of the relations

$$\int \|Bx\|^2 G(dx) = k \leqslant \left[\int \|Bx\|^s G(dx)\right]^{2/s}.$$

Q.E.D.

THEOREM 9.11. *Let* $\mathbf{X}_1, \ldots, \mathbf{X}_n$ *be* n *independent random vectors in* R^k *with zero means, covariance matrices* $\mathrm{V}_1, \ldots, \mathrm{V}_n$ *(at least one of which is nonsingular), and finite* s*th absolute moments for some integer* s *not smaller than* 3. *Then there exist two positive constants* $\mathrm{c}_{22}(\mathrm{s},\mathrm{k})$, $\mathrm{c}_{23}(\mathrm{s},\mathrm{k})$ *such that for all* t *in* R^k *satisfying*

$$\|t\| \leqslant c_{22}(s,k)\left(\frac{n^{1/2}}{\eta_s^{1/(s-2)}}\right)^{(s-2)/s},$$

one has for all α, $0 \leqslant |\alpha| \leqslant s$,

$$\left|D^{\alpha}\left[\prod_{j=1}^{n} E\left(\exp\left\{\left\langle \frac{iBt}{n^{1/2}}, \mathbf{X}_j\right\rangle\right\}\right) - \exp\left\{-\frac{\|t\|^2}{2}\right\}\sum_{r=0}^{s-3} n^{-r/2}\tilde{P}_r(iBt : \{\chi_\nu\})\right]\right|$$
$$\leqslant \frac{c_{23}(s,k)\eta_s}{n^{(s-2)/2}}\left[\|t\|^{s-|\alpha|} + \|t\|^{3(s-2)+|\alpha|}\right]\exp\left\{-\frac{\|t\|^2}{4}\right\},$$

where the notation is as in Theorem 9.9.

Proof. As in the proof of Theorem 9.9 (see the concluding observations), it is enough to prove this theorem for $B = V = I$. In this case, for all t satisfying

$$\|t\| \leqslant \left[\frac{n^{1/2}}{\eta_s^{1/(s-2)}}\right]^{(s-2)/s},$$

one has

$$\left|\hat{G}_j\left(\frac{t}{n^{1/2}}\right) - 1\right| \leqslant \frac{E\langle t, \mathbf{X}_j\rangle^2}{2n} \leqslant \frac{\|t\|^2 E\|\mathbf{X}_j\|^2}{2n}$$
$$\leqslant \frac{\|t\|^2 (E\|\mathbf{X}_j\|^s)^{2/s}}{2n} \leqslant \frac{\|t\|^2}{2n}(n \cdot \eta_s)^{2/s} \leqslant \tfrac{1}{2}.$$

Q.E.D.

Before concluding this section, we state another theorem that may be easily proved along the lines of Theorem 9.9, by taking one more term in the Taylor expansion.

THEOREM 9.12. *Under the hypothesis of Theorem* 9.10,

$$\left| D^{\alpha}\left[\hat{G}^{n}\left(\frac{Bt}{n^{1/2}}\right) - \exp\left\{-\frac{\|t\|^2}{2}\right\} \sum_{r=0}^{s-2} n^{-r/2}\tilde{P}_r(iBt:\{\chi_\nu\})\right]\right|$$

$$\leqslant \frac{\delta(n)}{n^{(s-2)/2}}\left[\|t\|^{s-|\alpha|} + \|t\|^{3(s-2)+|\alpha|}\right]\exp\left\{-\frac{\|t\|^2}{4}\right\} \tag{9.53}$$

where $\delta(n)\to 0$ *as* $n\to\infty$. *In fact*, (9.53) *holds even for* $s \geqslant 2$, $|\alpha| \leqslant s$.

Remark. Under the hypothesis of Theorem 9.11, one may also prove that if

$$\|t\| \leqslant c_{22}(s,k)\Lambda^{-1/2}\left(\frac{n^{1/2}}{\eta_s^{1/(s-2)}}\right)^{(s-2)/s}, \tag{9.54}$$

where Λ is the largest eigenvalue of V, then

$$\left| D^{\alpha}\left[\prod_{j=1}^{n} E\left(\exp\left\{\langle \frac{it}{n^{1/2}}, \mathbf{X}_j\rangle\right\}\right) - \exp\{-\tfrac{1}{2}\langle t, Vt\rangle\}\cdot \sum_{r=0}^{s-3} n^{-r/2}\tilde{P}_r(it:\{\chi_\nu\})\right]\right|$$

$$\leqslant \frac{c_{24}(s,k)\Lambda^{|\alpha|/2}\eta_s}{n^{(s-2)/2}}\left[\langle t, Vt\rangle^{(s-|\alpha|)/2}\right.$$

$$\left. + \langle t, Vt\rangle^{(3(s-2)+|\alpha|)/2}\right]\exp\{-\tfrac{1}{4}\langle t, Vt\rangle\}. \tag{9.55}$$

For $\alpha=0$ this follows by replacing Bt by t in Theorem 9.11. The general case follows by induction on $|\alpha|$. Note that the derivative D^{α} is with respect to t. Completely analogous modifications hold in the statements of Theorems 9.10 and 9.12.

10. A CLASS OF KERNELS

For $a>0$ let $U_{[-a,a]}$ denote the probability measure on R^1 with density

$$u_{[-a,a]}(x) = \frac{1}{2a} \qquad \text{for } -a \leqslant x \leqslant a$$

$$= 0 \qquad \text{for } |x| > a. \tag{10.1}$$

The measure $U_{[-a,a]}$ is called the *uniform distribution* on $[-a,a]$. One has

$$\hat{U}_{[-a,a]}(t)=\frac{1}{2a}\int_{-a}^{a}\cos tx\,dx=\frac{\sin at}{at}\qquad (t\in R^1). \tag{10.2}$$

The probability measure

$$T_a\equiv U^{*2}_{[-a/2,a/2]} \tag{10.3}$$

is called the *triangular distribution on* $[-a,a]$. It is easy to show that its density is

$$\begin{aligned} t_a(x)&=\frac{1}{a}\left(1-\frac{|x|}{a}\right) &&\text{for } |x|\leqslant a\\ &=0 &&\text{for } |x|>a, \end{aligned} \tag{10.4}$$

and that

$$\hat{T}_a(t)=\left(\frac{\sin\frac{at}{2}}{\frac{at}{2}}\right)^2\qquad (t\in R^1). \tag{10.5}$$

One can write

$$c(m)=\left(\int_{R^1}\left|\frac{\sin x}{x}\right|^m dx\right)^{-1}\qquad (m=2,\dots). \tag{10.6}$$

For $a>0$ and integer $m\geqslant 2$ let $G_{a,m}$ denote the probability measure on R^1 with density

$$g_{a,m}(x)=ac(m)\left|\frac{\sin ax}{ax}\right|^m\qquad (x\in R^1). \tag{10.7}$$

It follows from (10.2) that for even integers $m\geqslant 2$

$$\left|\frac{\sin at}{at}\right|^m=\widehat{U^{*m}_{[-a,a]}}(t)\qquad (t\in R^1), \tag{10.8}$$

so that by the Fourier inversion theorem [Theorem 4.1(iv)]

$$\begin{aligned} \hat{G}_{a,m}(t)&=2\pi ac(m)u^{*m}_{[-a,a]}(t)\qquad (t\in R^1),\\ &=0 \text{ if } |t|\geqslant ma. \end{aligned} \tag{10.9}$$

Let $\mathbf{Z}_1,\dots,\mathbf{Z}_k$ be independent random variables each with distribution $G_{1/2r,2r}$ and let $\mathbf{Z}=(\mathbf{Z}_1,\dots,\mathbf{Z}_k)$. Then for each $d>0$

$$\operatorname{Prob}(\|\mathbf{Z}\|\geqslant d)\leqslant k\operatorname{Prob}\left(|\mathbf{Z}_1|\geqslant\frac{d}{\sqrt{k}}\right). \tag{10.10}$$

Thus for any α in $(0,1)$, there exists a constant d depending only on k such that

$$\operatorname{Prob}(\|\mathbf{Z}\|\geqslant d)\leqslant 1-\alpha. \tag{10.11}$$

Note that the characteristic function of $\mathbf{Z}$ vanishes outside $[-1,1]^k$. Now let K_1 denote the distribution of $\mathbf{Z}/d$; then

$$\begin{aligned} &K_1(\{x:\|x\|\geqslant 1\})\leqslant 1-\alpha,\\ &\hat{K}_1(t)=0 \qquad \text{if } t\not\in[-d,d]^k. \end{aligned} \tag{10.12}$$

One thus has

THEOREM 10.1. *Let* r *be any given positive integer and let* $\alpha\in(0,1)$. *There exists a probability measure* K_1 *on* R^k *such that*

(i) $K_1(\{x:\|x\|\geqslant 1\})\leqslant 1-\alpha$
(ii) *For all nonnegative integral vectors* β *satisfying*

$$0\leqslant|\beta|\leqslant 2r-2,$$

one has

$$\int|x^\beta|K_1(dx)<\infty.$$

(iii) $\hat{K}_1(t)=0$ *for* $t\not\in[-d,d]^k$ *where* d *is a constant depending only on* α *and* r.

We also need as kernels certain probability measures having compact support and fast-decreasing characteristic functions. To this end we prove

THEOREM 10.2. *Let* u *be a real-valued, nonnegative, nonincreasing function on* $[1,\infty)$ *such that*

$$\int_1^\infty\frac{u(t)}{t}\,dt<\infty. \tag{10.13}$$

e.g. $u(t)=\frac{1}{\log^2 t}$

$\frac{u(t)}{t}=-\frac{d}{dt}\frac{1}{\log t}$

Then for every $l>0$ *there exists a probability measure* K *on* R^1 *satisfying*

(i) *Support of* $K\subset[-l,l]$
(ii) $|\hat{K}(t)|=0(\exp\{-|t|u(|t|)\})\qquad(|t|\to\infty).$

$e^{-|t|u(t)}=e^{-\frac{|t|}{\log^2 t}}$

Proof. Define a nonincreasing sequence of nonnegative numbers $\{a_r: r \geqslant 1\}$ by

$$a_r = \frac{eu(r_0)}{r_0}, \qquad 1 \leqslant r \leqslant r_0,$$

$$= \frac{eu(r)}{r}, \qquad r > r_0,$$

where r_0 is a positive integer chosen to satisfy

$$\sum_{r=1}^{\infty} a_r \equiv eu(r_0) + e \sum_{r=r_0+1}^{\infty} \frac{u(r)}{r} \leqslant l.$$

Let the probability measure K be defined as the infinite convolution

$$K = U_{[-a_1, a_1]} * U_{[-a_2, a_2]} * \cdots .$$

Clearly, support of K is

$$\left[-\sum_{r=1}^{\infty} a_r, \sum_{r=1}^{\infty} a_r \right] \subset [-l, l].$$

Also,

$$\hat{K}(t) = \prod_{r=1}^{\infty} \frac{\sin a_r t}{a_r t} \qquad (t \in R^1).$$

For every positive integer $s \geqslant r_0$, therefore,

$$|\hat{K}(t)| \leqslant \prod_{r=1}^{s} \frac{1}{|a_r t|} \leqslant \frac{1}{|a_s t|^s} = \left(\frac{s}{eu(s)|t|} \right)^s.$$

Now choose $s = [\,|t|u(|t|)\,]$, the *integer part* of $|t|u(|t|)$. We assume without loss of generality that

$$|t|u(|t|) \to \infty \qquad \text{as } |t| \to \infty.$$

[For otherwise one may replace $u(t)$ by $u_1(t) = u(t) + (t+1)^{-1/2}$ and note that (10.13) holds for u if and only if it holds for u_1, and that (ii) holds for u if it holds for u_1 (since $u_1 > u$).] Thus for sufficiently large $|t|$, one has

$r_0 \leqslant s \leqslant |t|$ so that

$$|\hat{K}(t)| \leqslant \left(\frac{[|t|u(|t|)]}{eu(s)|t|} \right)^{[|t|u(|t|)]}$$

$$\leqslant \left(\frac{|t|u(|t|)}{eu(|t|)|t|} \right)^{[|t|u(|t|)]} \leqslant e^{-|t|u(|t|)+1}.$$

Q.E.D.

Remark. The above result is due to Ingham [1], who has also shown that the above theorem provides the best possible rate of growth for the characteristic function of a probability measure with compact support. To be precise, he has shown that if u is a real-valued, nonnegative, nonincreasing function on $[1,\infty)$ such that $\int_1^\infty [u(t)/t]\,dt = \infty$, then there does not exist any probability measure with compact support whose characteristic function is $0(\exp\{-|t|u(|t|)\})$ as $|t| \to \infty$. Note that this also follows from the following result of Wiener and Paley†:

If $f \in L^2(a,\infty)$ then

$$\int_{-\infty}^{\infty} \frac{|\log|\hat{f}(t)||}{1+t^2}\,dt < \infty. \tag{10.14}$$

THEOREM 10.3. *Given a nonnegative, nonincreasing function* u *on* $[1,\infty)$ *satisfying* (10.13), *and a nonnegative integer* s, *there exists a probability measure* K *on* R^k *with support contained in the unit ball* $S = \{x : \|x\| \leqslant 1\}$ *such that*

$$|D^\alpha \hat{K}(t)| \leqslant c(s,k,u) \exp\left\{ -\sum_{i=1}^{k} |t_i| u(|t_i|) \right\} \tag{10.15}$$

for all nonnegative integral vectors $\alpha = (\alpha_1, \ldots, \alpha_k)$ *satisfying* $0 \leqslant |\alpha| \leqslant s$. *Here* c(s, k, u) *is a positive constant depending only on* s, k, *and* u.

Proof. By Theorem 10.2 there exists a measure M on R^1 satisfying

(i) $M(\{x : x \in R^1, |x| \leqslant k^{-1/2}(s+1)^{-1}\}) = 1,$
(ii) $|\hat{M}(t)| \leqslant c_1(s,k,u) \exp\{-|t|u(|t|)\} \qquad (t \in R^1).$

Define $M_1 = M^{*(s+1)}$. Then

$$M_1(\{x : x \in R^1, |x| \leqslant k^{-1/2}\}) = 1$$

$$\hat{M}_1(t) = (\hat{M}(t))^{s+1} \qquad (t \in R^1),$$

---

†See Paley, E. A. C. and Wiener, N. [1], Theorem XII, pp. 16, 17.

so that for $0 \leqslant r \leqslant s$,

$$\begin{aligned}\left|\frac{d^r}{dt^r}\hat{M}_1(t)\right| &\leqslant (s+1)^r E|\mathbf{X}|^r |\hat{M}(t)|^{s+1-r}\\ &\leqslant (s+1)^r k^{-r/2}(s+1)^{-r}|\hat{M}(t)|\\ &\leqslant c_2(s,k,u)\exp\{-|t|u(|t|)\},\end{aligned}$$

where $\mathbf{X}$ denotes a random variable with distribution M. Now define K on R^k as the product measure $M_1 \times M_1 \times \cdots \times M_1$. Then

$$\begin{aligned}&K\big(\{x : x \in R^k, \|x\| \leqslant 1\}\big)\\ &\qquad \geqslant K\big(\{x : x = (x_1,\dots,x_k), |x_i| \leqslant k^{-1/2} \text{ for } 1 \leqslant i \leqslant k\}\big)\\ &\qquad = \big[M_1(\{x : x \in R^1, |x| \leqslant k^{-1/2}\})\big]^k = 1,\end{aligned}$$

and

$$\begin{aligned}|D^\alpha \hat{K}(t)| &= \left|\frac{d^{\alpha_1}}{dt_1^{\alpha_1}}\hat{M}_1(t_1)\right| \cdots \left|\frac{d^{\alpha_k}}{dt_k^{\alpha_k}}\hat{M}_1(t_k)\right|\\ &\leqslant c_2^k(s,k,u)\exp\left\{-\sum_{i=1}^{k}|t_i|u(|t_i|)\right\} \qquad (t=(t_1,\dots,t_k)\in R^k),\end{aligned}$$

for $0 \leqslant |\alpha| \leqslant s$. Q.E.D.

COROLLARY 10.4. *For any positive integer* s, *there exists a probability measure* K *on* R^k *satisfying*

(i) $K(\{x : \|x\| \leqslant 1\}) = 1$,

(ii) *for all nonnegative integral vectors* α, $0 \leqslant |\alpha| \leqslant s$,

$$|D^\alpha \hat{K}(t)| \leqslant c(s,k)\exp\{-\|t\|^{1/2}\} \qquad (t \in R^k).$$

Proof. In Theorem 10.3 take $u(t) = t^{-1/2}$ on $[1,\infty)$ and note $\Sigma_{i=1}^k |t_i|^{1/2} \geqslant \|t\|^{1/2}$, for all $t = (t_1,\dots,t_k) \in R^k$. Q.E.D.

NOTES

Sections 4 and 5. The material on the Fourier transform and the Fourier–Stieltjes transform reviewed here is fairly standard and may be found in Cramér [3, 4], Chung [1], Feller [3], Katznelson [1], and Stein and Weiss [1].

Section 6. The best reference for the inequalities of this section is Hardy, Littlewood, and Polya [1].

Sections 7–9. The idea of expanding the distribution function F_n of the normalized sum of n independent random variables as $\Sigma_r n^{-r/2} P_r(-\Phi : \{\chi_\nu\})$ appears for the first time in Chebyshev [1]; later it was investigated independently by Edgeworth [1]. However the asymptotic expansion of the characteristic function of F_n was obtained by Cramér [1, 3] (Chapter VII), who used it to give the first rigorous derivation of the asymptotic expansion of F_n under the so-called *Cramér's condition* (20.1) (see Chapter IV). Theorems 8.4–8.6 are refinements (and extensions to R^k) of results of Cramér [3] (Chapter VII); analogs of Theorems 8.7 and 8 .9 for $k=1$ were obtained by Laipounov [2]. There are many such refinements available in the literature, for example, Esseen [1], Gnedenko and Kolmogorov [1], and Petrov [1] in one dimension; Rao [1], Bikjalis [3, 6], and Bhattacharya [3] in multidimension. Theorems 9.9–9.12 are generalizations of analogous results of Bikjalis [6].

Section 10. Kernels such as K_1 (Theorem 10.1) were used in one dimension by Berry [1] and Esseen [1]. Theorem 10.2 is due to Ingham [1].

CHAPTER 3

Bounds for Errors of Normal Approximation

Our goal in the present chapter is to estimate $|\int f dQ_n - \int f d\Phi|$ for a large class of functions f on R^k; here Q_n is the distribution of the normalized sum of n independent random vectors and Φ is the standard normal distribution on R^k. For bounded f, the error bound is computed in terms of the average modulus of oscillation of f with respect to Φ, or the supremum of this average over all translates of f. This is entirely appropriate in view of the characterization of Φ-uniformity classes proved in Chapter 1; also, in most cases of practical importance, these moduli can be estimated efficiently. The main tools used for obtaining these error bounds are expansions of the characteristic function $\hat{Q}_n$ and its derivatives as derived in Chapter 2, together with some smoothing inequalities proved in Section 11. In Section 12 the classical Berry–Esseen bound is obtained with an estimation of the constant involved. Section 13 is devoted to estimations of $|\int f dQ_n - \int f d\Phi|$ for bounded f under the simplifying assumption that fourth moments are finite; the proofs here are not only simpler than those of the general results of Sections 15–17, but also yield numerical values for constants involved in the bounds. In Section 14 we obtain truncation estimates that enable us to derive the main results of Section 15 on rates of convergence for unbounded f's under the assumption of finiteness of third moments. After a short section showing how to deal with different normalizations of the sum of n independent random vectors, in Section 17 we consider a number of important applications of the theorems in Section 15. A final section deals with rates of convergence under the sole assumption of finiteness of second moments.

To facilitate comprehension we briefly sketch the main ideas underlying the rather long route leading to the main results. For the sake of simplicity assume that $\{\mathbf{X}_n: n \geqslant 1\}$ is a sequence of independent and identically distributed (i.i.d.) random vectors with common distribution Q_1. Suppose that Q_1 has mean zero, covariance I (identity matrix), and a finite third absolute moment ρ_3. Let Q_n denote the distribution of $n^{-1/2}(\mathbf{X}_1 + \cdots + \mathbf{X}_n)$. If Q_1 has an integrable characteristic function (c.f.) $\hat{Q}_1$, then the c.f. $\hat{Q}_n$ of Q_n is integrable for all n. One can then use Fourier inversion to estimate the density h_n of the signed measure $Q_n - \Phi$ as

$$\|h_n\|_\infty \equiv \sup_{y \in R^k} |h_n(y)| \leqslant (2\pi)^{-k} \|\hat{Q}_n - \hat{\Phi}\|_1. \tag{1}$$

The fairly precise estimates of $\hat{Q}_n - \hat{\Phi}$ of Chapter 2 (e.g., Theorem 8.4) yield $\|h_n\| = 0(n^{-1/2})$. Since such a uniform estimate of h_n cannot be integrated over the unbounded domain R^k, one may estimate the variation norm $\|Q_n - \Phi\| = \|h_n\|_1$ by estimating the integral of $|h_n|$ over a sphere S of radius $0(\log^{1/2} n)$ as $\|h_n\|_\infty \cdot \mathrm{vol}(S)$, and the integral over the complement of S by the classical Berry–Esseen theorem and a Chebyshev-type inequality. To avoid the loss in precision arising from the factor $\mathrm{vol}(S) = 0(\log^{k/2} n)$, assume that Q_1 has a finite fourth moment and apply Theorem 8.5, adding one term $[n^{-1/2} P_1(-\Phi: \{\chi_\nu\})]$ to Φ and later subtracting the contribution from this term, which is of the order $n^{-1/2}$. In the general case (i.e., when $\hat{Q}_1$ may not be integrable) we smoothen Q_1 by convolving it with a smooth kernel K_ε (a probability measure with an integrable c.f., which for small ε assigns most of its mass near zero) and apply the above argument to $(Q_n - \Phi) * K_\varepsilon$, with a proper choice of ε (depending on n). The smoothing inequalities of Section 11 enable one to estimate the perturbation due to this convolution by K_ε, and one arrives at Theorems 13.2 and 13.3, which express bounds for $|\int f\, d(Q_n - \Phi)|$ in terms of the average moduli of oscillation $\bar{\omega}_f$ or ω_f^* and the range $\omega_f(R^k)$ of f. To estimate $|\int f\, d(Q_n - \Phi)|$ for unbounded f and at the same time relax the assumption of finiteness of fourth moments (in case of bounded f) we compare the measures $\|x\|^r Q_n(dx)$ and $\|x\|^r \Phi(dx)$ for nonnegative integers r in the same manner in which Q_n and Φ are compared above. Because of the fact that for odd integers r $\|x\|^r Q_n(dx)$ does not have Fourier–Stieltjes transforms as well behaved as those for even r, we replace r by r_0, where $r_0 = r+1$ if r is odd and $r_0 = r$ if r is even. As above, we smoothen $\nu_{r_0} \equiv \|x\|^{r_0}(Q_n - \Phi)(dx)$ as $\nu_{r_0} * K_\varepsilon$ and apply Lemma 11.6 to the density g of $\nu_{r_0} * K_\varepsilon$, thus obtaining

$$\|\nu_{r_0} * K_\varepsilon\| = \|g\|_1 \leqslant c(k) \max_{|\beta| = 0, k+1} \|D^\beta \hat{g}\|_1, \tag{2}$$

where $\hat{g}$ is the Fourier transform of g. The Fourier–Stieltjes transform of ν_{r_0} is estimated by Theorems 9.9–9.12 and a sharp estimate of $\|\nu_{r_0} * K_\varepsilon\|$ is obtained provided $\int \|x\|^{r_0+k+1} Q_1(dx)$ is finite (which ensures the existence of $D^{\beta}\hat{g}$ for $|\beta|=k+1$). To relax this last hypothesis, which is rather restrictive, one resorts to a truncation of the random vectors $\{\mathbf{X}_n: n \geqslant 1\}$ and applies the above procedure to these truncated vectors. The various lemmas in Section 14 allow one to take care of the perturbation due to truncation. As in the case $r_0=0$, for final accounting (i.e., to estimate the effect of smoothing by K_ε) one uses the smoothing inequalities of Section 11. The main theorems of Section 15 are obtained in this manner. A further truncation enables one to obtain corresponding analogs when only the finiteness of absolute moments of order $2+\delta$ is assumed for some δ, $0 \leqslant \delta < 1$, thus yielding generalizations and refinements of the classical one-dimensional theorems of Liapounov and Lindeberg.

11. SMOOTHING INEQUALITIES

Lemmas 11.1 and 11.4 show how the difference $\mu-\nu$ between a finite measure μ and a finite signed measure ν is perturbed by convolution with a probability measure K_ε that concentrates (for small ε) most of its mass near zero.

Let f be a real-valued, Borel-measurable function on R^k. Recall that in Chapter 1 we defined the following:

$$\omega_f(A)=\sup\{|f(x)-f(y)|: x,y \in A\} \qquad (A \subset R^k), \tag{11.1}$$

$$\omega_f(x:\varepsilon)=\omega_f(B(x:\varepsilon)) \qquad (x \in R^k, \quad \varepsilon>0).$$

Also define

$$M_f(x:\varepsilon)=\sup\{f(y): y \in B(x:\varepsilon)\}, \tag{11.2}$$

$$m_f(x:\varepsilon)=\inf\{f(y): y \in B(x:\varepsilon)\} \qquad (x \in R^k, \quad \varepsilon>0).$$

Note that

$$\omega_f(x:\varepsilon)=M_f(x:\varepsilon)-m_f(x:\varepsilon) \qquad (x \in R^k, \quad \varepsilon>0). \tag{11.3}$$

The functions $M_f(\cdot:\varepsilon)$, $m_f(\cdot:\varepsilon)$ are lower and upper semicontinuous, respectively, for every real-valued function f that is bounded on each compact subset of R^k. Also, $\omega_f(\cdot:\varepsilon)$ is lower semicontinuous. These follow from

$$\{x: M_f(x:\varepsilon)>c\} = \bigcup_x \{B(x:\varepsilon): f(x)>c\} \qquad (c \in R^1),$$

$$m_f(x:\varepsilon)=-M_{-f}(x:\varepsilon) \qquad (x \in R^k, \quad \varepsilon>0). \tag{11.4}$$

In particular, it follows that $M_f(\cdot:\varepsilon), m_f(\cdot:\varepsilon), \omega_f(\cdot:\varepsilon)$ are Borel-measurable for every real-valued function f on R^k that is bounded on compacts. Recall that the translate f_y of f by $y(\in R^k)$ is defined by

$$f_y(x)=f(x+y) \qquad (x\in R^k). \tag{11.5}$$

LEMMA 11.1 *Let μ be a finite measure and ν a finite signed measure on* R^k. *Let ε be a positive number and K_ε a probability measure on R^k satisfying*

$$K_\varepsilon(B(0:\varepsilon))=1. \tag{11.6}$$

Then for every real-valued, Borel-measurable function f *on* R^k *that is bounded on compacts,*

$$\left|\int f\,d(\mu-\nu)\right|\leqslant\gamma(f:\varepsilon)+\tau(f:2\varepsilon), \tag{11.7}$$

where

$$\gamma(f:\varepsilon)=\max\left\{\int M_f(\cdot:\varepsilon)\,d(\mu-\nu)*K_\varepsilon,\ -\int m_f(\cdot:\varepsilon)\,d(\mu-\nu)*K_\varepsilon\right\},$$

$$\tau(f:2\varepsilon)=\max\left\{\int(M_f(\cdot:2\varepsilon)-f)\,d\nu^+,\ \int(f-m_f(\cdot:2\varepsilon))\,d\nu^+\right\}, \tag{11.8}$$

provided that $|\mathrm{M_f}(\cdot:2\varepsilon)|$ *and* $|\mathrm{m_f}(\cdot:2\varepsilon)|$ *are integrable with respect to* μ *and* $|\nu|$.

Proof. One has

$$\begin{aligned}
\gamma(f:\varepsilon)&\geqslant\int M_f(\cdot:\varepsilon)\,d(\mu-\nu)*K_\varepsilon\\
&=\int_{B(0:\varepsilon)}\left[\int M_f(y+x:\varepsilon)(\mu-\nu)(dy)\right]K_\varepsilon(dx)\\
&=\int_{B(0:\varepsilon)}\left[\int M_f(y+x:\varepsilon)\,\mu(dy)-\int f(y)\nu(dy)\right.\\
&\qquad\left.-\int(M_f(y+x:\varepsilon)-f(y))\nu(dy)\right]K_\varepsilon(dx)\\
&\geqslant\int_{B(0:\varepsilon)}\left[\int f(y)\,\mu(dy)-\int f(y)\nu(dy)\right.\\
&\qquad\left.-\int(M_f(y+x:\varepsilon)-f(y))\nu^+(dy)\right]K_\varepsilon(dx)\\
&\geqslant\int f\,d(\mu-\nu)-\int(M_f(\cdot:2\varepsilon)-f)\,d\nu^+.
\end{aligned} \tag{11.9}$$

Similarly

$$-\gamma(f:\varepsilon) \leqslant \int m_f(\cdot:\varepsilon)\, d(\mu-\nu)*K_\varepsilon$$

$$= \int_{B(0:\varepsilon)} \left[\int m_f(y+x:\varepsilon)\,\mu(dy) - \int f(y)\nu(dy) \right.$$

$$\left. + \int (f(y) - m_f(y+x:\varepsilon))\nu(dy) \right] K_\varepsilon(dx)$$

$$\leqslant \int f\, d(\mu-\nu) + \int (f - m_f(\cdot:2\varepsilon))\, d\nu^+. \tag{11.10}$$

From (11.9), (11.10), one gets

$$-\gamma(f:\varepsilon) - \tau(f:2\varepsilon) \leqslant -\gamma(f:\varepsilon) - \int (f - m_f(\cdot:2\varepsilon))\, d\nu^+$$

$$\leqslant \int f\, d(\mu-\nu) \leqslant \gamma(f:\varepsilon) + \int (M_f(\cdot:2\varepsilon) - f)\, d\nu^+$$

$$\leqslant \gamma(f:\varepsilon) + \tau(f:2\varepsilon).$$

Q.E.D.

COROLLARY 11.2 *Under the hypothesis of Lemma* 11.1, *one has*

$$\left| \int f\, d(\mu-\nu) \right| \leqslant \gamma(f:\varepsilon) + \int \omega_f(\cdot:2\varepsilon)\, d\nu^+. \tag{11.11}$$

If, in addition, f *is bounded and*

$$\mu(R^k) = \nu(R^k), \tag{11.12}$$

then

$$\left| \int f\, d(\mu-\nu) \right| \leqslant \tfrac{1}{2}\omega_f(R^k) \|(\mu-\nu)*K_\varepsilon\| + \int \omega_f(\cdot:2\varepsilon)\, d\nu^+. \tag{11.13}$$

Proof. The inequality (11.11) follows from (11.7) and the relation (11.3). To prove (11.13) note that, in view of (11.12),

$$\int (f-c)\, d(\mu-\nu) = \int f\, d(\mu-\nu), \qquad \omega_{f-c}(\cdot:2\varepsilon) = \omega_f(\cdot:2\varepsilon),$$

$$\gamma(f-c:\varepsilon) = \max\left\{ \int M_{f-c}(\cdot:\varepsilon)\, d(\mu-\nu)*K_\varepsilon,\ -\int m_{f-c}(\cdot:\varepsilon)\, d(\mu-\nu)*K_\varepsilon \right\}$$

$$\leqslant \sup\{|f(x)-c| : x \in R^k\} \|(\mu-\nu)*K_\varepsilon\|$$

for all c in R^1. Letting

$$c=\tfrac{1}{2}\left(\sup\{f(x):x\in R^k\}+\inf\{f(x):x\in R^k\}\right),\qquad(11.14)$$

and observing that for this value of c

$$\omega_{f-c}(R^k)=\omega_f(R^k)=\sup\{f(x)-c:x\in R^k\}$$

$$-\inf\{f(x)-c:x\in R^k\}=2\sup\{|f(x)-c|:x\in R^k\},$$

one obtains (11.13). Q.E.D.

COROLLARY 11.13 *For every Borel subset* A *of* R^k *one has*

$$\mu(A)-\nu(A)\leqslant\tfrac{1}{2}\|(\mu-\nu)*K_\varepsilon\|+\nu^+(A^{2\varepsilon}\setminus A)$$

if μ, ν, K_ε *are as in Lemma* 11.1 *and if* (11.12) *holds.*

Proof. The inequality (11.15) follows from (11.9) with $f=I_A$ if one notes that

$$\int M_{f-1/2}(\cdot:2\varepsilon)\,d(\mu-\nu)*K_\varepsilon\leqslant\tfrac{1}{2}\|(\mu-\nu)*K_\varepsilon\|,$$

as in the proof of Corollary 11.2. Q.E.D.

LEMMA 11.4 *Let* μ *be a finite measure and* ν *a finite signed measure on* R^k. *Let* ε *be a positive number and* K_ε *a probability measure on* R^k *satisfying*

Götze for Stein

$$\alpha\equiv K_\varepsilon(B(0:\varepsilon))>\tfrac{1}{2}.\qquad(11.16)$$

Then for each real-valued, Borel-measurable, bounded function f *on* R^k, *one has*

$$\left|\int f\,d(\mu-\nu)\right|\leqslant(2\alpha-1)^{-1}\left[\gamma^*(f:\varepsilon)+\alpha\tau^*(f:2\varepsilon)+(1-\alpha)\tau^*(f:\varepsilon)\right]\qquad(11.17)$$

where

$$\gamma^*(f:\varepsilon)=\sup\{\gamma(f_y:\varepsilon):y\in R^k\},$$

$$\tau^*(f:2\varepsilon)=\sup\{\tau(f_y:2\varepsilon):y\in R^k\}.\qquad(11.18)$$

Proof. Let

$$\delta=\sup\left\{\left|\int f_y\,d(\mu-\nu)\right|:y\in R^k\right\}.$$

Assume first that

$$\delta = \sup\left\{ \int f_y \, d(\mu - \nu) : y \in R^k \right\}. \tag{11.19}$$

Then, given any positive number η, there exists y^0 in R^k such that

$$\int f_{y^0} d(\mu - \nu) \geq \delta - \eta. \tag{11.20}$$

In this case

$$\begin{aligned}
\gamma^*(f:\varepsilon) &\geq \int M_{f_{y^0}}(\cdot:\varepsilon)\, d(\mu-\nu)*K_\varepsilon \\
&= \int_{B(0:\varepsilon)} \left[\int M_{f_{y^0}}(y+x:\varepsilon)(\mu-\nu)(dy) \right] K_\varepsilon(dx) \\
&\quad + \int_{R^k \setminus B(0:\varepsilon)} \left[\int M_{f_{y^0}}(y+x:\varepsilon)(\mu-\nu)(dy) \right] K_\varepsilon(dx) \\
&\geq \int_{B(0:\varepsilon)} \left[\int f_{y^0}(y)\,\mu(dy) - \int f_{y^0}(y)\,\nu(dy) \right. \\
&\qquad \left. - \int \left(M_{f_{y^0}}(y+x:\varepsilon) - f_{y^0}(y)\right)\nu(dy) \right] K_\varepsilon(dx) \\
&\quad + \int_{R^k \setminus B(0:\varepsilon)} \left[\int f_{y^0}(y+x)\,\mu(dy) - \int f_{y^0}(y+x)\,\nu(dy) \right. \\
&\qquad \left. - \int \left(M_{f_{y^0}}(y+x:\varepsilon) - f_{y^0}(y+x)\right)\nu(dy) \right] K_\varepsilon(dx) \\
&\geq \int_{B(0:\varepsilon)} \left[\delta - \eta - \int \left(M_{f_{y^0}}(y+x:\varepsilon) - f_{y^0}(y)\right)\nu^+(dy) \right] K_\varepsilon(dx) \\
&\quad + \int_{R^k \setminus B(0:\varepsilon)} \left[-\delta - \int \left(M_{f_{y^0}}(y+x:\varepsilon) - f_{y^0}(y+x)\right)\nu^+(dy) \right] K_\varepsilon(dx) \\
&\geq \int_{B(0:\varepsilon)} \left[\delta - \eta - \int \left(M_{f_{y^0}}(y:2\varepsilon) - f_{y^0}(y)\right)\nu^+(dy) \right] K_\varepsilon(dx) \\
&\quad + \int_{R^k \setminus B(0:\varepsilon)} \left[-\delta - \tau^*(f:\varepsilon) \right] K_\varepsilon(dx) \\
&\geq \int_{B(0:\varepsilon)} \left[\delta - \eta - \tau^*(f:2\varepsilon) \right] K_\varepsilon(dx) + \left[-\delta - \tau^*(f:\varepsilon) \right](1-\alpha) \\
&= \left[\delta - \eta - \tau^*(f:2\varepsilon) \right]\alpha - \left[\delta + \tau^*(f:\varepsilon) \right](1-\alpha) \\
&= (2\alpha-1)\delta - \alpha\tau^*(f:2\varepsilon) - (1-\alpha)\tau^*(f:\varepsilon) - \alpha\eta.
\end{aligned} \tag{11.21}$$

Since η may be chosen arbitrarily close to zero,

$$\gamma^*(f:\varepsilon) \geqslant (2\alpha-1)\delta - \alpha\tau^*(f:2\varepsilon) - (1-\alpha)\tau^*(f:\varepsilon),$$

from which (11.17) follows. If instead of (11.19) one has

$$\delta = -\inf\left\{ \int f_y \, d(\mu-\nu) : y \in R^k \right\},$$

then, given any $\eta > 0$, find y^0 such that

$$-\int f_{y^0} \, d(\mu - \nu) \geqslant \delta - \eta.$$

Now look at $-f_{y^0}$ (instead of f_{y^0}) and note that

$$M_{-f_y}(\cdot:\varepsilon) = -m_{f_y}(\cdot:\varepsilon), \tag{11.22}$$

$$\int \left(f_y - m_{f_y}(\cdot:\varepsilon)\right) d\nu^+ = \int \left(M_{-f_y}(\cdot:\varepsilon) - (-f_y)\right) d\nu^+ \qquad (y \in R^k).$$

Proceeding exactly as in (11.21), one obtains

$$\begin{aligned} \gamma^*(f:\varepsilon) &\geqslant -\int m_{f_{y^0}}(\cdot:\varepsilon) \, d(\mu-\nu)*K_\varepsilon \geqslant (2\alpha-1)\delta \\ &\quad - \alpha\tau^*(f:2\varepsilon) - (1-\alpha)\tau^*(f:\varepsilon) - \alpha\eta. \end{aligned}$$

Q.E.D.

We define the *average modulus of oscillation* $\bar{\omega}_f(\varepsilon:\mu)$ *of* f *with respect to a finite measure* μ by

$$\bar{\omega}_f(\varepsilon:\mu) = \int \omega_f(x:\varepsilon) \, \mu(dx) \qquad (\varepsilon > 0). \tag{11.23}$$

Here f is a real-valued, Borel-measurable function on R^k. We also define $\omega_f^*(\varepsilon:\mu)$ as the supremum of the above average over all translates of f, that is,

$$\omega_f^*(\varepsilon:\mu) = \sup\left\{ \bar{\omega}_{f_y}(\varepsilon:\mu) : y \in R^k \right\} \qquad (\varepsilon > 0). \tag{11.24}$$

COROLLARY 11.5 *Under the hypothesis of Lemma* 11.4, *one has*

$$\begin{aligned} \left| \int f \, d(\mu-\nu) \right| &\leqslant (2\alpha-1)^{-1}\left[\gamma^*(f:\varepsilon) + \alpha\omega_f^*(2\varepsilon:\nu^+) + (1-\alpha)\omega_f^*(\varepsilon:\nu^+)\right] \\ &\leqslant (2\alpha-1)^{-1}\left[\gamma^*(f:\varepsilon) + \omega_f^*(2\varepsilon:\nu^+)\right]. \end{aligned} \tag{11.25}$$

If, in addition, $\mu(R^k)=\nu(R^k)$, *then*

$$\left|\int f\,d(\mu-\nu)\right| \leqslant (2\alpha-1)^{-1}\left[\tfrac{1}{2}\omega_f(R^k)\|(\mu-\nu)*K_\varepsilon\| + \alpha\omega_f^*(2\varepsilon:\nu^+)\right.$$
$$\left. +(1-\alpha)\omega_f^*(\varepsilon:\nu^+)\right]$$
$$\leqslant (2\alpha-1)^{-1}\left[\tfrac{1}{2}\omega_f(R^k)\|(\mu-\nu)*K_\varepsilon\| + \omega_f^*(2\varepsilon:\nu^+)\right]. \quad (11.26)$$

Proof. This corollary follows from Lemma 11.4 exactly as Corollary 11.2 follows from Lemma 11.1. Q.E.D.

The final result of this section relates the L_1-norm of an integrable function g to the L_1-norms of certain derivatives of its Fourier transform $\hat{g}$.

LEMMA 11.6 *Let* g *be a real-valued function in* $L^1(R^k)$ *satisfying*

$$\int \|x\|^{k+1}|g(x)|\,dx < \infty. \quad (11.27)$$

Then there exists a positive constant c(k) *depending only on* k (*and not on* g) *such that*

$$\|g\|_1 \leqslant c(k) \max_{|\beta|=0,k+1} \int |D^\beta \hat{g}(t)|\,dt. \quad (11.28)$$

Proof. We assume that $D^\beta\hat{g}$ is integrable with respect to Lebesgue measure for $0\leqslant|\beta|\leqslant k+1$ [else, the right side of (11.28) is $+\infty$]. Define the set

$$A=\{x:\ x\in R^k, g(x)\geqslant 0\},$$

and the 2^k quadrants E_α, one for each vector $\alpha=(\alpha_1,\dots,\alpha_k)$ (in R^k) having zeros or ones as coordinates, by

$$E_\alpha=\{x=(x_1,\dots,x_k):\ x_j\geqslant 0 \text{ if the } j\text{th coordinate of } \alpha \text{ is zero},$$
$$x_j<0 \text{ if the } j\text{th coordinate of } \alpha \text{ is one, } 1\leqslant j\leqslant k\}.$$

Since the Fourier transform of the function

$$x\to\left[1+\left(\sum_{j=1}^{k}(-1)^{\alpha_j}x_j\right)^{k+1}\right]g(x) \qquad (x\in R^k)$$

is of the form

$$\sum_{|\beta|=0,k+1} c(\alpha,\beta,k)D^\beta\hat{g},$$

where $c(\alpha,\beta,k)$ depends only on its arguments (and not on g), one has

$$\|g\|_1 = \sum_{\alpha}\left[\int_{A\cap E_\alpha} g(x)\,dx - \int_{(R^k\setminus A)\cap E_\alpha} g(x)\,dx\right]$$

$$= \sum_{\alpha}\left(\int_{A\cap E_\alpha} - \int_{(R^k\setminus A)\cap E_\alpha}\right)(1+|x|^{k+1})^{-1}(1+|x|^{k+1})\,g(x)\,dx$$

$$= \sum_{\alpha}\left(\int_{A\cap E_\alpha} - \int_{(R^k\setminus A)\cap E_\alpha}\right)(1+|x|^{k+1})^{-1}$$

$$\times\left[1+\left(\sum_{j=1}^{k}(-1)^{\alpha_j}x_j\right)^{k+1}\right]g(x)\,dx$$

$$= \sum_{\alpha}\left(\int_{A\cap E_\alpha} - \int_{(R^k\setminus A)\cap E_\alpha}\right)(1+|x|^{k+1})^{-1}(2\pi)^{-k}$$

$$\times\left[\int \exp\{-i\langle t,x\rangle\}\left(\sum_{|\beta|=0,k+1} c(\alpha,\beta,k)D^{\beta}\hat{g}(t)\right)dt\right]dx$$

$$\leqslant\left(\int(1+|x|^{k+1})^{-1}\,dx\right)c'(k)\max_{|\beta|=0,k+1}\int|D^{\beta}\hat{g}(t)|\,dt.$$

Q.E.D.

12. BERRY–ESSEEN THEOREM

Let P be a finite measure and Q a finite signed measure on R^1 with distribution functions F and G, respectively,

$$F(x)=P((-\infty,x]),\qquad G(x)=Q((-\infty,x])\qquad (x\in R^1). \tag{12.1}$$

Let K_ε be a probability measure on R^1 such that

$$\alpha\equiv K_\varepsilon((-\varepsilon,\varepsilon))>\tfrac{1}{2} \tag{12.2}$$

for a given $\varepsilon>0$. It follows easily from Lemma 11.4 that

$$\sup_{x\in R^1}|F(x)-G(x)|\leqslant(2\alpha-1)^{-1}\Big[\sup\{|(P-Q)*K_\varepsilon((-\infty,x])|:x\in R^1\}$$

$$+\sup\{|G^+(x)-G^+(y)|:|x-y|\leqslant 2\varepsilon\}\Big],$$

where G^+ is the distribution function of Q^+.

However with a more restricted choice of Q and the kernel K_ε this inequality can be sharpened.

LEMMA 12.1 *Let* P *be a finite measure and* Q *a finite signed measure on* R^1 *with distribution functions* F *and* G, *respectively. If* Q *has a density bounded above in magnitude by* m, *and if* K_ε *is a probability measure on* R^1 *that is symmetric and satisfies* (12.2), *then*

$$\sup_{x\in R^1}|F(x)-G(x)|\leqslant(2\alpha-1)^{-1}\Big[\sup_{x\in R^1}|(P-Q)*K_\varepsilon((-\infty,x])|+\alpha m\varepsilon\Big].$$

Proof. First assume that

$$\delta\equiv\sup_{x\in R^1}|F(x)-G(x)|=-\inf_{x\in R^1}(F(x)-G(x)). \tag{12.3}$$

Given $\eta>0$, there exists x_0 such that

$$F(x_0)-G(x_0)<-\delta+\eta.$$

Then

$$\begin{aligned}
&(P-Q)*K_\varepsilon((-\infty,x_0-\varepsilon])\\
&\quad=\int[F(x_0-\varepsilon-y)-G(x_0-\varepsilon-y)]K_\varepsilon(dy)\\
&\quad=\int_{B(0:\varepsilon)}[F(x_0-\varepsilon-y)-G(x_0-\varepsilon-y)]K_\varepsilon(dy)\\
&\qquad+\int_{R^1\setminus B(0:\varepsilon)}[F(x_0-\varepsilon-y)-G(x_0-\varepsilon-y)]K_\varepsilon(dy)\\
&\quad\leqslant\int_{B(0:\varepsilon)}[F(x_0)-G(x_0-\varepsilon-y)]K_\varepsilon(dy)+\delta(1-\alpha)\\
&\quad\leqslant\int_{B(0:\varepsilon)}[F(x_0)-G(x_0)+m(\varepsilon+y)]K_\varepsilon(dy)+\delta(1-\alpha)\\
&\quad\leqslant\int_{B(0:\varepsilon)}[-\delta+\eta+m(\varepsilon+y)]K_\varepsilon(dy)+\delta(1-\alpha)\\
&\quad=(-\delta+\eta+m\varepsilon)\alpha+m\int_{B(0:\varepsilon)}yK_\varepsilon(dy)+\delta(1-\alpha)\\
&\quad=-\delta(2\alpha-1)+m\alpha\varepsilon+\eta\alpha.
\end{aligned}$$

Hence

$$\delta(2\alpha-1) \leqslant \sup_{x \in R^1} \left|(P-Q)*K_\varepsilon((-\infty,x])\right| + m\alpha\varepsilon, \tag{12.4}$$

which proves the lemma if (12.3) holds. If, on the other hand,

$$\delta = \sup_{x \in R^1} (F(x) - G(x)),$$

then given $\eta > 0$, there exists x_0 such that

$$F(x_0) - G(x_0) > \delta - \eta.$$

Then

$$\begin{aligned}
&(P-Q)*K_\varepsilon((-\infty, x_0+\varepsilon]) \\
&\quad = \int_{B(0:\varepsilon)} [F(x_0+\varepsilon-y) - G(x_0+\varepsilon-y)] K_\varepsilon(dy) \\
&\quad\quad + \int_{R^1 \setminus B(0:\varepsilon)} [F(x_0+\varepsilon-y) - G(x_0+\varepsilon-y)] K_\varepsilon(dy) \\
&\quad \geqslant \int_{B(0:\varepsilon)} [F(x_0) - G(x_0) - (\varepsilon - y)m] K_\varepsilon(dy) - \delta(1-\alpha) \\
&\quad \geqslant (\delta-\eta)\alpha - \alpha m\varepsilon - \delta(1-\alpha) = \delta(2\alpha-1) - \alpha m\varepsilon - \eta\alpha,
\end{aligned}$$

so that

$$(2\alpha-1)\delta \leqslant \sup_{x \in R^1} (P-Q)*K_\varepsilon((-\infty,x]) + \alpha m\varepsilon.$$

Q.E.D.

LEMMA 12.2 *Let* P *be a finite measure and* Q *a finite signed measure on* R^1 *with distribution functions* F *and* G, *respectively. Assume that*

$$\int |x| P(dx) < \infty, \qquad \int |x| |Q|(dx) < \infty, \qquad P(R^1) = Q(R^1).$$

If Q *has a density bounded above in magnitude by* m, *and if* K_ε *is a symmetric probability measure on* R^1 *satisfying*

$$\alpha \equiv K_\varepsilon((-\varepsilon,\varepsilon)) > \tfrac{1}{2}, \qquad \int |\hat{K}_\varepsilon(t)| \, dt < \infty$$

for some $\varepsilon > 0$, *then*

$$\sup_{x \in R^1} |F(x) - G(x)| \leqslant (2\alpha - 1)^{-1} \times \left[(2\pi)^{-1} \int |t|^{-1} |(\hat{P}(t) - \hat{Q}(t)) \hat{K}_\varepsilon(t)| \, dt + \alpha m \varepsilon \right]. \quad (12.5)$$

Proof. By Fourier inversion, the density of the signed measure $(P - Q) * K_\varepsilon$ is

$$(2\pi)^{-1} \int \exp\{-itx\} (\hat{P}(t) - \hat{Q}(t)) \hat{K}_\varepsilon(t) \, dt \qquad (x \in R^1). \quad (12.6)$$

It is simple to check that the function (12.6) is the derivative of the function

$$(2\pi)^{-1} \int \exp\{-itx\} (-it)^{-1} (\hat{P}(t) - \hat{Q}(t)) \hat{K}_\varepsilon(t) \, dt \qquad (x \in R^1). \quad (12.7)$$

Note that

$$|t|^{-1} |\hat{P}(t) - \hat{Q}(t)| = \left| \int t^{-1} (\exp\{itx\} - 1)(P - Q)(dx) \right| \leqslant \int |x| |P - Q|(dx) < \infty.$$

By the Riemann–Lebesgue lemma [Theorem 4.1(iii)], the function (12.7) goes to zero as $|x| \to \infty$. Thus

$$(P - Q) * K_\varepsilon((-\infty, x]) = (2\pi)^{-1} \int \exp\{-itx\} (-it)^{-1} (\hat{P}(t) - \hat{Q}(t)) \hat{K}_\varepsilon(t) \, dt,$$

the constant of integration being zero. The inequality (12.5) now follows from Lemma 12.1. Q.E.D.

Remark. If, in addition to the assumptions on P and Q in Lemma 12.2, one also assumes that they have integrable Fourier–Stieltjes transforms and that

$$\int |t|^{-1} |\hat{P}(t) - \hat{Q}(t)| \, dt < \infty, \quad (12.8)$$

then the above argument yields

$$\sup_{x \in R^1} |F(x) - G(x)| \leqslant (2\pi)^{-1} \int |t|^{-1} |\hat{P}(t) - \hat{Q}(t)| \, dt. \quad (12.9)$$

We shall use Lemma 12.2 to prove the Berry–Esseen theorem below. The kernel K_ε is the distribution of $\varepsilon \mathbf{Z}/3.25$, where $\mathbf{Z}$ is a random variable whose distribution has the density $g_{1/2,2}$ given by

$$g_{1/2,2}(x)=(2\pi)^{-1}\left[\left(\frac{x}{2}\right)^{-1}\sin\frac{x}{2}\right]^2 \qquad (x\in R^1). \tag{12.10}$$

Recall [see (10.9)] that

$$\hat{K}_\varepsilon(t)=\hat{g}_{1/2,2}\left(\frac{\varepsilon t}{3.25}\right)=1-\frac{\varepsilon|t|}{3.25} \qquad \text{if } |t|\leqslant\frac{3.25}{\varepsilon},$$

$$=0 \qquad \text{if } |t|>\frac{3.25}{\varepsilon}. \tag{12.11}$$

By a careful but straightforward numerical integration one also obtains

$$\alpha\equiv K_\varepsilon((-\varepsilon,\varepsilon))=\int_{\{|x|\leqslant 3.25\}} g_{1/2,2}(x)\,dx\geqslant 0.79. \tag{12.12}$$

The following lemma will be useful in estimating the constant in the Berry–Esseen theorem.

LEMMA 12.3 *Let* P *be a probability measure on* R^1 *with zero mean and variance one. Let* F *denote the distribution function of* P *and* Φ *that of the standard normal distribution on* R^1. *Then*

$$\sup_{x\in R^1}|F(x)-\Phi(x)|\leqslant 0.5416.$$

Proof. We first prove the so-called *one-sided Chebyshev inequality*

$$F(-x)\leqslant(1+x^2)^{-1}, \qquad 1-F(x)\leqslant(1+x^2)^{-1} \qquad (x\geqslant 0). \tag{12.13}$$

Fix $x>0$. For every $b\geqslant 0$, one has

$$1+b^2=\int(y-b)^2P(dy)\geqslant\int_{(-\infty,-x]}(y-b)^2P(dy)\geqslant(x+b)^2F(-x),$$

so that

$$g(b)\equiv(1+b^2)(x+b)^{-2}\geqslant F(-x).$$

The minimum of g in $[0,\infty)$ occurs at $b=x^{-1}$, and

$$g(x^{-1})=(1+x^2)^{-1},$$

which gives the first inequality in (12.13) [note that for $x=0$, (12.13) is trivial]. The second is obtained similarly (or, by looking at $\tilde{P}$). This gives

$$F(-x)-\Phi(-x)\leqslant(1+x^2)^{-1}-\Phi(-x)\equiv h(x),$$

say, for $x\geqslant 0$. The supremum of h in $[0,\infty)$ is attained at a point x_0 satisfying

$$h'(x_0)=0, \qquad \text{or} \qquad 2x_0(1+x_0^2)^{-2}=(2\pi)^{-1/2}e^{-x_0^2/2}.$$

A numerical computation yields $x_0=0.2135$ and $h(x_0)=0.5416$, thus proving

$$|F(x)-\Phi(x)|\leqslant 0.5416 \tag{12.14}$$

for all $x\leqslant 0$ [note that $\Phi(x)-F(x)\leqslant .5$ for all $x\leqslant 0$]. The inequality (12.14) for $x\geqslant 0$ follows similarly (or, by looking at $\tilde{P}$). Q.E.D.

THEOREM 12.4 (Berry–Esseen Theorem) *Let* $\mathbf{X}_1,\ldots,\mathbf{X}_n$ *be* n *independent random variables each with zero mean and a finite absolute third moment. If*

$$\rho_2\equiv n^{-1}\sum_{j=1}^{n} E\mathbf{X}_j^2>0,$$

then

$$\sup_{x\in R^1}|F_n(x)-\Phi(x)|\leqslant(2.75)l_{3,n}, \tag{12.15}$$

where F_n *is the distribution function of* $(n\rho_2)^{-1/2}(\mathbf{X}_1+\cdots+\mathbf{X}_n)$, Φ *is the standard normal distribution function, and* $l_{3,n}$ *is the Liapounov ratio*

$$l_{3,n}=\left(\frac{\rho_3}{\rho_2^{3/2}}\right)n^{-1/2}, \qquad \rho_3=n^{-1}\sum_{j=1}^{n} E|\mathbf{X}_j|^3.$$

If, in addition to the above hypothesis, $\mathbf{X}_1,\ldots,\mathbf{X}_n$ *are identically distributed, then*

$$\sup_{x\in R^1}|F_n(x)-\Phi(x)|\leqslant(1.6)l_{3,n}. \tag{12.16}$$

Proof. To prove (12.15) first assume that $\rho_2=1$. For convenience write

$$l_n=l_{3,n}. \tag{12.17}$$

In view of Lemma 12.3,

$$\sup_{x\in R^1}|F_n(x)-\Phi(x)|\leqslant 0.5416, \tag{12.18}$$

so that we may assume that

$$l_n \leqslant \frac{0.5416}{2.75} \leqslant 0.196. \tag{12.19}$$

For, if (12.19) is violated, (12.15) reduces to (12.18). Let P_n denote the distribution of $n^{-1/2}(\mathbf{X}_1 + \cdots + \mathbf{X}_n)$. Take

$$\varepsilon = \tfrac{2}{3}(3.25)l_n, \tag{12.20}$$

and let K_ε be the distribution specified before the statement of Lemma 12.3.

It follows from Lemma 12.2 and inequality (12.12) that

$$\sup_{x \in R^1} |F_n(x) - \Phi(x)| \leqslant (0.58)^{-1} \Bigg[(2\pi)^{-1} \int_{\{|t| \leqslant (3/2) l_n^{-1}\}} \left| \frac{\left(\hat{P}_n(t) - e^{-t^2/2}\right)}{t} \right| \times \left(1 - \tfrac{2}{3} l_n |t|\right) dt + (2\pi)^{-1/2} (3.25)(0.79) \tfrac{2}{3} l_n \Bigg]. \tag{12.21}$$

Write

$$I \equiv \int_{\{|t| \leqslant (3/2) l_n^{-1}\}} \left| \frac{\left(\hat{P}_n(t) - e^{-t^2/2}\right)}{t} \right| \left(1 - \tfrac{2}{3} l_n |t|\right) dt \leqslant I_1 + I_2 + I_3 + I_4 + I_5, \tag{12.22}$$

where

$$\begin{aligned}
I_1 &= \int_{\{|t| \leqslant l_n^{-1/3}\}} \left| \frac{\left(\hat{P}_n(t) - e^{-t^2/2}\right)}{t} \right| dt, \\
I_2 &= \int_{\{l_n^{-1/3} < |t| \leqslant (1/2) l_n^{-1}\}} \left| \frac{\hat{P}_n(t)}{t} \right| dt, \\
I_3 &= \int_{\{(1/2) l_n^{-1} < |t| \leqslant l_n^{-1}\}} \left| \frac{\hat{P}_n(t)}{t} \right| \left(1 - \tfrac{2}{3} l_n |t|\right) dt, \\
I_4 &= \int_{\{l_n^{-1} < |t| \leqslant (3/2) l_n^{-1}\}} \left| \frac{\hat{P}_n(t)}{t} \right| \left(1 - \tfrac{2}{3} l_n |t|\right) dt, \\
I_5 &= \int_{\{|t| > l_n^{-1/3}\}} \left| \frac{e^{-t^2/2}}{t} \right| dt.
\end{aligned} \tag{12.23}$$

Applying Theorem 8.4 with $d=1$ and using (12.19), one gets

$$I_1 \leqslant (0.36) l_n \int_{R^1} t^2 \exp\{-0.3779 t^2\}\, dt \leqslant (1.3118) l_n. \tag{12.24}$$

By Theorem 8.9 (with $\delta = 1$),

$$\begin{aligned} I_2 &\leqslant l_n^{1/3} \int_{\{|t| > l_n^{-1/3}\}} e^{-t^2/3}\, dt \leqslant l_n^{2/3} \int_{\{|t| > l_n^{-1/3}\}} |t| e^{-t^2/3}\, dt \\ &= 3 l_n^{2/3} \exp\{-\tfrac{1}{3} l_n^{-2/3}\} = 3 l_n \left(l_n^{-1/3} \exp\{-\tfrac{1}{3} l_n^{-2/3}\}\right) \\ &\leqslant (1.9320) l_n. \end{aligned} \tag{12.25}$$

Again using Theorem 8.8, this time with $\delta = \frac{1}{2}$, we get

$$\begin{aligned} I_3 &= \int_{\{(1/2) l_n^{-1} < |t| \leqslant l_n^{-1}\}} \left| \frac{\hat{P}_n(t)}{t} \right| \left(1 - \tfrac{2}{3} l_n |t|\right) dt \\ &\leqslant \tfrac{2}{3} 2 l_n \int_{\{|t| > (1/2) l_n^{-1}\}} e^{-t^2/6}\, dt \leqslant \tfrac{2}{3} (2 l_n)^2 \int_{\{|t| > (1/2) l_n^{-1}\}} |t| e^{-t^2/6}\, dt \\ &= 16 l_n^2 \exp\left\{ \frac{-l_n^{-2}}{24} \right\} \leqslant (1.06) l_n. \end{aligned} \tag{12.26}$$

Noting now that Theorem 8.9 was derived from the inequality [see (8.48)]

$$|\hat{P}_n(t)| \leqslant \exp\left\{ -\frac{t^2}{2} + \tfrac{1}{3} |t|^3 l_n \right\},$$

which holds for $|t| \leqslant \frac{3}{2} l_n^{-1}$, one has

$$\begin{aligned} I_4 &\leqslant \tfrac{1}{3} l_n \int_{\{l_n^{-1} < |t| \leqslant (3/2) l_n^{-1}\}} \exp\left\{ -\frac{t^2}{2} + \tfrac{1}{3} |t|^3 l_n \right\} dt \\ &= \tfrac{1}{3} l_n \int_{\{l_n^{-1} < |t| \leqslant (3/2) l_n^{-1}\}} \exp\left\{ -\tfrac{1}{3} l_n t^2 \left(\tfrac{3}{2} l_n^{-1} - |t|\right) \right\} dt \\ &\leqslant \tfrac{1}{3} l_n \int_{\{l_n^{-1} < |t| \leqslant (3/2) l_n^{-1}\}} \exp\left\{ -\left(\frac{l_n^{-1}}{3}\right) \left(\tfrac{3}{2} l_n^{-1} - |t|\right) \right\} dt \\ &\leqslant \tfrac{2}{3} l_n \int_0^\infty \exp\left\{ -\left(\frac{l_n^{-1}}{3}\right) u \right\} du \\ &= 2 l_n^2 \leqslant (0.3920) l_n. \end{aligned} \tag{12.27}$$

Finally,

$$I_5=\int_{\{|t|>l_n^{-1/3}\}}\left|\frac{e^{-t^2/2}}{t}\right|dt$$

$$\leqslant l_n^{2/3}\int_{\{|t|>l_n^{-1/3}\}}|t|e^{-t^2/2}\,dt$$

$$=2l_n^{2/3}\exp\{-\tfrac{1}{2}l_n^{-2/3}\}$$

$$\leqslant(0.8096)l_n. \tag{12.28}$$

The estimates (12.24)–(12.28) are now used in (12.22) to give

$$I\leqslant(5.5054)l_n. \tag{12.29}$$

Using this in (12.21) one obtains

$$\sup_{x\in R^1}|F_n(x)-\Phi(x)|\leqslant(0.58)^{-1}\left[(2\pi)^{-1}(5.5054)l_n+(2\pi)^{-1/2}(3.25)(0.79)\tfrac{2}{3}l_n\right]$$

$$\leqslant(2.676)l_n. \tag{12.30}$$

This proves (12.15) under the assumption $\rho_2=1$. In the general case, look at random variables $\mathbf{Y}_j=\mathbf{X}_j/\rho_2^{1/2}$, $1\leqslant j\leqslant n$.

We now prove (12.16). Again assume that $E\mathbf{X}_1^2=1$ and

$$l_n\leqslant\frac{0.5416}{1.6}=0.3385. \tag{12.31}$$

As before, we proceed to estimate I and write

$$I\leqslant I_1'+I_2'+I_3', \tag{12.32}$$

where

$$I_1'=\int_{\{|t|\leqslant 2^{1/2}l_n^{-1}\}}\left|\frac{\left(\hat{P}_n(t)-e^{-t^2/2}\right)}{t}\right|dt,$$

$$I_2'=\int_{\{2^{1/2}l_n^{-1}<|t|\leqslant(3/2)l_n^{-1}\}}\left|\frac{\hat{P}_n(t)}{t}\right|\left(1-\tfrac{2}{3}l_n|t|\right), \tag{12.33}$$

$$I_3'=\int_{\{2^{1/2}l_n^{-1}<|t|\leqslant(3/2)l_n^{-1}\}}\left|\frac{e^{-t^2/2}}{t}\right|\left(1-\tfrac{2}{3}l_n|t|\right)dt.$$

Writing

$$a = E\left(\exp\left\{\frac{it\mathbf{X}_1}{n^{1/2}}\right\}\right), \qquad b = \exp\left\{-\tfrac{1}{2}\frac{t^2}{n}\right\}, \tag{12.34}$$

one has

$$\left|\hat{P}_n(t) - e^{-t^2/2}\right| = |a^n - b^n| = |a-b|\cdot|a^{n-1} + a^{n-2}b + \cdots + b^{n-1}|. \tag{12.35}$$

But, for all t, writing $\rho_3 = E|\mathbf{X}_1|^3$ yields

$$|a-b| \leqslant \left|a - 1 + \frac{t^2}{2n}\right| + \left|e^{-t^2/2n} - 1 + \frac{t^2}{2n}\right| \leqslant \tfrac{1}{6}\rho_3|t|^3 n^{-3/2} + \tfrac{1}{8}t^4 n^{-2}, \tag{12.36}$$

and for $|t| \leqslant 2^{1/2} l_n^{-1}$ the quantity $1 - \frac{1}{2}t^2/n$ is nonnegative, so that

$$|a| \leqslant 1 - \frac{t^2}{2n} + \frac{\rho_3|t|^3}{6n^{3/2}} \leqslant 1 - \frac{t^2}{2n} + \frac{0.2358}{n}t^2 = 1 - \frac{0.2642t^2}{n}$$

$$\leqslant \exp\left\{-\frac{(0.2642)t^2}{n}\right\}. \tag{12.37}$$

Hence, writing

$$c = 0.2642, \tag{12.38}$$

one has, using (12.36), (12.37) in (12.35),

$$\left|\hat{P}_n(t) - e^{-t^2/2}\right| \leqslant \left(\frac{\rho_3|t|^3}{6n^{3/2}} + \frac{t^4}{8n^2}\right)\sum_{r=0}^{n-1}\exp\left\{-\frac{n-1-r}{n}ct^2 - \frac{r}{2n}t^2\right\} \tag{12.39}$$

Now

$$\sum_{r=0}^{n-1}\exp\left\{-\frac{n-1-r}{n}ct^2 - \frac{r}{2n}t^2\right\}$$

$$= \exp\left\{\left(\frac{1}{2n} - c\right)t^2\right\}\sum_{r=1}^{n}\exp\left\{-\left(\tfrac{1}{2} - c\right)t^2\frac{r}{n}\right\}$$

$$\leqslant n\exp\left\{\left(\frac{1}{2n} - c\right)t^2\right\}\int_0^1 e^{-(\frac{1}{2}-c)t^2x}\,dx$$

$$= n\exp\left\{\left(\frac{1}{2n} - c\right)t^2\right\}\frac{1 - \exp\{-(\frac{1}{2}-c)t^2\}}{(\frac{1}{2}-c)t^2}. \tag{12.40}$$

The estimate (12.40) is used in (12.39) to yield

$$\left|\frac{\hat{P}_n(t)-e^{-t^2/2}}{|t|}\right| \leqslant \left(\tfrac{1}{2}-c\right)^{-1}\left(\frac{\rho_3}{6n^{1/2}}+\frac{|t|}{8n}\right)$$

$$\times\left(\exp\left\{-\left(c-\frac{1}{2n}\right)t^2\right\}-\exp\left\{-\frac{n-1}{2n}t^2\right\}\right). \tag{12.41}$$

Recalling from (12.31) that $\rho_3 n^{-1/2} \leqslant 0.3385$, so that $n \geqslant 9$, one obtains

$$I_1' \leqslant \left(\tfrac{1}{2}-c\right)^{-1} l_n \int_{R^1}\left(\frac{1}{6}+\frac{|t|}{24}\right)\left(\exp\left\{-\left(c-\frac{1}{2n}\right)t^2\right\}-\exp\left\{-\frac{n-1}{2n}t^2\right\}\right)dt$$

$$=\left(\tfrac{1}{2}-c\right)^{-1} l_n\left[\frac{(2\pi)^{1/2}}{6}\left\{\left(2c-\frac{1}{n}\right)^{-1/2}-\left(\frac{n-1}{n}\right)^{-1/2}\right\}\right.$$

$$\left.+\frac{1}{24}\left\{\left(c-\frac{1}{2n}\right)^{-1}-\left(\frac{n-1}{2n}\right)^{-1}\right\}\right]$$

$$\leqslant \left(\tfrac{1}{2}-c\right)^{-1} l_n\left[\frac{(2\pi)^{1/2}}{6}\left\{\left(2c-\tfrac{1}{9}\right)^{-1/2}-1\right\}+\tfrac{1}{24}\left\{\left(c-\tfrac{1}{18}\right)^{-1}-2\right\}\right]$$

$$\leqslant (1.465)l_n. \tag{12.42}$$

Next, proceeding as in the estimation of I_4 above, one has

$$I_2' \leqslant 2^{-1/2}\left(1-\frac{2^{3/2}}{3}\right) l_n \int_{\{2^{1/2}l_n^{-1}<|t|\leqslant(3/2)l_n^{-1}\}} \exp\left\{-\tfrac{2}{3}l_n^{-1}\left(\tfrac{3}{2}l_n^{-1}-|t|\right)\right\}dt$$

$$\leqslant (0.0404)l_n. \tag{12.43}$$

Finally,

$$I_3' \leqslant \left(2^{-1/2}l_n\right)^2\left(1-\frac{2^{3/2}}{3}\right)\int_{\{2^{1/2}l_n^{-1}<|t|\leqslant(3/2)l_n^{-1}\}} |t|e^{-t^2/2}\,dt$$

$$\leqslant (0.00002)l_n. \tag{12.44}$$

Therefore

$$I \leqslant (1.506)l_n, \tag{12.45}$$

and, using (12.45) in (12.21),

$$\sup_{x\in R^1}|F_n(x)-\Phi(x)| \leqslant (0.58)^{-1}\left[(2\pi)^{-1}(1.506)+0.6829\right]l_n$$

$$\leqslant (1.6)l_n. \tag{12.46}$$

Q.E.D.

Some of the computations above may be sharpened to yield somewhat better bounds in (12.15), (12.16). In Chapter 5 we see that in the i.i.d. case, under the hypothesis of the Berry–Esseen theorem, the finite limit

$$d(P,\Phi) \equiv \lim_{n\to\infty} \sqrt{n} \sup_{x\in R^1} |F_n(x)-\Phi(x)| \tag{12.47}$$

exists, where P is the distribution of $\mathbf{X}_1$, and that

$$\sup_P \frac{\sigma^3}{\rho_3} d(P,\Phi) = \frac{\sqrt{10}+3}{6\sqrt{2\pi}} \cong 0.409, \tag{12.48}$$

where $\sigma^2 = E\mathbf{X}_1^2$, and the supremum is over all P having mean zero, positive variance and finite third absolute moment. Thus $(\sqrt{10}+3)/(6\sqrt{2\pi})$ is the *asympototically best constant* for the Berry–Esseen theorem. It is also known that in the general non-i.i.d. case one has (see Notes at the end of this chapter)

$$\sup_{x\in R^1} |F_n(x)-\Phi(x)| \leqslant (0.7975) l_{3,n}. \tag{12.49}$$

13. RATES OF CONVERGENCE ASSUMING FINITE FOURTH MOMENTS

We begin with a lemma that is used in computing error bounds.

LEMMA 13.1 *Let* $\mathbf{X}_1,\ldots,\mathbf{X}_n$ *be* n *independent random vectors with values in* R^k *having finite third absolute moments* $\rho_{3,j}(1\leqslant j\leqslant n)$ *and satisfying*

$$E\mathbf{X}_j = 0, \qquad n^{-1}\sum_{j=1}^{n} \mathrm{Cov}(\mathbf{X}_j) = I, \tag{13.1}$$

where I *is the* $k\times k$ *identity matrix. As usual, let*

$$\begin{aligned} \chi_{\nu,j} &= \nu\text{th cumulant of } \mathbf{X}_j \qquad (1\leqslant j\leqslant n),\\ \chi_\nu &= n^{-1}\sum_{j=1}^{n} \chi_{\nu,j} \qquad (|\nu|=3),\\ \rho_3 &= n^{-1}\sum_{j=1}^{n}\rho_{3,j}. \end{aligned} \tag{13.2}$$

Then the variation norm of the signed measure $P_1(-\Phi: \{\chi_\nu\})$ *defined in Section 7 satisfies*

$$\|P_1(-\Phi:\{\chi_\nu\})\| \leqslant \left[\tfrac{1}{3}(2\pi)^{-1/2}(4e^{-3/2}+1) + (\pi e^{1/2})^{-1} k^{1/2}\left(1-\frac{1}{k}\right) + \left(\frac{2}{\pi}\right)^{3/2}(k^{3/2}-3k^{1/2}+2)\right]\rho_3. \tag{13.3}$$

In the special case k = 1,

$$\|P_1(-\Phi:\{\chi_\nu\})\| = \tfrac{1}{3}(2\pi)^{-1/2}(4e^{-3/2}+1)|\mu_3|$$
$$\leqslant \tfrac{1}{3}(2\pi)^{-1/2}(4e^{-3/2}+1)\rho_3. \tag{13.4}$$

where

$$\mu_3 = n^{-1}\sum_{j=1}^{n} E\mathbf{X}_j^3.$$

Proof. We first prove (13.4). By (7.21),

$$\begin{aligned}
&\|P_1(-\Phi:\{\chi_\nu\})\| \\
&= \tfrac{1}{6}|\mu_3|\int_{R^1}|x^3-3x|\phi(x)\,dx \\
&= \tfrac{1}{3}|\mu_3|\left[\int_{[0,3^{1/2}]}(3x-x^3)\phi(x)\,dx + \int_{(3^{1/2},\infty)}(x^3-3x)\phi(x)\,dx\right] \\
&= \tfrac{1}{3}(2\pi)^{-1/2}\left\{\left[(x^2-1)e^{-x^2/2}\right]_0^{3^{1/2}} + \left[(1-x^2)e^{-x^2/2}\right]_{3^{1/2}}^{\infty}\right\}|\mu_3| \\
&= \tfrac{1}{3}(2\pi)^{-1/2}(4e^{-3/2}+1)|\mu_3| \leqslant \tfrac{1}{3}(2\pi)^{-1/2}(4e^{-3/2}+1)\rho_3,
\end{aligned} \tag{13.5}$$

where ϕ is the standard normal density on R^1.

For arbitrary k, we use (7.20) to get

$$\begin{aligned}
\|P_1(-\Phi:\{\chi_\nu\})\| &\leqslant \tfrac{1}{6}\left[|\chi_{(3,0,\dots,0)}| + \cdots \right.\\
&\quad \left. + |\chi_{(0,\dots,0,3)}|\right]\int_{R^1}|x^3-3x|\phi(x)\,dx \\
&\quad + \tfrac{1}{2}\left[|\chi_{(2,1,0,\dots,0)}| + |\chi_{(2,0,1,0,\dots,0)}| + \cdots \right.\\
&\quad\quad \left. + |\chi_{(0,\dots,0,1,2)}|\right]\int_{R^2}\left|x_2(1-x_1^2)\right|\phi(x_1)\phi(x_2)\,dx_1\,dx_2 \\
&\quad + \left[|\chi_{(1,1,1,0,\dots,0)}| + \cdots \right.\\
&\quad\quad \left. + |\chi_{(0,\dots,0,1,1,1)}|\right]\int_{R^3}|x_1x_2x_3|\phi(x_1)\phi(x_2)\phi(x_3)\,dx_1\,dx_2\,dx_3 \\
&\leqslant \tfrac{1}{3}(2\pi)^{-1/2}(4e^{-3/2}+1)\theta_{n,1} + (\pi e^{1/2})^{-1}\theta_{n,2} + \left(\frac{2}{\pi}\right)^{3/2}\theta_{n,3},
\end{aligned} \tag{13.6}$$

where, writing $\mathbf{X}_j = (\mathbf{X}_{j,1}, \ldots, \mathbf{X}_{j,k})$ and using (6.21) give

$$\theta_{n,1} = |\chi_{(3,0,\ldots,0)}| + \cdots + |\chi_{(0,\ldots,0,3)}|$$

$$= \sum_{i=1}^{k} \left| n^{-1} \sum_{j=1}^{n} E\mathbf{X}_{j,i}^3 \right|$$

$$\leqslant n^{-1} \sum_{j=1}^{n} \sum_{i=1}^{k} E|\mathbf{X}_{j,i}|^3 \equiv \beta_3 \leqslant \rho_3,$$

$$\theta_{n,2} = |\chi_{(2,1,0,\ldots,0)}| + \cdots + |\chi_{(0,\ldots,0,1,2)}|$$

$$= \sum_{\substack{1 \leqslant i,i' \leqslant k,\\ i \neq i'}} \left| n^{-1} \sum_{j=1}^{n} E\left(\mathbf{X}_{j,i}^2 \mathbf{X}_{j,i'}\right) \right|$$

$$\leqslant n^{-1} \sum_{j=1}^{n} E\left(\sum_{i=1}^{k} \mathbf{X}_{j,i}^2 \cdot \sum_{i'=1}^{k} |\mathbf{X}_{j,i'}| - \sum_{i=1}^{k} |\mathbf{X}_{j,i}|^3 \right)$$

$$\leqslant n^{-1} \sum_{j=1}^{n} \left[E\left(\|\mathbf{X}_j\|^2 k^{1/2} \|\mathbf{X}_j\|\right) - \sum_{i=1}^{k} E|\mathbf{X}_{j,i}|^3 \right]$$

$$\leqslant k^{1/2}\rho_3 - k^{-1/2}\rho_3 = k^{1/2}\left(1 - \frac{1}{k}\right)\rho_3,$$

$$\theta_{n,3} = |\chi_{(1,1,1,0,\ldots,0)}| + \cdots + |\chi_{(0,\ldots,0,1,1,1)}|$$

$$= \sum_{\substack{1 \leqslant i,i',i'' \leqslant k\\ i \neq i', i' \neq i'', i'' \neq i}} \left| n^{-1} \sum_{j=1}^{n} E\left(\mathbf{X}_{j,i}\mathbf{X}_{j,i'}\mathbf{X}_{j,i''}\right) \right|$$

$$\leqslant n^{-1} \sum_{j=1}^{n} E\left[\left(\sum_{i=1}^{k} |\mathbf{X}_{j,i}| \right)^3 - 3\sum_{i=1}^{k} \mathbf{X}_{j,i}^2 \left(\sum_{i'=1}^{k} |\mathbf{X}_{j,i'}| - |\mathbf{X}_{j,i}| \right) - \sum_{i=1}^{k} |\mathbf{X}_{j,i}|^3 \right]$$

$$\leqslant k^{3/2}\rho_3 - 3k^{1/2}\rho_3 + 2\beta_3$$

$$\leqslant (k^{3/2} - 3k^{1/2} + 2)\rho_3. \tag{13.7}$$

Q.E.D.

Let us now introduce a kernel probability measure K on R^k whose density ψ is given by

$$\psi(x)=\prod_{i=1}^{k} g_{a,4}(x_i) \qquad \left[x=(x_1,\dots,x_k)\in R^k\right], \tag{13.8}$$

where $a>0$, and $g_{a,4}$ is the probability density on R^1 defined by (10.7). In fact, in this case we may compute the constant in the expression for $g_{a,4}$ and get

$$g_{a,4}(y)=\frac{3a}{2\pi}\left(\frac{\sin ay}{ay}\right)^4 \qquad (y\in R^1). \tag{13.9}$$

We shall choose

$$a=\tfrac{1}{4}\left(\frac{512}{\pi}\right)^{1/3}k^{5/6}=2\pi^{-1/3}k^{5/6}. \tag{13.10}$$

By (10.9),

$$\hat{K}(t)=\prod_{i=1}^{k}\hat{g}_{a,4}(t_i)=0 \qquad \text{if } t\notin[-4a,4a]^k=\{t: |t_i|\leqslant 4a \quad \text{for } 1\leqslant i\leqslant k\},$$

$$\left[t=(t_1,\dots,t_k)\in R^k\right]. \tag{13.11}$$

Note that

$$K(\{x: \|x\|\geqslant 1\})\leqslant\frac{3ak}{\pi}\int_{[k^{-1/2},\infty)}\left(\frac{\sin ay}{ay}\right)^4 dy$$

$$\leqslant\frac{3k}{\pi}\int_{[ak^{-1/2},\infty)}u^{-4}\,dy=\frac{k}{\pi}(ak^{-1/2})^{-3}=\tfrac{1}{8}. \tag{13.12}$$

Hence

$$K(\{x: \|x\|<1\})\geqslant\tfrac{7}{8}. \tag{13.13}$$

In the proofs of the theorems below we use some theorems of Section 8, substituting $n^{-(s-2)/2}\rho_s$ for $l_{s,n}$ $(s>2)$. This is justified in view of (8.12).

THEOREM 13.2 *Let* $\mathbf{X}_1,\dots,\mathbf{X}_n$ *be* n *independent and identically distributed random vectors with values in* R^k *satisfying*

$$E\mathbf{X}_1=0, \qquad \mathrm{Cov}(\mathbf{X}_1)=I, \qquad \rho_4\equiv E\|\mathbf{X}_1\|^4<\infty. \tag{13.14}$$

Let Q_n *denote the distribution of* $n^{-1/2}(\mathbf{X}_1+\cdots+\mathbf{X}_n)$ *and let* Φ *be the standard normal distribution on* R^k. *Then for every real-valued, bounded, Borel-measurable function* f *on* R^k,

$$\left|\int f\,d(Q_n-\Phi)\right|$$

$$\leqslant \omega_f(R^k)\Big[\big(\tfrac{8}{3}c_0+a_1(k)\big)\rho_3 n^{-1/2}$$

$$+a_2(k,\rho_3,\rho_4)n^{-1}(\log n)^{k/2}+\tfrac{4}{3}\pi^{-1/2}kn^{-1}(\log n)^{-1/2}$$

$$+\tfrac{32}{3}\pi^{-1}k^{5/2}\rho_3^3(n\log n)^{-3/2}+\tfrac{32}{3}(2k)^{(k-2)/2}\left(\Gamma\left(\frac{k}{2}\right)\right)^{-2}\delta_n\Big]$$

$$+\tfrac{4}{3}\omega_f^*\left(2^{7/2}\pi^{-1/3}k^{4/3}\rho_3 n^{-1/2}:\Phi\right), \tag{13.15}$$

where ω_f^* *is defined by* (11.24), c_0 *is the universal constant appearing in the Berry–Esseen bound* (12.49), *and*

a₁(1) ≤ .168

$$a_1(k)=\tfrac{2}{9}(2\pi)^{-1/2}(4e^{-3/2}+1)+\tfrac{2}{3}(\pi e^{1/2})^{-1}k^{1/2}(1-k^{-1})$$

$$+\tfrac{2}{3}\left(\frac{2}{\pi}\right)^{3/2}(k^{3/2}-3k^{1/2}+2),$$

$$a_2(k,\rho_3,\rho_4)=\tfrac{4}{3}k^{k/2}(k+2)2^{k+1}\left(\Gamma\left(\frac{k}{2}\right)\right)^{-1}\left(0.14+\frac{\rho_4}{24}\right)$$

$$+(0.19)(2.3)^k k^{k/2}(k+2)(k+4)\left(\Gamma\left(\frac{k}{2}\right)\right)^{-1}\rho_3^2,$$

$$\delta_n=(\log n)^{k/2}\int_{n^{1/2}/(2\rho_4^{1/2})}^{\infty}\Big[r^{k-1}e^{-0.264r^2}+r^{k-1}e^{-(1/2)r^2}$$

$$+\tfrac{1}{6}\rho_3 n^{-1/2}r^{k+2}e^{-(1/2)r^2}\Big]\,dr. \tag{13.16}$$

Proof. Let $\mathbf{Z}$ denote a random vector whose distribution is K [see (13.8)–(13.13)]. By Corollary 11.5 [more specifically, inequality (11.26)],

$$\left|\int f\,d(Q_n-\Phi)\right|\leqslant\tfrac{4}{3}\left[\tfrac{1}{2}\omega_f(R^k)\|(Q_n-\Phi)*K_\varepsilon\|+\omega_f^*(2\varepsilon:\Phi)\right] \tag{13.17}$$

for all $\varepsilon > 0$. Choose

$$\varepsilon = 4ak^{1/2}\rho_3(2n)^{-1/2}. \tag{13.18}$$

Then

$$\hat{K}_\varepsilon(t) = 0 \qquad \text{if } \|t\| \geqslant \frac{(2n)^{1/2}}{\rho_3}. \tag{13.19}$$

For every Borel set B and $r \geqslant 0$, write

$$B_1 = B \cap B(0:r), \qquad B_2 = B \setminus B_1. \tag{13.20}$$

One has

$$|(Q_n - \Phi) * K_\varepsilon(B_1)| \leqslant |H_n * K_\varepsilon(B_1)| + n^{-1/2}|P_1(-\Phi: \{\chi_\nu\}) * K_\varepsilon(B_1)|, \tag{13.21}$$

where $H_n = Q_n - \Phi - n^{-1/2}P_1(-\Phi: \{\chi_\nu\})$ and χ_ν is the νth cumulant of $\mathbf{X}_1$ (only cumulants with $|\nu| = 3$ enter into the expression for P_1). By Fourier inversion and (13.19),

$$|H_n * K_\varepsilon(B_1)| \leqslant (2\pi)^{-k}\lambda_k(B_1)\int |\hat{H}_n(t)\hat{K}_\varepsilon(t)|\,dt \leqslant (2\pi)^{-k}\lambda_k(B_1)(I_1 + I_2), \tag{13.22}$$

where λ_k is Lebesgue measure on R^k, and (by Theorem 8.5)

$$\begin{aligned} I_1 &\equiv \int_{\{\|t\| \leqslant n^{1/2}/(2\rho_4^{1/2})\}} |\hat{H}_n(t)|\,dt \\ &\leqslant \left(\frac{0.14}{n} + \frac{\rho_4}{24n}\right)\int \|t\|^4 \exp\{-\tfrac{1}{2}\|t\|^2\}\,dt \\ &\quad + \frac{0.03\rho_3^2}{n}\int \|t\|^6 \exp\{-0.383\|t\|^2\}\,dt \\ &\leqslant \left(\frac{0.14}{n} + \frac{\rho_4}{24n}\right)k(k+2)(2\pi)^{k/2} \\ &\quad + \frac{0.07\rho_3^2}{n}\left(\frac{\pi}{0.383}\right)^{k/2} k(k+2)(k+4), \end{aligned} \tag{13.23}$$

and (by Theorem 8.7)

$$I_2 \equiv \int_{\{n^{1/2}/(2\rho_4^{1/2}) < \|t\| < (2n)^{1/2}/\rho_3\}} |\hat{H}_n(t)|\,dt$$

$$\leqslant \int_{\{\|t\| > n^{1/2}/(2\rho_4^{1/2})\}} \Big[\exp\{-0.264\|t\|^2\}$$

$$+\left(1+\tfrac{1}{6}n^{-1/2}\rho_3\|t\|^3\right)\exp\{-\tfrac{1}{2}\|t\|^2\}\Big]\,dt$$

$$=2\pi^{k/2}\left(\Gamma\left(\frac{k}{2}\right)\right)^{-1}\int_{n^{1/2}/(2\rho_4^{1/2})}^{\infty}\Big[r^{k-1}e^{-0.264r^2}+r^{k-1}e^{-(1/2)r^2}$$

$$+\tfrac{1}{6}\rho_3 n^{-1/2} r^{k+2} e^{-(1/2)r^2}\Big]\,dr$$

$$=2\pi^{k/2}\left(\Gamma\left(\frac{k}{2}\right)\right)^{-1}\delta_n'. \tag{13.24}$$

Thus

$$|H_n * K_\varepsilon(B_1)| \leqslant 2^{-(k-2)/2}\left(\Gamma\left(\frac{k}{2}\right)\right)^{-1}(k+2)\left(\frac{0.14}{n}+\frac{\rho_4}{24n}\right)r^k$$

$$+(0.14)(1.532)^{-k/2}\left(\Gamma\left(\frac{k}{2}\right)\right)^{-1}(k+2)(k+4)\rho_3^2 n^{-1} r^k$$

$$+2^{-(k-2)}k^{-1}\left(\Gamma\left(\frac{k}{2}\right)\right)^{-2}\delta_n' r^k. \tag{13.25}$$

Also, by Lemma 13.1,

$$n^{-1/2}|P_1(-\Phi:\{\chi_\nu\}) * K_\varepsilon(B_1)|$$

$$\leqslant \tfrac{1}{2}n^{-1/2}\|P_1(-\Phi:\{\chi_\nu\})\|$$

$$\leqslant \tfrac{1}{2}\Big[\tfrac{1}{3}(2\pi)^{-1/2}(4e^{-3/2}+1)+(\pi e^{1/2})^{-1}k^{1/2}(1-k^{-1})$$

$$+\left(\frac{2}{\pi}\right)^{3/2}(k^{3/2}-3k^{1/2}+2)\Big]\rho_3 n^{-1/2}$$

$$=a_1'(k)\rho_3 n^{-1/2}, \tag{13.26}$$

say. Next,

$$|(Q_n-\Phi)*K_\varepsilon(B_2)| \leqslant \max\{Q_n*K_\varepsilon(B_2),\ \Phi*K_\varepsilon(B_2)\}$$

$$\leqslant \max\left\{\operatorname{Prob}\left(\|n^{-1/2}(\mathbf{X}_1+\cdots+\mathbf{X}_n)\| \geqslant \frac{r}{2}\right),\right.$$

$$\left.\int_{\{\|x\|\geqslant r/2\}} \phi(x)\,dx\right\} + \operatorname{Prob}\left(\|\varepsilon\mathbf{Z}\| \geqslant \frac{r}{2}\right). \quad (13.27)$$

By the Berry-Esseen theorem,

$$\operatorname{Prob}\left(\|n^{-1/2}(\mathbf{X}_1+\cdots+\mathbf{X}_n)\| \geqslant \frac{r}{2}\right)$$

$$\leqslant \sum_{i=1}^{k} \operatorname{Prob}\left(|n^{-1/2}(\mathbf{X}_{1,i}+\cdots+\mathbf{X}_{n,i})| \geqslant \frac{r}{2k^{1/2}}\right)$$

$$\leqslant 2c_0\rho_3 n^{-1/2} + 4k^{3/2}(2\pi)^{-1/2}r^{-1}e^{-r^2/8k},$$

$$\int_{\{\|x\|\geqslant r/2\}} \phi(x)\,dx \leqslant 4k^{3/2}(2\pi)^{-1/2}r^{-1}e^{-r^2/8k},$$

$$\operatorname{Prob}\left(\|\varepsilon\mathbf{Z}\| \geqslant \frac{r}{2}\right) \leqslant \frac{512k^4\rho_3^3}{2^{3/2}\pi r^3 n^{3/2}}, \quad (13.28)$$

where $\mathbf{X}_j = (\mathbf{X}_{j,1},\ldots,\mathbf{X}_{j,k})$, $1 \leqslant j \leqslant n$. Thus

$$|(Q_n-\Phi)*K_\varepsilon(B_2)| \leqslant 2c_0\rho_3 n^{-1/2} + 4k^{3/2}(2\pi)^{-1/2}r^{-1}e^{-r^2/8k}$$

$$+512\frac{k^4\rho_3^3}{2^{3/2}\pi r^3 n^{3/2}}. \quad (13.29)$$

Now take

$$r = (8k\log n)^{1/2}. \quad (13.30)$$

Then from (13.25), (13.26), and (13.29),

$$|(Q_n-\Phi)*K_\varepsilon(B_1)|$$

$$\leqslant 2^{(k+1)}\left(\Gamma\left(\frac{k}{2}\right)\right)^{-1}k^{k/2}(k+2)\left(\frac{0.14}{n}+\frac{\rho_4}{24n}\right)(\log n)^{k/2}$$

$$+(0.14)(2.3)^k\left(\Gamma\left(\frac{k}{2}\right)\right)^{-1}k^{k/2}(k+2)(k+4)\rho_3^2n^{-1}(\log n)^{k/2}$$

$$+2^{(k+4)/2}k^{(k-2)/2}\left(\Gamma\left(\frac{k}{2}\right)\right)^{-2}\delta_n'(\log n)^{k/2}+a_1'(k)\rho_3n^{-1/2}, \quad (13.31)$$

$$|(Q_n-\Phi)*K_\varepsilon(B_2)|\leqslant 2c_0\rho_3n^{-1/2}+k\pi^{-1/2}(n\log^{1/2}n)^{-1}$$

$$+8\pi^{-1}k^{5/2}\rho_3^3(n\log n)^{-3/2}.$$

Now since

$$\|(Q_n-\Phi)*K_\varepsilon\|=2\sup\{|(Q_n-\Phi)*K_\varepsilon(B)|: B\in\mathcal{B}^k\}, \quad (13.32)$$

by using (13.31) in (13.17) one obtains the desired inequality (13.15). Q.E.D.

One can easily deduce from (13.15) that there exists a constant $a_3(k)$ (depending only on k) such that

$$\left|\int fd(Q_n-\Phi)\right|\leqslant \omega_f(R^k)a_3(k)\rho_4n^{-1/2}+\tfrac{4}{3}\omega_f^*(2^{7/2}\pi^{-1/3}k^{4/3}\rho_3n^{-1/2}:\Phi). \quad (13.33)$$

We prove a generalization of (13.33) to the non-i.i.d. case.

THEOREM 13.3 *Let* $\mathbf{X}_1,\ldots,\mathbf{X}_n$ *be* n *independent random vectors with values in* R^k *satisfying*

$$E\mathbf{X}_j=0 \qquad (1\leqslant j\leqslant n),$$

$$n^{-1}\sum_{j=1}^{n}\mathrm{Cov}(\mathbf{X}_j)=I, \quad (13.34)$$

$$\rho_4\equiv n^{-1}\sum_{j=1}^{n}E\|\mathbf{X}_j\|^4<\infty.$$

Let Q_n *denote the distribution of* $\mathrm{n}^{-1/2}(\mathbf{X}_1+\cdots+\mathbf{X}_n)$. *Then there exists a constant* $a_3(k)$ *depending only on* k *such that for every real-valued, bounded, Borel-measurable function* f *on* R^k *one has*

$$\left|\int f d(Q_n - \Phi)\right| \leqslant \omega_f(R^k) a_3(k)\rho_4 n^{-1/2} + \tfrac{4}{3}\omega_f^*\left(2^{7/2}\pi^{-1/3}k^{4/3}\rho_3 n^{-1/2} : \Phi\right). \tag{13.35}$$

Proof. We use the same notation as in the preceding proof. Then we have [in place of (13.22)–(13.24)]

$$|H_n * K_\varepsilon(B_1)| \leqslant (2\pi)^{-k}\lambda_k(B_1)(J_1 + J_2). \tag{13.36}$$

Theorem 8.6 yields

$$J_1 \equiv \int_{\{\|t\| \leqslant \frac{1}{2} n^{1/4}/\rho_4^{1/4}\}} |\hat{H}_n(t)|\, dt \leqslant a_4(k)\rho_4 n^{-1} \tag{13.37}$$

if [see (8.30)]

$$\rho_4 n^{-1} \leqslant 1, \tag{13.38}$$

and Theorem 8.9, with $\delta = \frac{3}{2} - 2^{1/2}$, leads to the estimate

$$\begin{aligned} J_2 &\equiv \int_{\{\frac{1}{2} n^{1/4}/\rho_4^{1/4} < \|t\| < (2n)^{1/2}/\rho_3\}} |\hat{H}_n(t)|\, dt \\ &\leqslant \int_{\{\|t\| > \frac{1}{2}(n/\rho_4)^{1/4}\}} \left[\exp\left\{\frac{-\delta}{3}\|t\|^2\right\} \right. \\ &\qquad \left. + \left(1 + \tfrac{1}{6}\rho_3 n^{-1/2}\|t\|^3\right)\exp\left\{-\tfrac{1}{2}\|t\|^2\right\}\right] dt \\ &\leqslant a_5(k)\rho_4 n^{-1}. \end{aligned} \tag{13.39}$$

Note that we used (13.38) in the last step of (13.39). Now choose $r = (8k\log n)^{1/2}$ and obtain

$$|H_n * K_\varepsilon(B_1)| \leqslant a_6(k)\rho_4 n^{-1}(\log n)^{k/2} \leqslant a_7(k)\rho_4 n^{-1/2}, \tag{13.40}$$

and

$$\begin{aligned} |H_n * K_\varepsilon(B_2)| &\leqslant a_8(k)\rho_3 n^{-1/2} + k\pi^{-1/2}\left(n\log^{1/2} n\right)^{-1} \\ &\quad + 8\pi^{-1}k^{5/2}\rho_3^3(n\log n)^{-3/2} \leqslant a_9(k)\rho_4 n^{-1/2}. \end{aligned} \tag{13.41}$$

The rest follows exactly as in the preceding proof. It is simple to check that (13.35) holds with $a_3(k)=2$ if (13.38) is violated. Thus (13.35) is proved for all cases. Q.E.D.

One may compute an explicit error bound, with numerical values for the constants involved, in Theorem 13.3 much the same way as in the i.i.d. case (Theorem 13.2).

A significant application is to the class $\mathcal{C}$ of all Borel-measurable convex subsets of R^k. By Corollary 3.2 (with $s=0$)

$$\sup_{C\in\mathcal{C}} \omega^*_{I_C}(2\varepsilon:\Phi)=\sup_{C\in\mathcal{C}}\Phi\big((\partial C)^{2\varepsilon}\big)$$
$$\leqslant \frac{2^{5/2}\Gamma((k+1)/2)}{\Gamma(k/2)}\varepsilon. \tag{13.42}$$

Using this in Theorem 13.3 one obtains

$$\sup_{C\in\mathcal{C}}|Q_n(C)-\Phi(C)|\leqslant a_{10}(k)\rho_4 n^{-1/2}. \tag{13.43}$$

In the i.i.d. case one also has

$$\overline{\lim_{n\to\infty}}\ \sup_{C\in\mathcal{C}}\sqrt{n}\,|Q_n(C)-\Phi(C)|$$
$$\leqslant\left[a_1(k)+\frac{128}{3}\pi^{-1/3}k^{4/3}\frac{\Gamma((k+1)/2)}{\Gamma(k/2)}\right]\rho_3, \tag{13.44}$$

where $a_1(k)$ is given by (13.16). This follows from Theorem 13.2 and the estimate (13.42), provided that one replaces the first estimate in (13.28) by

$$Q_n\big(\{x:\ \|x\|\geqslant\tfrac{1}{2}(8k\log n)^{1/2}\}\big)=o(n^{-1/2})\qquad (n\to\infty), \tag{13.45}$$

which holds (see Corollary 17.12) if $\rho_3<\infty$.

14. TRUNCATION

We need some truncation estimates for relaxing the moment condition (namely, finiteness of fourth moments) of the last section to obtain rates of convergence for integrals of unbounded functions and (in Chapters 4 and 5) to derive precise asymptotic expansions.

Throughout this section $\mathbf{X}_1,\ldots,\mathbf{X}_n$ *are* n *independent random vectors with values in* R^k *having zero means. We write, as usual,*

$$\mu_{\alpha,j}=E\mathbf{X}_j^{\alpha}, \qquad \chi_{\alpha,j}=\alpha\text{th cumulant of } \mathbf{X}_j,$$

$$\rho_{s,j}=E\|\mathbf{X}_j\|^s \qquad (1\leqslant j\leqslant n),$$

$$\mu_{\alpha}=n^{-1}\sum_{j=1}^{n}\mu_{\alpha,j}, \qquad \chi_{\alpha}=n^{-1}\sum_{j=1}^{n}\chi_{\alpha,j}, \tag{14.1}$$

$$\rho_s=n^{-1}\sum_{j=1}^{n}\rho_{s,j} \qquad \left[\alpha\in(\mathbf{Z}^+)^k,\quad s\geqslant 0\right].$$

Define *truncated random vectors*

$$\mathbf{Y}_j=\begin{cases}\mathbf{X}_j & \text{if } \|\mathbf{X}_j\|\leqslant n^{1/2},\\ 0 & \text{if } \|\mathbf{X}_j\|>n^{1/2},\end{cases} \tag{14.2}$$

$$\mathbf{Z}_j=\mathbf{Y}_j-E\mathbf{Y}_j \qquad (1\leqslant j\leqslant n),$$

and write

$$\mu'_{\alpha,j}=E\mathbf{Z}_j^{\alpha}, \qquad \chi'_{\alpha,j}=\alpha\text{th cumulant of } \mathbf{Z}_j,$$

$$\rho'_{s,j}=E\|\mathbf{Z}_j\|^s, \tag{14.3}$$

$$\rho'_s=n^{-1}\sum_{j=1}^{n}E\|\mathbf{Z}_j\|^s, \qquad \chi'_{\alpha}=n^{-1}\sum_{j=1}^{n}\chi'_{\alpha,j}.$$

Also introduce

$$\Delta_{n,j,s}=\int_{\{\|\mathbf{X}_j\|>n^{1/2}\}}\|\mathbf{X}_j\|^s, \quad \bar{\Delta}_{n,s}=n^{-1}\sum_{j=1}^{n}\Delta_{n,j,s}. \tag{14.4}$$

Finally, write

$$V=n^{-1}\sum_{j=1}^{n}\operatorname{Cov}(\mathbf{X}_j)=((v_{ij})),$$

$$D=n^{-1}\sum_{j=1}^{n}\operatorname{Cov}(\mathbf{Z}_j)=((d_{ij})). \tag{14.5}$$

The constants c'*s depend only on their arguments.*

LEMMA 14.1 *Let* $\rho_s < \infty$ *for some* $s \geqslant 2$.

(i) *One has*

$$\rho_{s,j} = E\|\mathbf{Y}_j\|^s + \Delta_{n,j,s}, \quad \overline{\Delta}_{n,s} \leqslant \rho_s. \tag{14.6}$$

(ii) *If* α *is a nonnegative integral vector satisfying* $1 \leqslant |\alpha| \leqslant s$, *then*

$$\begin{aligned} |E\mathbf{X}_j^\alpha - E\mathbf{Y}_j^\alpha| &\leqslant n^{-(s-|\alpha|)/2}\Delta_{n,j,s}, \\ |E\mathbf{Y}_j^\alpha - E\mathbf{Z}_j^\alpha| &\leqslant |\alpha|(2^{|\alpha|}+1)n^{-(s-|\alpha|)/2}\Delta_{n,j,s}. \end{aligned} \tag{14.7}$$

(iii) *One has*

$$|d_{il} - v_{il}| \leqslant 2n^{-(s-2)/2}\overline{\Delta}_{n,s} \qquad (1 \leqslant i, l \leqslant k). \tag{14.8}$$

(iv) *If* $2 \leqslant s' \leqslant s$, *then*

$$\begin{gathered} E\|\mathbf{Y}_j\|^{s'} \leqslant \rho_{s',j}, \qquad \rho'_{s',j} = E\|\mathbf{Z}_j\|^{s'} \leqslant 2^{s'}\rho_{s',j}, \\ \rho'_{s'} \leqslant 2^{s'}\rho_{s'}. \end{gathered} \tag{14.9}$$

(v) *If* $s' \geqslant s$, *then*

$$\begin{aligned} E\|\mathbf{Y}_j\|^{s'} &\leqslant (\varepsilon n^{1/2})^{(s'-s)} \int_{\{\|\mathbf{X}_j\| \leqslant \varepsilon n^{1/2}\}} \|\mathbf{X}_j\|^s \\ &\quad + n^{(s'-s)/2} \int_{\{\varepsilon n^{1/2} < \|\mathbf{X}_j\| \leqslant n^{1/2}\}} \|\mathbf{X}_j\|^s \leqslant n^{(s'-s)/2}\rho_{s,j} \qquad (0 \leqslant \varepsilon \leqslant 1), \\ \rho'_{s',j} &= E\|\mathbf{Z}_j\|^{s'} \leqslant 2^{s'}E\|\mathbf{Y}_j\|^{s'}. \end{aligned} \tag{14.10}$$

Proof. The relations (14.6) follow from definition. To prove the first inequality in (14.7), use the inequality (6.29) to obtain

$$\begin{aligned} |E\mathbf{X}_j^\alpha - E\mathbf{Y}_j^\alpha| &= \left| \int_{\{\|\mathbf{X}_j\| > n^{1/2}\}} \mathbf{X}_j^\alpha \right| \leqslant \int_{\{\|\mathbf{X}_j\| > n^{1/2}\}} \|\mathbf{X}_j\|^{|\alpha|} \\ &\leqslant n^{-(s-|\alpha|)/2} \int_{\{\|\mathbf{X}_j\| > n^{1/2}\}} \|\mathbf{X}_j\|^s = n^{-(s-|\alpha|)/2}\Delta_{n,j,s}. \end{aligned} \tag{14.11}$$

To derive the second inequality in (14.7), first observe that

$$|x^{\alpha}-y^{\alpha}|=|x_1^{\alpha_1}\cdots x_k^{\alpha_k}-y_1^{\alpha_1}\cdots y_k^{\alpha_k}|$$

$$=|x_1^{\alpha_1}\cdots x_{k-1}^{\alpha_{k-1}}(x_k^{\alpha_k}-y_k^{\alpha_k})$$

$$+x_1^{\alpha_1}\cdots x_{k-2}^{\alpha_{k-2}}(x_{k-1}^{\alpha_{k-1}}-y_{k-1}^{\alpha_{k-1}})y_k^{\alpha_k}+\cdots+(x_1^{\alpha_1}-y_1^{\alpha_1})y_2^{\alpha_2}\cdots y_k^{\alpha_k}|$$

$$\leqslant|x_1^{\alpha_1}\cdots x_{k-1}^{\alpha_{k-1}}(x_k-y_k)|\alpha_k\max\left(|x_k|^{\alpha_k-1},|y_k|^{\alpha_k-1}\right)$$

$$+\cdots+\alpha_1\max\left(|x_1|^{\alpha_1-1},|y_1|^{\alpha_1-1}\right)|y_2^{\alpha_2}\cdots y_k^{\alpha_k}(x_1-y_1)|$$

$$\leqslant|\alpha|(\|x\|^{|\alpha|-1}+\|y\|^{|\alpha|-1})\max\{|x_i-y_i|:1\leqslant i\leqslant k\}, \tag{14.12}$$

$$\left[x=(x_1,\ldots,x_k),y=(y_1,\ldots,y_k)\in R^k\right].$$

Since by (14.11)

$$|\mathbf{Y}_{j,i}-\mathbf{Z}_{j,i}|=|E\mathbf{Y}_{j,i}|\leqslant n^{-(s-1)/2}\Delta_{n,j,s} \qquad (1\leqslant i\leqslant k), \tag{14.13}$$

it follows from (14.12) that

$$|E\mathbf{Y}_j^{\alpha}-E\mathbf{Z}_j^{\alpha}|\leqslant|\alpha|n^{-(s-1)/2}\Delta_{n,j,s}\left(E\|\mathbf{Y}_j\|^{|\alpha|-1}+E\|\mathbf{Z}_j\|^{|\alpha|-1}\right). \tag{14.14}$$

Now

$$E\|\mathbf{Y}_j\|^{|\alpha|-1}=\int_{\{\|\mathbf{X}_j\|\leqslant n^{1/2}\}}\|\mathbf{X}_j\|^{|\alpha|-1}\leqslant n^{(|\alpha|-1)/2}, \tag{14.15}$$

$$\|E\mathbf{Y}_j\|\leqslant\left(E\|\mathbf{Y}_j\|^2\right)^{1/2},$$

$$E\|\mathbf{Z}_j\|^{|\alpha|-1}\leqslant 2^{|\alpha|-1}\left(E\|\mathbf{Y}_j\|^{|\alpha|-1}+\|E\mathbf{Y}_j\|^{|\alpha|-1}\right)\leqslant 2^{|\alpha|}n^{(|\alpha|-1)/2}.$$

Hence

$$|E\mathbf{Y}_j^{\alpha}-E\mathbf{Z}_j^{\alpha}|\leqslant|\alpha|(2^{|\alpha|}+1)n^{-(s-|\alpha|)/2}\Delta_{n,j,s}.$$

Next we have

$$|d_{il}-v_{il}|\leqslant n^{-1}\sum_{j=1}^{n}|\text{cov}(\mathbf{X}_{j,i},\mathbf{X}_{j,l})-\text{cov}(\mathbf{Y}_{j,i},\mathbf{Y}_{j,l})|$$

$$=n^{-1}\sum_{j=1}^{n}|E(\mathbf{X}_{j,i}\mathbf{X}_{j,l}-\mathbf{Y}_{j,i}\mathbf{Y}_{j,l})+E(\mathbf{Y}_{j,i})E(\mathbf{Y}_{j,l})|$$

$$\leqslant n^{-1}\sum_{j=1}^{n}\left(n^{-(s-2)/2}\Delta_{n,j,s}+n^{1/2}n^{-(s-1)/2}\Delta_{n,j,s}\right)$$

$$=2n^{-(s-2)/2}\overline{\Delta}_{n,s}. \tag{14.16}$$

To prove (iv) note that [using (6.26)]

$$\|\mathbf{Z}_j\|^{s'}\leqslant(\|\mathbf{Y}_j\|+\|E\mathbf{Y}_j\|)^{s'}\leqslant 2^{s'-1}\left(\|\mathbf{Y}_j\|^{s'}+\left(E\|\mathbf{Y}_j\|^2\right)^{s'/2}\right)$$

$$\leqslant 2^{s'-1}\left(\|\mathbf{Y}_j\|^{s'}+E\|\mathbf{Y}_j\|^{s'}\right).$$

Proof of (v) is clear. Q.E.D.

Recall that the *norm of a* k×k *matrix* T, *denoted* ||T||, is defined by

$$\|T\|=\sup_{x\in R^k,\,\|x\|\leqslant 1}\|Tx\|. \tag{14.17}$$

COROLLARY 14.2 *Under the hypothesis of Lemma* 14.1, *one has*

$$|\langle t,Dt\rangle-\langle t,Vt\rangle|\leqslant 2kn^{-(s-2)/2}\overline{\Delta}_{n,s}\|t\|^2$$

$$\leqslant 2kn^{-(s-2)/2}\rho_s\|t\|^2\qquad (t\in R^k), \tag{14.18}$$

so that

$$\|D-V\|\leqslant 2kn^{-(s-2)/2}\overline{\Delta}_{n,s}\leqslant 2kn^{-(s-2)/2}\rho_s. \tag{14.19}$$

In particular, if V = I *and*

$$\overline{\Delta}_{n,s}\leqslant\frac{n^{(s-2)/2}}{8k}, \tag{14.20}$$

then D *is nonsingular and*

$$\tfrac{3}{4}\leqslant\|D\|\leqslant\tfrac{5}{4},\qquad \|D^{-1}\|\leqslant\tfrac{4}{3}. \tag{14.21}$$

Proof. By (14.8) and using (6.26) in the last step, we have

$$|\langle t,Dt\rangle-\langle t,Vt\rangle|=\left|\sum_{i,l=1}^{k}t_it_l(d_{il}-v_{il})\right|$$

$$\leqslant 2n^{-(s-2)/2}\bar{\Delta}_{n,s}\left(\sum_{i=1}^{k}|t_i|\right)^2\leqslant 2kn^{-(s-2)/2}\bar{\Delta}_{n,s}\|t\|^2. \tag{14.22}$$

Since the matrices D, V are symmetric,

$$\|D-V\|=\sup_{\|t\|\leqslant 1}|\langle t,(D-V)t\rangle|\leqslant 2kn^{-(s-2)/2}\bar{\Delta}_{n,s}. \tag{14.23}$$

Also, if $V=I$ and (14.20) holds, then

$$\|D-I\|\leqslant\tfrac{1}{4},\qquad \tfrac{3}{4}\leqslant\|D\|\leqslant\tfrac{5}{4}, \tag{14.24}$$

and

$$\langle t,Dt\rangle\geqslant\|t\|^2-\tfrac{1}{4}\|t\|^2=\tfrac{3}{4}\|t\|^2. \tag{14.25}$$

The last inequality implies that D is nonsingular and that $\|D^{-1}\|\leqslant\frac{4}{3}$. Q.E.D.

The next lemma concerns the growth of derivatives of the characteristic function of $n^{-1/2}(\mathbf{Z}_1+\cdots+\mathbf{Z}_n)$.

LEMMA 14.3 *Suppose* $\rho_s<\infty$ *for some* $s\geqslant 3$, *and that* $V=I$. *Let* g_j *denote the characteristic function of* $\mathbf{Z}_j (1\leqslant j\leqslant n)$. *Then if*

$$\|t\|\leqslant\frac{n^{1/2}}{16\rho_3}, \tag{14.26}$$

one has, for every nonnegative integral vector α,

$$\left|\left(D^\alpha\prod_{j=1}^{n}g_j\right)\left(\frac{t}{n^{1/2}}\right)\right|$$

$$\leqslant c_1(\alpha,k)(1+\|t\|^{|\alpha|})\exp\left\{-\tfrac{1}{2}\langle t,Dt\rangle+\tfrac{1}{6}\|t\|^2\right\}. \tag{14.27}$$

Proof. By inequality (8.46) one has

$$\left|g_j\left(\frac{t}{n^{1/2}}\right)\right|^2\leqslant\exp\left\{-n^{-1}E\langle t,\mathbf{Z}_j\rangle^2+\tfrac{2}{3}n^{-3/2}E|\langle t,\mathbf{Z}_j\rangle|^3\right\}. \tag{14.28}$$

Observe that

$$\exp\left\{n^{-1}E\langle t,\mathbf{Z}_j\rangle^2-\tfrac{2}{3}n^{-3/2}E|\langle t,\mathbf{Z}_j\rangle|^3\right\}$$
$$\leqslant \exp\left\{n^{-1}\left(E|\langle t,\mathbf{Z}_j\rangle|^3\right)^{2/3}-\tfrac{2}{3}n^{-3/2}E|\langle t,\mathbf{Z}_j\rangle|^3\right\}$$
$$\leqslant \sup_{a\geqslant 0}\exp\left\{a^2-\tfrac{2}{3}a^3\right\}=e^{1/3}. \tag{14.29}$$

Now let N_r be a subset of $N=\{1,2,\dots,n\}$ containing r elements. Then in the given range (14.26) one has

$$\prod_{j\in N\setminus N_r}\left|g_j\left(\frac{t}{n^{1/2}}\right)\right|^2\leqslant \prod_{j\in N\setminus N_r}\exp\left\{-n^{-1}E\langle t,\mathbf{Z}_j\rangle^2+\tfrac{2}{3}n^{-3/2}E|\langle t,\mathbf{Z}_j\rangle|^3\right\}$$
$$=\exp\left\{-\langle t,Dt\rangle+\tfrac{2}{3}n^{-3/2}\sum_{j=1}^{n}E|\langle t,\mathbf{Z}_j\rangle|^3\right\}$$
$$\times\prod_{j\in N_r}\exp\left\{n^{-1}E\langle t,\mathbf{Z}_j\rangle^2-\tfrac{2}{3}n^{-3/2}E|\langle t,\mathbf{Z}_j\rangle|^3\right\}$$
$$\leqslant \exp\left\{-\langle t,Dt\rangle+\tfrac{2}{3}n^{-1/2}\rho_3'\|t\|^3\right\}e^{r/3}$$
$$\leqslant \exp\left\{-\langle t,Dt\rangle+\tfrac{1}{3}\|t\|^2\right\}e^{r/3}, \tag{14.30}$$

using (14.9) (with $s'=3$) and (14.26) in the last step. It follows that

$$\prod_{j\in N\setminus N_r}\left|g_j\left(\frac{t}{n^{1/2}}\right)\right|\leqslant \exp\left\{\frac{r}{6}-\tfrac{1}{2}\langle t,Dt\rangle+\tfrac{1}{6}\|t\|^2\right\}, \tag{14.31}$$

which proves the lemma for $\alpha=0$. To prove it for nonzero α, note that

$$\left|(D_m g_j)\left(\frac{t}{n^{1/2}}\right)\right|=n^{-1/2}\left|E\left(\mathbf{Z}_{j,m}\exp\left\{i\langle n^{-1/2}t,\mathbf{Z}_j\rangle\right\}\right)\right|$$
$$=n^{-1/2}\left|E\left[\mathbf{Z}_{j,m}\left(\exp\left\{i\langle n^{-1/2}t,\mathbf{Z}_j\rangle\right\}-1\right)\right]\right|$$
$$\leqslant n^{-1/2}E\left|\mathbf{Z}_{j,m}\langle n^{-1/2}t,\mathbf{Z}_j\rangle\right|\leqslant n^{-1}\|t\|\,E\,\|\mathbf{Z}_j\|^2$$
$$\leqslant 4n^{-1}\rho_{2,j}\|t\| \qquad \left[\mathbf{Z}_j=(\mathbf{Z}_{j,1},\dots,\mathbf{Z}_{j,k})\right], \tag{14.32}$$

using (14.9) in the last step (with $s'=2$). For a nonnegative integral vector

β satisfying $|\beta| \geqslant 2$,

$$\left|(D^{\beta}g_j)\left(\frac{t}{n^{1/2}}\right)\right| \leqslant n^{-|\beta|/2}E|\mathbf{Z}_j^{\beta}| \leqslant n^{-|\beta|/2}\rho'_{|\beta|,j}$$

$$\leqslant 2^{|\beta|}n^{-1}\rho_{2,j}, \tag{14.33}$$

by (6.29) and (14.10). Thus

$$\left|(D^{\beta}g_j)\left(\frac{t}{n^{1/2}}\right)\right| \leqslant c_2(\alpha,k)n^{-1}\rho_{2,j}\,\overline{\|t\|} \qquad (1 \leqslant |\beta| \leqslant |\alpha|), \tag{14.34}$$

where

$$c_2(\alpha,k) = \max\{4, 2^{|\alpha|}\}, \qquad \overline{\|t\|} = \max\{1, \|t\|\}. \tag{14.35}$$

By Leibniz' rule for differentiation of the product of n functions, $(D^{\alpha}\prod_{j=1}^{n} g_j)(t/n^{1/2})$ is the sum of $n^{|\alpha|}$ terms, a typical term being

$$\left[\prod_{j \notin N_r} g_j\left(\frac{t}{n^{1/2}}\right)\right]\left[D^{\beta_1}g_{j_1}\left(\frac{t}{n^{1/2}}\right)\right]\cdots\left[D^{\beta_r}g_{j_r}\left(\frac{t}{n^{1/2}}\right)\right], \tag{14.36}$$

where $N_r = \{j_1,\dots,j_r\}$, $1 \leqslant r \leqslant |\alpha|$, and $\beta_1,\dots,\beta_r$ are nonnegative integral vectors satisfying

$$|\beta_i| \geqslant 1 \qquad (1 \leqslant i \leqslant r), \qquad \sum_{i=1}^{r} \beta_i = \alpha. \tag{14.37}$$

The number of times the expression (14.36) is repeated among the $n^{|\alpha|}$ summands is given by

$$\frac{\alpha_1!\cdots\alpha_k!}{\prod_{i=1}^{r}\prod_{j=1}^{k}\beta_{ij}!}, \tag{14.38}$$

where

$$\alpha = (\alpha_1,\dots,\alpha_k), \qquad \beta_i = (\beta_{i1},\dots,\beta_{ik}) \qquad (1 \leqslant i \leqslant r). \tag{14.39}$$

In view of (14.31) and (14.34), the expression (14.36) is bounded in magnitude by

$$\exp\left\{\frac{r}{6} - \tfrac{1}{2}\langle t, Dt\rangle + \tfrac{1}{6}\|t\|^2\right\}\prod_{j \in N_r} b_j, \tag{14.40}$$

where

$$b_j = n^{-1}c_2(\alpha,k)\rho_{2,j}\,\overline{\|t\|} \qquad (1 \leqslant j \leqslant n). \tag{14.41}$$

Therefore

$$\left|\left(D^\alpha \prod_{j=1}^{n} g_j\right)\left(\frac{t}{n^{1/2}}\right)\right|$$

$$\leqslant \sum_{1 \leqslant r \leqslant |\alpha|} c_3(\alpha,r)\exp\left\{\frac{r}{6} - \tfrac{1}{2}\langle t, Dt\rangle + \tfrac{1}{6}\|t\|^2\right\}\cdot \sum{}^*\left(\sum_{j\in N_r} b_j\right), \tag{14.42}$$

where, for each r, Σ^* denotes summation over all choices of r indices from $N = \{1,\dots,n\}$. Hence

$$\sum{}^*\left(\prod_{j\in N_r} b_j\right) \leqslant \left(\sum_{j=1}^{n} b_j\right)^r = \left(c_2(\alpha,k)\rho_2\,\overline{\|t\|}\,\right)^r$$

$$= \left(c_2(\alpha,k)k\,\overline{\|t\|}\,\right)^r. \tag{14.43}$$

The inequality (14.27) follows on using (14.43) in (14.42). Q.E.D.

COROLLARY 14.4 *If in addition to the hypothesis of Lemma* 14.3 *the inequality* (14.20) *holds, then*

$$\left|\left(D^\alpha \prod_{j=1}^{n} g_j\right)\left(\frac{t}{n^{1/2}}\right)\right| \leqslant c_1(\alpha,k)(1+\|t\|^{|\alpha|})\exp\left\{-\tfrac{5}{24}\|t\|^2\right\} \tag{14.44}$$

for all t *satisfying* (14.26).

Proof. Use (14.25) in (14.27). Q.E.D.

We also need an estimate for the difference between a polynomial expression in the cumulants of $n^{-1/2}(\mathbf{X}_1 + \cdots + \mathbf{X}_n)$ and the corresponding expression in the cumulants of $n^{-1/2}(\mathbf{Z}_1 + \cdots + \mathbf{Z}_n)$.

LEMMA 14.5 *Suppose that* V = I *and* (14.20) *holds for some* s ⩾ 3. *Let* $\nu_1,\dots,\nu_m$ *be nonnegative integral vectors satisfying*

$$|\nu_i| \geqslant 3 \qquad (1 \leqslant i \leqslant m), \qquad \sum_{i=1}^{m}(|\nu_i|-2) = r, \tag{14.45}$$

for some positive integer r, 1 ⩽ r ⩽ s − 2. *Then*

$$|\chi_{\nu_1}\cdots\chi_{\nu_m} - \chi'_{\nu_1}\cdots\chi'_{\nu_m}| \leqslant c_4(s,k)n^{-(s-r-2)/2}\bar{\Delta}_{n,s}$$

$$\leqslant c_4(s,k)n^{-(s-r-2)/2}\rho_s. \tag{14.46}$$

Proof. First recall [see (6.14), (6.15)] that for each nonnegative integral vector ν, $\chi_{\nu,j}$ is a linear combination of terms like

$$\mu_{\alpha_1,j}^{r_1}\cdots\mu_{\alpha_p,j}^{r_p}, \tag{14.47}$$

where $\alpha_1,\ldots,\alpha_p$ are nonnegative integral vectors and $r_1,\ldots,r_p$ are positive integers satisfying

$$\sum_{i=1}^{p} r_i\alpha_i=\nu \qquad (\alpha_i>0, 1\leqslant i\leqslant p). \tag{14.48}$$

To estimate the difference $\chi_{\nu,j}-\chi'_{\nu,j}$ it is therefore enough to estimate

$$\mu_{\alpha_1,j}^{r_1}\cdots\mu_{\alpha_p,j}^{r_p}-\mu'^{r_1}_{\alpha_1,j}\cdots\mu'^{r_p}_{\alpha_p,j} \tag{14.49}$$

subject to (14.48) and $3\leqslant|\nu|\leqslant s$. By Lemma 14.1(ii),

$$|\mu_{\alpha,j}-\mu'_{\alpha,j}|\leqslant c_5(\alpha)n^{-(s-|\alpha|)/2}\Delta_{n,j,s} \qquad (1\leqslant|\alpha|\leqslant s). \tag{14.50}$$

We may use (14.50) and (14.12) to get

$$\begin{aligned}
&|\mu_{\alpha_1,j}^{r_1}\cdots\mu_{\alpha_p,j}^{r_p}-\mu'^{r_1}_{\alpha_1,j}\cdots\mu'^{r_p}_{\alpha_p,j}|\\
&\qquad\leqslant\sum_{i=1}^{p} r_ic_6(\alpha_i)n^{-(s-|\alpha_i|)/2}\Delta_{n,j,s}\big(|\mu_{\alpha_i,j}^{r_i-1}|+|\mu'^{r_i-1}_{\alpha_i,j}|\big)\\
&\qquad\qquad\times|\mu_{\alpha_1,j}^{r_1}\cdots\mu_{\alpha_{i-1},j}^{r_i}\mu'^{r_{i+1}}_{\alpha_{i+1},j}\cdots\mu'^{r_p}_{\alpha_p,j}|\\
&\qquad\leqslant\sum_{i=1}^{p} c_7n^{-(s-|\alpha_i|)/2}\Delta_{n,j,s}\rho_{s,j}^{m_i/s},
\end{aligned} \tag{14.51}$$

where c_7 depends on i, α's, and r's, and

$$m_i=\sum_{i'=1}^{p} r_{i'}|\alpha_{i'}|-|\alpha_i|=|\nu|-|\alpha_i|, \tag{14.52}$$

by (14.48). In the last step of (14.51) we have used the inequalities [see (6.27), Lemma 6.2(ii), and (14.9)]

$$\begin{aligned}
&|\mu_{\alpha_i,j}|\leqslant\rho_{|\alpha_i|,j}\leqslant\rho_{s,j}^{|\alpha_i|/s},\\
&|\mu'_{\alpha_i,j}|\leqslant2^{|\alpha_i|}\rho_{|\alpha_i|,j}\leqslant2^{|\alpha_i|}\rho_{s,j}^{|\alpha_i|/s}.
\end{aligned} \tag{14.53}$$

In view of (14.20) one has

$$\rho_{s,j}=\int_{\{\|\mathbf{X}_j\|\leqslant n^{1/2}\}}\|\mathbf{X}_j\|^s+\Delta_{n,j,s}\leqslant n^{s/2}+\Delta_{n,j,s}$$

$$\leqslant n^{s/2}+n\bar{\Delta}_{n,s}\leqslant n^{s/2}+\frac{n^{s/2}}{8k}, \tag{14.54}$$

so that

$$n^{-(s-|\alpha_i|)/2}\rho_{s,j}^{m_i/s}=n^{-(s-|\alpha_i|)/2}\rho_{s,j}^{(|\nu|-|\alpha_i|)/s}$$

$$\leqslant n^{-(s-|\alpha_i|)/2}n^{(|\nu|-|\alpha_i|)/2}\left(1+\frac{1}{8k}\right)^{(|\nu|-|\alpha_i|)/s}$$

$$\leqslant c_8(\alpha_i,k)n^{-(s-|\nu|)/2}. \tag{14.55}$$

It now follows from (14.51) that

$$|\chi_{\nu,j}-\chi'_{\nu,j}|\leqslant c_9(\nu,k)n^{-(s-|\nu|)/2}\Delta_{n,j,s} \tag{14.56}$$

and, therefore, averaging over $j=1,\ldots,n$,

$$|\chi_\nu-\chi'_\nu|\leqslant c_9(\nu,k)\bar{\Delta}_{n,s}n^{-(s-|\nu|)/2}\qquad(3\leqslant|\nu|\leqslant s). \tag{14.57}$$

Let now $\nu_1,\ldots,\nu_m$ be nonnegative integral vectors satisfying (14.45). By (14.57) one has

$$|\chi_{\nu_1}\cdots\chi_{\nu_m}-\chi'_{\nu_1}\cdots\chi'_{\nu_m}|$$

$$\leqslant\sum_{i=1}^{m}c_9(\nu_i,k)\bar{\Delta}_{n,s}n^{-(s-|\nu_i|)/2}|\chi_{\nu_1}\cdots\chi_{\nu_{i-1}}\chi'_{\nu_{i+1}}\cdots\chi'_{\nu_m}|. \tag{14.58}$$

By Lemma 6.3 (and averaging), Lemma 6.2(iii), and Lemma 14.1(iv) one gets (recall that $\rho_2=k$)

$$|\chi_{\nu_1}\cdots\chi_{\nu_{i-1}}\chi'_{\nu_{i+1}}\cdots\chi'_{\nu_m}|$$

$$\leqslant c'_9\rho_{|\nu_1|}\cdots\rho_{|\nu_{i-1}|}\rho'_{|\nu_{i+1}|}\cdots\rho'_{|\nu_m|}$$

$$\leqslant c_{10}(s,k)\rho_{|\nu_1|}\cdots\rho_{|\nu_{i-1}|}\rho_{|\nu_{i+1}|}\cdots\rho_{|\nu_m|}$$

$$\leqslant c_{11}(s,k)\rho_s^{(r-|\nu_i|+2)/(s-2)}. \tag{14.59}$$

But because of (14.20), one has

$$\rho_s = n^{-1}\sum_{j=1}^{n}\rho_{s,j}$$

$$= n^{-1}\sum_{j=1}^{n}\left[\int_{\{\|\mathbf{X}_j\|\leqslant n^{1/2}\}}\|\mathbf{X}_j\|^s + \Delta_{n,j,s}\right]$$

$$\leqslant n^{-1}\sum_{j=1}^{n}\left[n^{(s-2)/2}\int_{\{\|\mathbf{X}_j\|\leqslant n^{1/2}\}}\|\mathbf{X}_j\|^2 + \Delta_{n,j,s}\right]$$

$$\leqslant kn^{(s-2)/2} + \bar{\Delta}_{n,s} \leqslant \left(1+\frac{1}{8k}\right)n^{(s-2)/2}. \tag{14.60}$$

Using (14.60) in (14.59), one gets

$$|\chi_{\nu_1}\cdots\chi_{\nu_{i-1}}\chi'_{\nu_{i+1}}\cdots\chi'_{\nu_m}| \leqslant c_{12}(s,k)n^{(r-|\nu_i|+2)/2}. \tag{14.61}$$

Substituting this in (14.58) we obtain (14.46). Q.E.D.

The above estimates also lead to

LEMMA 14.6 *Assume that* V = I, *and that* (14.20) *holds for some* s ≥ 3. *Then for every integer* r, 0 ≤ r ≤ s − 2, *one has*

$$n^{-r/2}|P_r(-\phi:\{\chi_\nu\})(x) - P_r(-\phi_{0,D}:\{\chi'_\nu\})(x)|$$

$$\leqslant c_{13}(r,k,s)\bar{\Delta}_{n,s}n^{-(s-2)/2}(1+\|x\|^{3r+2})$$

$$\times \exp\left\{-\frac{\|x\|^2}{6}+\|x\|\right\} \qquad (x\in R^k), \tag{14.62}$$

and

$$n^{-r/2}|P_r(-\phi:\{\chi_\nu\})(x+a_n) - P_r(-\phi:\{\chi_\nu\})(x)|$$

$$\leqslant c_{14}(r,k,s)\bar{\Delta}_{n,s}n^{-(s-2)/2}(1+\|x\|^{3r+1})$$

$$\times \exp\left\{-\tfrac{1}{2}\|x\|^2+\frac{\|x\|}{8k^{1/2}}\right\} \qquad (x\in R^k), \tag{14.63}$$

where

$$a_n = n^{-1/2}\sum_{j=1}^{n}E\mathbf{Y}_j. \tag{14.64}$$

Proof. We first obtain an auxiliary estimate. For $z \in \mathbf{C}^k$ write

$$g(z) = 1 - (\operatorname{Det} D)^{-1/2} \exp\left\{ \tfrac{1}{2} \sum_{i=1}^{k} z_i^2 - \tfrac{1}{2} \sum_{1 \leqslant i, i' \leqslant k} d^{ii'} z_i z_{i'} \right\}$$

$$[z = (z_1, \ldots, z_k)]. \tag{14.65}$$

Then

$$\begin{aligned} |g(z)| &= \left|(\operatorname{Det} D)^{-1/2} \exp\left\{ \tfrac{1}{2} \sum_{i=1}^{k} z_i^2 - \tfrac{1}{2} \sum_{1 \leqslant i, i' \leqslant k} d^{ii'} z_i z_{i'} \right\} - 1\right| \\ &\leqslant |(\operatorname{Det} D)^{-1/2} - 1| \exp\{ \|D^{-1} - I\| (\|z\|^2) \} \\ &\quad + \tfrac{1}{2} \|D^{-1} - I\| (\|z\|^2) \exp\{ \|D^{-1} - I\| (\|z\|^2) \} \\ &\qquad [z = (z_1, \ldots, z_k) \in \mathbf{C}^k], \end{aligned} \tag{14.66}$$

where Det denotes *determinant* and $d^{ii'}$ is the (i,i') element of D^{-1}.

By Corollary 14.2 [replacing t by Bt in (14.18), where $B^2 = D^{-1}$],

$$\begin{aligned} |\langle t, D^{-1} t\rangle - \|t\|^2| &\leqslant 2kn^{-(s-2)/2} \bar{\Delta}_{n,s} \|D^{-1}\| (\|t\|^2) \\ &\leqslant \tfrac{8}{3} kn^{-(s-2)/2} \bar{\Delta}_{n,s} \|t\|^2, \end{aligned} \tag{14.67}$$

which implies

$$\|D^{-1} - I\| \leqslant \tfrac{8}{3} kn^{-(s-2)/2} \bar{\Delta}_{n,s} \leqslant \tfrac{1}{3}. \tag{14.68}$$

Also, expanding $\operatorname{Det} D$ and making use of (14.8) and (14.20), we get

$$\begin{aligned} \left|(\operatorname{Det} D)^{-1/2} - 1\right| &= (\operatorname{Det} D)^{-1/2} \left|1 - (\operatorname{Det} D)^{1/2}\right| \\ &\leqslant (\operatorname{Det} D)^{-1/2} \left|1 - (\operatorname{Det} D)^{1/2}\right| \left(1 + (\operatorname{Det} D)^{1/2}\right) \\ &= (\operatorname{Det} D)^{-1/2} |1 - \operatorname{Det} D| \leqslant \|D^{-1}\|^{k/2} |1 - \operatorname{Det} D| \\ &\leqslant c_{15}(s,k) \bar{\Delta}_{n,s} n^{-(s-2)/2}. \end{aligned} \tag{14.69}$$

The estimates (14.68), (14.69) are now substituted in (14.66) to yield

$$|g(z)| \leqslant c_{16}(s,k) \bar{\Delta}_{n,s} n^{-(s-2)/2} (1 + \|z\|^2) \exp\left\{ \frac{\|z\|^2}{3} \right\} \qquad (z \in \mathbf{C}^k). \tag{14.70}$$

Hence

$$\max_{\{z'\in \mathbf{C}^k,\,\|z'-z\|\leqslant 1\}} |g(z)| \leqslant c_{16}(s,k)\bar{\Delta}_{n,s} n^{-(s-2)/2}(2+\|z\|)^2 \exp\left\{\frac{(\|z\|+1)^2}{3}\right\}$$

$$\leqslant c_{17}(s,k)\bar{\Delta}_{n,s} n^{-(s-2)/2}(1+\|z\|^2)\exp\left\{\frac{\|z\|^2}{3}+\|z\|\right\}. \tag{14.71}$$

By Cauchy's estimate (Lemma 9.2) one then obtains

$$|D^{\nu-\alpha} g(z)| \leqslant c_{18}(s,k,\nu-\alpha)\bar{\Delta}_{n,s} n^{-(s-2)/2}(1+\|z\|^2)\exp\left\{\frac{\|z\|^2}{3}+\|z\|\right\}. \tag{14.72}$$

From this follows

$$|D^{\nu}(\phi-\phi_{0,D})(x)| = \left|\sum_{0\leqslant\alpha\leqslant\nu} c_{19}(\alpha)(D^{\alpha}\phi)(x)(D^{\nu-\alpha}g)(x)\right|$$

$$\leqslant c_{20}(s,k)\bar{\Delta}_{n,s} n^{-(s-2)/2}(1+\|x\|^{|\nu|+2})\exp\left\{-\frac{\|x\|^2}{6}+\|x\|\right\}$$

$$(x\in R^k). \tag{14.73}$$

Now recall from Section 7 that [using (7.3) and Lemma 7.2]

$$P_r(-\phi:\{\chi_\nu\})$$

$$=\sum_{m=1}^{r}\frac{1}{m!}\left\{\sum_{j_1,\ldots,j_m}^{*}\left(\sum_{\nu_1,\ldots,\nu_m}^{**}\frac{\chi_{\nu_1}\cdots\chi_{\nu_m}}{\nu_1!\cdots\nu_m!}(-1)^{r+2m}D^{\nu_1+\cdots+\nu_m}\phi\right)\right\} \tag{14.74}$$

where Σ^* denotes summation over all m-tuples of positive integers $(j_1,\ldots,j_m)$ satisfying

$$\sum_{i=1}^{m} j_i = r, \tag{14.75}$$

and Σ^{**} denotes summation [for fixed $(j_1,\ldots,j_m)$] over all m-tuples of nonnegative integral vectors $(\nu_1,\ldots,\nu_m)$ satisfying

$$|\nu_i| = j_i + 2 \qquad (1 \leqslant i \leqslant m). \tag{14.76}$$

The function $P_r(-\phi_{0,D}\colon \{\chi'_\nu\})$ is similarly defined by replacing χ_ν by χ'_ν and ϕ by $\phi_{0,D}$ in (14.74). Next observe that

$$\begin{aligned}&|\chi_{\nu_1}\cdots\chi_{\nu_m} D^{\nu_1+\cdots+\nu_m}\phi(x) - \chi'_{\nu_1}\cdots\chi'_{\nu_m} D^{\nu_1+\cdots+\nu_m}\phi_{0,D}(x)|\\ &\quad \leqslant |\chi'_{\nu_1}\cdots\chi'_{\nu_m} D^{\nu_1+\cdots+\nu_m}(\phi(x)-\phi_{0,D}(x))|\\ &\qquad + |(\chi_{\nu_1}\cdots\chi_{\nu_m} - \chi'_{\nu_1}\cdots\chi'_{\nu_m}) D^{\nu_1+\cdots+\nu_m}\phi(x)| \qquad (x\in R^k).\end{aligned} \tag{14.77}$$

By Lemma 14.5

$$|\chi_{\nu_1}\cdots\chi_{\nu_m} - \chi'_{\nu_1}\cdots\chi'_{\nu_m}| \leqslant c_4(s,k) n^{-(s-r-2)/2}\bar{\Delta}_{n,s}. \tag{14.78}$$

Also, as in (14.61), one has

$$|\chi'_{\nu_1}\cdots\chi'_{\nu_m}| \leqslant c_{21}(s,k) n^{r/2}. \tag{14.79}$$

Using (14.78), (14.79), and (14.73) in (14.77), we have

$$\begin{aligned}&|\chi_{\nu_1}\cdots\chi_{\nu_m} D^{\nu_1+\cdots+\nu_m}\phi(x) - \chi'_{\nu_1}\cdots\chi'_{\nu_m} D^{\nu_1+\cdots+\nu_m}\phi_{0,D}(x)|\\ &\quad \leqslant c_{22}(s,k)\bar{\Delta}_{n,s} n^{-(s-r-2)/2}(1+\|x\|^{3r+2})\exp\left\{-\frac{\|x\|^2}{6}+\|x\|\right\}\\ &\qquad (x\in R^k).\end{aligned} \tag{14.80}$$

The inequality (14.62) follows from this and the expression P_r given by (14.74).

To prove (14.63), first note that by Lemma 14.1(ii)

$$\|a_n\| \leqslant n^{-1/2}\sum_{j=1}^{n}\|E\mathbf{Y}_j\| \leqslant k^{1/2}\bar{\Delta}_{n,s} n^{-(s-2)/2} \leqslant (8k^{1/2})^{-1}. \tag{14.81}$$

From the expression (14.74) and the inequality (14.79) (which also holds if the primes are deleted), we get

$$\begin{aligned}&|P_r(-\phi\colon \{\chi_\nu\})(x+a_n) - P_r(-\phi\colon \{\chi_\nu\})(x)|\\ &\quad \leqslant c_{23}(s,k) n^{r/2}\max\{|D^\nu\phi(x+a_n) - D^\nu\phi(x)|\colon 0\leqslant|\nu|\leqslant 3r\}.\end{aligned} \tag{14.82}$$

L14.7 : $X_1,\ldots,X_n$ ind, $\Sigma\sigma_k^2 = 1$, $\sigma_k^2 = \mathrm{Var}\,X_k$. $X_k' = X_k I[|X_k| \leq 1$
$S_n = \Sigma_1^n X_k = \Sigma X_k' + \Sigma X_k'' = S_n' + S_n''$ $X_k'' = X_k - X_k'$, con
$\Rightarrow E|S_n|^r \leq 2^{r-1}\{E|S_n'|^r + E|S_n''|^r\} \leq$ Rosenthal
EX_k'
$= EX$

But $D^{\nu}\phi = q_{\nu}\phi$, where q_{ν} is a polynomial of degree $|\nu|$ with coefficients depending only on ν and k, so that

$$|D^{\nu}\phi(x+a_n) - D^{\nu}\phi(x)|$$
$$\leqslant |q_{\nu}(x+a_n) - q_{\nu}(x)|\phi(x) + |q_{\nu}(x+a_n)||\phi(x+a_n) - \phi(x)|. \tag{14.83}$$

But by (14.12) and (14.81),

$$|(x+a_n)^{\alpha} - x^{\alpha}| \leqslant |\alpha|(\|x+a_n\|^{|\alpha|-1} + \|x\|^{|\alpha|-1})\|a_n\|$$
$$\leqslant c_{24}(\alpha,k)\bar{\Delta}_{n,s} n^{-(s-2)/2}(1+\|x\|^{|\alpha|-1}) \qquad (x \in R^k), \tag{14.84}$$

and

$$|(x+a_n)^{\alpha}[\phi(x+a_n) - \phi(x)]|$$
$$= |(x+a_n)^{\alpha}\phi(x)|\,|\exp\{-\tfrac{1}{2}\|x+a_n\|^2 + \tfrac{1}{2}\|x\|^2\} - 1|$$
$$\leqslant |(x+a_n)^{\alpha}\phi(x)\tfrac{1}{2}(\|x+a_n\|^2 - \|x\|^2)\exp\{-\tfrac{1}{2}\|x+a_n\|^2 + \tfrac{1}{2}\|x\|^2\}|$$
$$\leqslant |(x+a_n)^{\alpha}|\phi(x)\|a_n\|(1+\|x\|)\exp\{\|a_n\|(1+\|x\|)\}$$
$$\leqslant c_{25}(\alpha,k)\bar{\Delta}_{n,s} n^{-(s-2)/2}(1+\|x\|^{|\alpha|+1})\exp\left\{-\tfrac{1}{2}\|x\|^2 + \frac{\|x\|}{8k^{1/2}}\right\}. \tag{14.85}$$

Relations (14.84), (14.85) are used in (14.83) to yield

$$|D^{\nu}\phi(x+a_n) - D^{\nu}\phi(x)| \leqslant c_{26}(\nu,k)\bar{\Delta}_{n,s} n^{-(s-2)/2}(1+\|x\|^{|\nu|+1})$$
$$\times \exp\left\{-\tfrac{1}{2}\|x\|^2 + \frac{\|x\|}{8k^{1/2}}\right\}. \tag{14.86}$$

Finally, (14.86) is used in (14.82) to get (14.62). Q.E.D.

A bound for absolute moments of sums of independent random vectors is needed to prove the important Lemma 14.8. This is provided by

LEMMA 14.7 *Suppose that* $\mathrm{V} = \mathrm{I}$ *and* (14.20) *holds for an integer* $\mathrm{s} \geqslant 2$. *Then there exists a constant* $c_{27}(\mathrm{r,s,k})$ *such that*

$$E\|\mathbf{X}_1 + \cdots + \mathbf{X}_n\|^r \leqslant c_{27}(r,s,k)n^{r/2} \tag{14.87}$$

for $1 \leqslant \mathrm{r} \leqslant \mathrm{s}$.

Rosenthal:
dim = 1: LHS ≤ K(Σ E|X_k|^r) ∨ (Σ EX_k^2)^{r/2}

Proof. In view of Lemma 6.2(ii) it is enough to prove the lemma for $r=s$. For $r=s=2$, (14.87) is trivially true. Let us then assume that $s\geqslant 3$. Also by (6.26),

$$E\|\mathbf{X}_1+\cdots+\mathbf{X}_n\|^s=E\left[\sum_{i=1}^{k}\left(\sum_{j=1}^{n}\mathbf{X}_{j,i}\right)^2\right]^{s/2}$$
$$\leqslant k^{(s-2)/2}\sum_{i=1}^{k}E\left|\sum_{j=1}^{n}\mathbf{X}_{j,i}\right|^s$$
$$\left[\mathbf{X}_j=(\mathbf{X}_{j,1},\ldots,\mathbf{X}_{j,k})\right],\tag{14.88}$$

so that it is enough to prove the lemma for $k=1$. If s is an even integer and $k=1$, then by Theorem 9.11,

$$n^{-s/2}E(\mathbf{X}_1+\cdots+\mathbf{X}_n)^s\leqslant\left|\sum_{m=0}^{s-3}n^{-m/2}D^s g_m(0)\right|+c_{28}(s)n^{-(s-2)/2}\rho_s,\tag{14.89}$$

where

$$g_m(t)=\tilde{P}_m(it:\{\chi_\nu\})\exp\left\{\frac{-t^2}{2}\right\}\qquad(t\in R^1).\tag{14.90}$$

By Lemma 9.5 (with $k=1$, $\alpha=s$, $z=0$) and inequality (14.60), one has

$$|D^s g_m(0)|\leqslant c_{29}(m,s)\rho_s^{m/(s-2)}\leqslant c_{30}(m,s)n^{m/2},$$
$$\rho_s\leqslant\tfrac{9}{8}n^{(s-2)/2}\qquad(0\leqslant m\leqslant s-3).\tag{14.91}$$

The lemma is proved for even integers s.

Next assume that s is an odd integer, $s\geqslant 3$, and $k=1$. Clearly,

$$E|\mathbf{X}_1+\cdots+\mathbf{X}_n|^s\leqslant E|\mathbf{Z}_1+\cdots+\mathbf{Z}_n|^s$$
$$+E|(\mathbf{X}_1+\cdots+\mathbf{X}_n)^s-(\mathbf{Z}_1+\cdots+\mathbf{Z}_n)^s|.\tag{14.92}$$

By Lemma 14.3 [inequality (14.27) with $k=1$, $\alpha=s+1$, $t=0$],

$$E|\mathbf{Z}_1+\cdots+\mathbf{Z}_n|^s\leqslant\left(E(\mathbf{Z}_1+\cdots+\mathbf{Z}_n)^{s+1}\right)^{s/(s+1)}$$
$$\leqslant c_{31}(s)n^{s/2}.\tag{14.93}$$

Next,

$$E|(\mathbf{X}_1+\cdots+\mathbf{X}_n)^s-(\mathbf{Z}_1+\cdots+\mathbf{Z}_n)^s|$$

$$\leqslant \sum_{j=1}^{n} E|(\mathbf{X}_1+\cdots+\mathbf{X}_j\ +\mathbf{Z}_{j+1}+\cdots+\mathbf{Z}_n)^s$$

$$-(\mathbf{X}_1+\cdots+\mathbf{X}_{j-1}+\mathbf{Z}_j+\cdots+\mathbf{Z}_n)^s|$$

$$\leqslant \sum_{j=1}^{n}\sum_{m=1}^{s}\binom{s}{m}E|(\mathbf{X}_j^m-\mathbf{Z}_j^m)(\mathbf{X}_1+\cdots+\mathbf{X}_{j-1}+\mathbf{Z}_{j+1}+\cdots+\mathbf{Z}_n)^{s-m}|$$

$$=\sum_{j=1}^{n}\sum_{m=1}^{s}\binom{s}{m}E|\mathbf{X}_j^m-\mathbf{Z}_j^m|E|\mathbf{X}_1+\cdots+\mathbf{X}_{j-1}+\mathbf{Z}_{j+1}+\cdots+\mathbf{Z}_n|^{s-m}. \tag{14.94}$$

By (6.26) and (14.54) we have

$$E|\mathbf{X}_1+\cdots+\mathbf{X}_{j-1}+\mathbf{Z}_{j+1}+\cdots+\mathbf{Z}_n|^{s-m}$$

$$\leqslant 2^{s-m-1}(E|\mathbf{X}_1+\cdots+\mathbf{X}_j+\mathbf{Z}_{j+1}+\cdots+\mathbf{Z}_n|^{s-m}+E|\mathbf{X}_j|^{s-m})$$

$$\leqslant 2^{s-m-1}E|\mathbf{X}_1+\cdots+\mathbf{X}_j+\mathbf{Z}_{j+1}+\cdots+\mathbf{Z}_n|^{s-m}$$

$$+2^{s-m-1}\tfrac{9}{8}n^{(s-m)/2}. \tag{14.95}$$

Writing

$$\tau_{r,j}=n^{-1}(E|\mathbf{X}_1|^r+\cdots+E|\mathbf{X}_j|^r+E|\mathbf{Z}_{j+1}|^r+\cdots+E|\mathbf{Z}_n|^r), \tag{14.96}$$

one has, using Lemma 14.1 and inequality (14.20),

$$\tau_{2,j}\leqslant n^{-1}\sum_{j=1}^{n}E\mathbf{X}_j^2=1,\qquad E\mathbf{Z}_{j+i}^2\leqslant E\mathbf{Y}_{j+i}^2\leqslant E\mathbf{X}_{j+i}^2, \tag{14.97}$$

$$\tau_{2,j}=n^{-1}\left[E(\mathbf{X}_1^2+\cdots+\mathbf{X}_n^2)-E(\mathbf{X}_{j+1}^2-E\mathbf{Z}_{j+1}^2)-\cdots-(E\mathbf{X}_n^2-E\mathbf{Z}_n^2)\right]$$

$$=n^{-1}\Big[n-E(\mathbf{X}_{j+1}^2-\mathbf{Y}_{j+1}^2)-\cdots-E(\mathbf{X}_n^2-\mathbf{Y}_n^2)$$

$$-(E\mathbf{Y}_{j+1})^2-\cdots-(E\mathbf{Y}_n)^2\Big]$$

$$\geqslant 1-n^{-1}\sum_{j=1}^{n} n^{-(s-2)/2}\Delta_{n,j,s}-n^{-1}\sum_{j=1}^{n} n^{-(s-1)}\Delta_{n,j,s}^{2}$$

$$\geqslant 1-2\bar{\Delta}_{n,s}n^{-(s-2)/2}\geqslant \tfrac{3}{4},$$

$$\tau_{r,j}\leqslant n^{-1}\left[E|\mathbf{X}_1|^r+\cdots+E|\mathbf{X}_j|^r+2^r\left(E|\mathbf{X}_{j+1}|^r+\cdots+E|\mathbf{X}_n|^r\right)\right]$$

$$\leqslant 2^r\phi_r.$$

Look at the random variables

$$\mathbf{W}_i=\begin{cases}\dfrac{\mathbf{X}_i}{\tau_{2,j}^{1/2}}, & 1\leqslant i\leqslant j,\\[2ex] \dfrac{\mathbf{Z}_i}{\tau_{2,j}^{1/2}}, & j+1\leqslant i\leqslant n.\end{cases}\tag{14.98}$$

Applying Theorem 9.11 (with $k=1$, $\alpha=s-1$, $t=0$) to these random variables, we have

$$E\left|\sum_{i=1}^{n}\mathbf{W}_i\right|^{s-m}\leqslant\left(E\left(\sum_{i=1}^{n}\mathbf{W}_i\right)^{s-1}\right)^{(s-m)/(s-1)}$$

$$\leqslant\left(c_{32}(s)n^{(s-1)/2}\right)^{(s-m)/(s-1)}$$

$$=c_{33}(s,m)n^{(s-m)/2}\qquad(1\leqslant m\leqslant s).\tag{14.99}$$

In deriving this we have proceeded exactly as in the proof for even integers s. More explicitly, we have used Lemma 9.5, inequalities (14.89), (14.91) with $\mathbf{W}$'s replacing $\mathbf{X}$'s, $\tau_{s,j}/\tau_{2,j}^{s/2}$ replacing ρ_s, and inequality (14.97) and the modified polynomials $\tilde{P}_m$ corresponding to the cumulants of $\mathbf{W}_i$, $1\leqslant i\leqslant n$. Hence

$$E|\mathbf{X}_1+\cdots+\mathbf{X}_j+\mathbf{Z}_{j+1}+\cdots+\mathbf{Z}_n|^{s-m}$$

$$=\tau_{2,j}^{(s-m)/2}E\left|\sum_{i=1}^{n}\mathbf{W}_i\right|^{s-m}\leqslant c_{34}(s,m)n^{(s-m)/2}\qquad(1\leqslant m\leqslant s).\tag{14.100}$$

Using this in (14.95), we get

$$E|\mathbf{X}_1+\cdots+\mathbf{X}_{j-1}+\mathbf{Z}_{j+1}+\cdots+\mathbf{Z}_n|^{s-m}$$

$$\leqslant c_{35}(s,m)n^{(s-m)/2}\qquad(1\leqslant m\leqslant s).\tag{14.101}$$

Also, by Lemma 14.1 and inequality (14.60),

$$\sum_{j=1}^{n} E|\mathbf{X}_j-\mathbf{Z}_j| \leqslant \sum_{j=1}^{n} (E|\mathbf{X}_j-\mathbf{Y}_j|+|E\mathbf{Y}_j|)$$

$$=\sum_{j=1}^{n}\left(\int_{\{|\mathbf{X}_j|>n^{1/2}\}} |\mathbf{X}_j|+|E\mathbf{Y}_j|\right)$$

$$\leqslant \sum_{j=1}^{n} \left(n^{-1/2}E\mathbf{X}_j^2+n^{-1/2}\right)=2n^{1/2},$$

$$\sum_{j=1}^{n} E|\mathbf{X}_j^2-\mathbf{Z}_j^2| \leqslant 2\sum_{j=1}^{n} E\mathbf{X}_j^2=2n,$$

$$\sum_{j=1}^{n} E|\mathbf{X}_j^m-\mathbf{Z}_j^m| \leqslant \sum_{j=1}^{n} (E|\mathbf{X}_j|^m+2^mE|\mathbf{X}_j|^m)=(1+2^m)n\rho_m$$

$$\leqslant (1+2^m)n\rho_s^{(m-2)/(s-2)}$$

$$\leqslant (1+2^m)n\left(\tfrac{9}{8}n^{(s-2)/2}\right)^{(m-2)/(s-2)}$$

$$=c_{36}(s,m)n^{m/2} \qquad (3\leqslant m\leqslant s). \tag{14.102}$$

Use (14.101), (14.102) in (14.94) to get

$$E|(\mathbf{X}_1+\cdots+\mathbf{X}_n)^s-(\mathbf{Z}_1+\cdots+\mathbf{Z}_n)^s| \leqslant c_{37}(s)n^{s/2}. \tag{14.103}$$

Together with (14.93) this implies

$$E|\mathbf{X}_1+\cdots+\mathbf{X}_n|^s \leqslant c_{38}(s)n^{s/2}, \tag{14.104}$$

completing the proof of the lemma. Q.E.D.

Before we state the next lemma, we define

$$\bar{\Delta}_{n,s}(\varepsilon)=n^{-1}\sum_{j=1}^{n}\int_{\{\|\mathbf{X}_j\|>\varepsilon n^{1/2}\}} \|\mathbf{X}_j\|^s \qquad (\varepsilon>0). \tag{14.105}$$

In this notation

$$\bar{\Delta}_{n,s}=\bar{\Delta}_{n,s}(1). \tag{14.106}$$

Let Q_n denote the distribution of $n^{-1/2}(\mathbf{X}_1+\cdots+\mathbf{X}_n)$ and Q_n'' that of $n^{-1/2}(\mathbf{Y}_1+\cdots+\mathbf{Y}_n)$.

LEMMA 14.8. *Let* $\mathrm{V=I}$. *If* $\rho_{\mathrm{s}}<\infty$ *for some* $\mathrm{s}\geqslant 0$, *then there exists a positive constant* $\mathrm{c}_{39}(\mathrm{s,k})$ *such that*

$$\|Q_n - Q_n''\| \leqslant c_{39}(s,k)\bar{\Delta}_{n,s} n^{-(s-2)/2}. \tag{14.107}$$

Also, there exist two positive constants $\mathrm{c}_{40}(\mathrm{s,k})$, $\mathrm{c}_{41}(\mathrm{s,k})$ *such that whenever*

$$\bar{\Delta}_{n,s}\left(\tfrac{2}{3}\right) \leqslant c_{40}(s,k) n^{(s-2)/2} \tag{14.108}$$

for some integer $\mathrm{s}\geqslant 2$,

$$\int \|x\|^r |Q_n - Q_n''|(dx) \leqslant c_{41}(s,k)\bar{\Delta}_{n,s} n^{-(s-2)/2} \tag{14.109}$$

for all $\mathrm{r}\in(0,\mathrm{s}]$.

Proof. Let G_j denote the distribution of $n^{-1/2}\mathbf{X}_j$, G_j'' that of $n^{-1/2}\mathbf{Y}_j$, $1\leqslant j\leqslant n$. Then

$$Q_n = G_1 * G_2 * \cdots * G_n, \qquad Q_n'' = G_1'' * G_2'' * \cdots * G_n'', \tag{14.110}$$

and

$$\begin{aligned}
\|Q_n - Q_n''\| &= \left\| \sum_{j=1}^{n} G_1 * \cdots * G_{j-1} * (G_j - G_j'') * G_{j+1}'' * \cdots * G_n'' \right\| \\
&\leqslant \sum_{j=1}^{n} \|G_j - G_j''\| = 2\sum_{j=1}^{n} P\left(\|\mathbf{X}_j\| > n^{1/2}\right) \\
&\leqslant 2\sum_{j=1}^{n} n^{-s/2} \int_{\{\|\mathbf{X}_j\| > n^{1/2}\}} \|\mathbf{X}_j\|^s = 2\bar{\Delta}_{n,s} n^{-(s-2)/2},
\end{aligned} \tag{14.111}$$

which proves (14.107). Now assume that s is an integer, $s\geqslant 2$. Since $\|x\|^r \leqslant 1 + \|x\|^s$ for $0\leqslant r\leqslant s$, it is enough to prove (14.109) for $r=0$ and $r=s$. The case $r=0$ is precisely (14.107). We therefore need only to prove (14.109) for $r=s$. One has

$$\begin{aligned}
&\int \|x\|^s |Q_n - Q_n''|(dx) \\
&\quad\leqslant \sum_{j=1}^{n} \int \|x\|^s |G_1 * \cdots * G_{j-1} * (G_j - G_j'') * G_{j+1}'' * \cdots * G_n''|(dx) \\
&\quad\leqslant \sum_{j=1}^{n} \int \left(\int \|u+v\|^s |G_j - G_j''|(dv) \right) G_1 * \cdots * G_{j-1} * G_{j+1}'' * \cdots * G_n''\,(du) \\
&\quad\leqslant 2^{s-1} \sum_{j=1}^{n} \left(\|G_j - G_j''\| \int \|u\|^s G_1 * \cdots * G_{j-1} * G_{j+1}'' * \cdots * G_n''\,(du) \right. \\
&\qquad\qquad \left. + \int \|v\|^s |G_j - G_j''|(dv) \right).
\end{aligned} \tag{14.112}$$

Now

$$\int \|v\|^s |G_j - G_j''|(dv) = \int_{\{\|\mathbf{X}_j\| > n^{1/2}\}} \|n^{-1/2}\mathbf{X}_j\|^s = n^{-s/2}\Delta_{n,j,s}. \quad (14.113)$$

Also,

$$\int \|u\|^s G_1 * \cdots * G_{j-1} * G_{j+1}'' * \cdots * G_n''\,(du)$$

$$= E\|n^{-1/2}(\mathbf{X}_1 + \cdots + \mathbf{X}_{j-1} + \mathbf{Y}_{j+1} + \cdots + \mathbf{Y}_n)\|^s$$

$$\leqslant 2^{s-1}\big(E\|n^{-1/2}(\mathbf{X}_1 + \cdots + \mathbf{X}_j + \mathbf{Y}_{j+1} + \cdots + \mathbf{Y}_n)\|^s$$

$$+ E\|n^{-1/2}\mathbf{X}_j\|^s\big)$$

$$\leqslant 2^{2(s-1)}\big(E\|n^{-1/2}(\mathbf{X}_1 + \cdots + \mathbf{X}_j + \mathbf{Z}_{j+1} + \cdots + \mathbf{Z}_n)\|^s$$

$$+ \|n^{-1/2}(E\mathbf{Y}_{j+1} + \cdots + E\mathbf{Y}_n)\|^s\big) + 2^{s-1}E\|n^{-1/2}\mathbf{X}_j\|^s. \quad (14.114)$$

By Lemma 14.1,

$$\|n^{-1/2}(E\mathbf{Y}_{j+1} + \cdots + E\mathbf{Y}_n)\|^s$$

$$\leqslant \left(n^{-1/2}k^{1/2}n^{-(s-1)/2}\sum_{j=1}^{n}\Delta_{n,j,s}\right)^s = \left(k^{1/2}n^{-(s-2)/2}\bar{\Delta}_{n,s}\right)^s. \quad (14.115)$$

By inequality (14.54),

$$E\|n^{-1/2}\mathbf{X}_j\|^s \leqslant n^{-s/2}\left(n^{s/2} + \Delta_{n,j,s}\right) \leqslant 1 + n^{-(s-2)/2}\bar{\Delta}_{n,s}. \quad (14.116)$$

We next estimate $E\|\mathbf{X}_1 + \cdots + \mathbf{X}_j + \mathbf{Z}_{j+1} + \cdots + \mathbf{Z}_n\|^s$ by using the preceding lemma. For this assume first that $k=1$, and define $\tau_{r,j}$ and random variables $\mathbf{W}_i$ as in (14.96), (14.98). Note that

$$\int_{\{|\mathbf{W}_i| > n^{1/2}\}} |\mathbf{W}_i|^s = \tau_{2,j}^{-s/2}\int_{\{|\mathbf{X}_i| > \tau_{2,j}^{1/2}n^{1/2}\}} |\mathbf{X}_i|^s \qquad (1 \leqslant i \leqslant j),$$

$$\int_{\{|\mathbf{W}_i| > n^{1/2}\}} |\mathbf{W}_i|^s = \tau_{2,j}^{-s/2}\int_{\{|\mathbf{Z}_i| > \tau_{2,j}^{1/2}n^{1/2}\}} |\mathbf{Z}_i|^s$$

$$\leqslant \tau_{2,j}^{-s/2}\left[\int_{\{|\mathbf{Y}_i| > \tau_{2,j}^{1/2}n^{1/2} - |E\mathbf{Y}_i|\}} |\mathbf{Y}_i|^s\right.$$

$$\left. + |E\mathbf{Y}_i|^s P\left(|\mathbf{Y}_i| > \tau_{2,j}^{1/2}n^{1/2} - |E\mathbf{Y}_i|\right)\right]$$

$$(j+1 \leqslant i \leqslant n). \quad (14.117)$$

Taking $c_{40}(s,k) \leqslant 1/(8k)$, one has [by (14.97) and Lemma 14.1]

$$\tau_{2,j}^{1/2} \geqslant \left(\tfrac{3}{4}\right)^{1/2},$$

$$|E\mathbf{Y}_i| \leqslant n^{-(s-1)/2}\Delta_{n,j,s} \leqslant n^{-(s-3)/2}\overline{\Delta}_{n,s} \leqslant \frac{n^{1/2}}{8},$$

$$\tau_{2,j}^{1/2}n^{1/2} - |E\mathbf{Y}_i| \geqslant \tfrac{2}{3}n^{1/2},$$

$$\begin{aligned}
|E\mathbf{Y}_i|^s P\left(|\mathbf{Y}_i| > \tau_{2,j}^{1/2}n^{1/2} - |E\mathbf{Y}_i|\right) &\leqslant \left(\frac{n^{1/2}}{8}\right)^s P\left(|\mathbf{Y}_i| > \frac{2n^{1/2}}{3}\right)\\
&\leqslant \left(\frac{n^{1/2}}{8}\right)^s P\left(|\mathbf{X}_i| > \frac{2n^{1/2}}{3}\right)\\
&\leqslant \left(\frac{n^{1/2}}{8}\right)^s \left(\frac{2n^{1/2}}{3}\right)^{-s} \int_{\{|\mathbf{X}_i| > 2n^{1/2}/3\}} |\mathbf{X}_i|^s\\
&\leqslant \left(\tfrac{3}{16}\right)^s \int_{\{|\mathbf{X}_i| > 2n^{1/2}/3\}} |\mathbf{X}_i|^s,
\end{aligned}$$

$$\int_{\{|\mathbf{Y}_i| > \tau_{2,j}^{1/2}n^{1/2} - |E\mathbf{Y}_i|\}} |\mathbf{Y}_i|^s \leqslant \int_{\{|\mathbf{X}_i| > 2n^{1/2}/3\}} |\mathbf{X}_i|^s. \tag{14.118}$$

Therefore, choosing $c_{40}(s,k) = \left(\tfrac{3}{4}\right)^{s/2}(16\mathrm{k})^{-1}$, one has

$$n^{-1}\sum_{i=1}^{n} \int_{\{|\mathbf{W}_i| > n^{1/2}\}} |\mathbf{W}_i|^s \leqslant \tau_{2,j}^{-s/2} 2\overline{\Delta}_{n,s}\left(\tfrac{2}{3}\right) \leqslant \frac{n^{(s-2)/2}}{8}, \tag{14.119}$$

so that Lemma 14.7 may be applied to yield

$$\begin{aligned}
E|\mathbf{X}_1 + \cdots + \mathbf{X}_j + \mathbf{Z}_{j+1} + \cdots + \mathbf{Z}_n|^s&\\
&= \tau_{2,j}^{s/2} E|\mathbf{W}_1 + \cdots + \mathbf{W}_n|^s\\
&\leqslant E|\mathbf{W}_1 + \cdots + \mathbf{W}_n|^s \leqslant c_{42}(s)n^{s/2}.
\end{aligned} \tag{14.120}$$

For arbitrary k, apply this estimate coordinatewise to get

$$E\|n^{-1/2}(\mathbf{X}_1 + \cdots + \mathbf{X}_j + \mathbf{Z}_{j+1} + \cdots + \mathbf{Z}_n)\|^s \leqslant c_{43}(s,k). \tag{14.121}$$

The inequalities (14.115), (14.116), and (14.121) now show that the left side of (14.114) is bounded by a constant (depending only on s and k). This fact, together with the estimates (14.113) and (14.111), now gives [on substitution in (14.112)]

$$\int \|x\|^s |Q_n - Q_n''|(dx) \leqslant c_{44}(s,k)\bar{\Delta}_{n,s} n^{-(s-2)/2}. \qquad (14.122)$$

Q.E.D.

15. MAIN THEOREMS

We are finally ready to prove the two major theorems of this chapter. These provide bounds for $|\int f dQ_n - \int f d\Phi|$ over (essentially) all Φ-continuous functions f that are integrable with respect to Q_n under given moment conditions. The bound provided by the first theorem, which is much more useful than the second, cannot in general be improved upon, and the hypothesis in it cannot be relaxed any further (at least if we are to have bounds involving moments alone). The utility of the second theorem is explained later.

We continue to use the same notation as in the preceding section. For ease of reference, recall that $\mathbf{X}_1, \ldots, \mathbf{X}_n$ are n independent random vectors with values in R^k satisfying

$$E\mathbf{X}_j = 0 \qquad (1 \leqslant j \leqslant n), \qquad n^{-1}\sum_{j=1}^{n} \operatorname{Cov}(\mathbf{X}_j) = I, \qquad (15.1)$$

I being the $k \times k$ identity matrix. The moments and cumulants of $\mathbf{X}_j$'s are denoted as in (14.1). In particular,

$$\rho_s = n^{-1}\sum_{j=1}^{n} E\|\mathbf{X}_j\|^s \qquad (s>0),$$

$$\chi_\nu = n^{-1}\sum_{j=1}^{n} \chi_{\nu,j} \qquad \left[\nu \in (\mathbf{Z}^+)^k\right], \qquad (15.2)$$

where $\chi_{\nu,j}$ is the νth cumulant of $\mathbf{X}_j$. The corresponding quantities for the centered truncated random vectors $\mathbf{Z}_j$ $(1 \leqslant j \leqslant n)$, defined by (14.2), are denoted by ρ_s', χ_ν'. Recall also the notation

$$D = n^{-1}\sum_{j=1}^{n} \operatorname{Cov}(\mathbf{Z}_j). \qquad (15.3)$$

We let Q_n, Q_n', Q_n'' denote the distributions of $n^{-1/2}\sum_{j=1}^n \mathbf{X}_j$, $n^{-1/2}\sum_{j=1}^n \mathbf{Z}_j$, $n^{-1/2}\sum_{j=1}^n \mathbf{Y}_j$, respectively. As usual, Φ denotes the standard normal distribution on R^k, and $\Phi_{a,V}$ is the normal distribution with mean a and covariance matrix V.

For a real-valued function f on R^k define

$$M_r(f) = \sup_{x\in R^k} (1+\|x\|^r)^{-1}|f(x)| \qquad \text{if } r>0,$$

$$M_0(f) = \sup_{x,y\in R^k} |f(x)-f(y)| = \omega_f(R^k). \tag{15.4}$$

If ν is a finite (signed) measure on R^k, define a new (signed) measure ν_r by

$$\nu_r(dx) = (1+\|x\|^r)\nu(dx) \qquad \text{if } r>0,$$

$$\nu_0 = \nu. \tag{15.5}$$

As in the preceding section, write

$$\bar{\Delta}_{n,s}(\epsilon) = n^{-1}\sum_{j=1}^n \int_{\{\|\mathbf{X}_j\|>\epsilon n^{1/2}\}} \|\mathbf{X}_j\|^s, \qquad \bar{\Delta}_{n,s} = \bar{\Delta}_{n,s}(1). \tag{15.6}$$

Then define

$$\Delta_{n,s}^* = \inf_{0\leqslant\epsilon\leqslant 1}\left[\bar{\Delta}_{n,s}(\epsilon) + \epsilon n^{-1}\sum_{j=1}^n \int_{\{\|\mathbf{X}_j\|\leqslant\epsilon n^{1/2}\}} \|\mathbf{X}_j\|^s\right]. \tag{15.7}$$

Note that

$$\Delta_{n,s}^* \leqslant \rho_s, \qquad \bar{\Delta}_{n,s} \equiv \bar{\Delta}_{n,s}(1) \leqslant \Delta_{n,s}^*. \tag{15.8}$$

If one takes $\epsilon = n^{-1/4}$ in the expression within brackets in (15.7), then one gets

$$\Delta_{n,s}^* \leqslant \bar{\Delta}_{n,s}(n^{-1/4}) + n^{-1/4}\rho_s. \tag{15.9}$$

Thus if $\mathbf{X}_1,\ldots,\mathbf{X}_n$ are the first n terms of a sequence of independent and identically distributed random vectors, then

$$n^{-(s-2)/2}\Delta_{n,s}^* = o(n^{-(s-2)/2}) \qquad (n\to\infty). \tag{15.10}$$

Finally recall the average moduli of oscillation $\bar{\omega}_f$ and ω_f^* defined by (11.23), (11.24).

THEOREM 15.1. *There exist positive constants* $c_i (i=1,2,3,4)$ *depending only on* k *and* s *such that if*

$$\bar{\Delta}_{n,s}\left(\tfrac{2}{3}\right) \leqslant c_1 n^{(s-2)/2} \tag{15.11}$$

holds for some integer $s \geqslant 3$, *then for every real-valued Borel-measurable function* f *satisfying*

$$M_r(f) < \infty \tag{15.12}$$

for some integer r, $0 \leqslant r \leqslant s$, *one has*

$$\left|\int f d(Q_n - \psi)\right| \leqslant c_2 M_r(f) n^{-(s-2)/2} \Delta_{n,s}^* + 2\omega_g^*\left(c_4 \rho_3 n^{-1/2} : (\psi^+)_{r_0}\right), \tag{15.13}$$

where

$$\psi = \sum_{m=0}^{s-2} n^{-m/2} P_m(-\Phi : \{\chi_\nu\}),$$

and

$$r_0 = \begin{cases} r & \text{if } r \text{ is even,} \\ r+1 & \text{if } r \text{ is odd,} \end{cases}$$

$$g(x) = \begin{cases} (1 + \|x\|^{r_0})^{-1} f(x) & \text{if } r > 0, \\ f(x) & \text{if } r = 0. \end{cases} \tag{15.14}$$

Proof. The constants c's appearing in the proof do not depend on anything other than r, k, s. We write

$$\psi' = \sum_{m=0}^{s-2} n^{-m/2} P_m(-\Phi_{0,D} : \{\chi'_\nu\}),$$

$$\psi''(B) = \psi(B + a_n) \qquad (B \in \mathfrak{B}^k), \tag{15.15}$$

where

$$a_n = n^{-1/2} \sum_{j=1}^{n} E\mathbf{Y}_j. \tag{15.16}$$

Thus ψ'' is the signed measure induced from ψ by the map $x \to x - a_n$. Now

$$\left|\int f d(Q_n - \psi)\right| \leqslant \left|\int f d(Q_n - Q_n'')\right| + \left|\int f d(Q_n'' - \psi)\right|. \tag{15.17}$$

If $r > 0$, then

$$|f(x)| \leqslant M_r(f)(1 + \|x\|^r) \qquad (x \in R^k), \tag{15.18}$$

so that using Lemma 14.8 one gets

$$\left|\int f d(Q_n - Q_n'')\right| \leqslant c_5 M_r(f) \bar{\Delta}_{n,s} n^{-(s-2)/2}. \tag{15.19}$$

Since the left side does not change when f is replaced by $f - c$, where c is the midpoint of the range of f, (15.19) also holds for $r = 0$.

Next

$$\left|\int f d(Q_n'' - \psi)\right| = \left|\int f_{a_n} d(Q_n' - \psi'')\right| \leqslant \left|\int f_{a_n} d(Q_n' - \psi')\right|$$

$$+ \left|f_{a_n} d(\psi' - \psi)\right| + \left|\int f_{a_n} d(\psi - \psi'')\right|, \tag{15.20}$$

where f_y denotes the translate of f by y; that is,

$$f_y(x) = f(x + y) \qquad (x \in R^k). \tag{15.21}$$

By Lemma 14.6,

$$\left|\int f_{a_n} d(\psi' - \psi)\right|$$

$$\leqslant M_r(f) \int (1 + \|x + a_n\|^r) |\psi' - \psi|(dx)$$

$$\leqslant M_r(f) \int (1 + 2^r \|a_n\|^r + 2^r \|x\|^r) |\psi' - \psi|(dx)$$

$$\leqslant M_r(f)\left[\|\psi' - \psi\| + 2^r \|a_n\|^r \|\psi' - \psi\| + 2^r \int \|x\|^r |\psi' - \psi|(dx)\right]$$

$$\leqslant c_6 M_r(f) \bar{\Delta}_{n,s} n^{-(s-2)/2}. \tag{15.22}$$

In the last step we also used inequality (14.81). Similarly,

$$\left|\int f_{a_n}\,d(\psi-\psi'')\right|$$
$$\leqslant M_r(f)\left[\,\|\psi-\psi''\|+2^r\|a_n\|^r\|\psi-\psi''\|+2^r\int\|x\|^r|\psi-\psi''|(dx)\right]$$
$$\leqslant c_7 M_r(f)\bar{\Delta}_{n,s}n^{-(s-2)/2}. \tag{15.23}$$

Therefore (15.17) reduces to [in view of (15.19), (15.20), (15.22), and (15.23)]

$$\left|\int f\,d(Q_n-\psi)\right|\leqslant c_8 M_r(f)\bar{\Delta}_{n,s}n^{-(s-2)/2}+\left|\int f_{a_n}\,d(Q_n'-\psi')\right|. \tag{15.24}$$

To estimate the second term on the right of (15.24) we introduce a kernel probability measure K on R^k satisfying

$$K(\{x:\|x\|<1\})\geqslant\tfrac{3}{4},\qquad \int\|x\|^{k+s+2}K(dx)<\infty,$$
$$\hat{K}(t)=0\qquad\text{if }\|t\|\geqslant c_9\qquad(t\in R^k). \tag{15.25}$$

One construction of such a probability measure is provided by Theorem 10.1. For $\epsilon>0$ define the probability measure K_ϵ by

$$K_\epsilon(B)=K(\epsilon^{-1}B)\qquad\left(B\in\mathfrak{B}^k,\quad \epsilon^{-1}B=\{\epsilon^{-1}x:x\in B\}\right). \tag{15.26}$$

Then one has

$$K_\epsilon(\{x:\|x\|<\epsilon\})\geqslant\tfrac{3}{4},\qquad \hat{K}_\epsilon(t)=0\qquad\text{if }\|t\|\geqslant\frac{c_9}{\epsilon}. \tag{15.27}$$

Now

$$\left|\int f_{a_n}\,d(Q_n'-\psi')\right|=\left|\int(1+\|x+a_n\|^{r_0})^{-1}f(x+a_n)\right.$$
$$\left.\times(1+\|x+a_n\|^{r_0})(Q_n'-\psi')(dx)\right|$$
$$\leqslant\left|\int(1+\|x+a_n\|^{r_0})^{-1}f(x+a_n)(1+\|x\|^{r_0})(Q_n'-\psi')(dx)\right|$$
$$+M_{r_0}(f)\int\big|\|x+a_n\|^{r_0}-\|x\|^{r_0}\big|(Q_n'+|\psi'|)(dx),$$

$$\int \left| \|x+a_n\|^{r_0} - \|x\|^{r_0} \right| (Q_n' + |\psi'|)(dx)$$

$$\leqslant r_0 \|a_n\| \int (\|x\|^{r_0-1} + \|a_n\|^{r_0-1})(Q_n' + |\psi'|)(dx)$$

$$\leqslant r_0 k^{1/2} \bar{\Delta}_{n,s} n^{-(s-2)/2} \Big[E\|n^{-1/2}(\mathbf{Z}_1 + \cdots + \mathbf{Z}_n)\|^{r_0-1} + (8k^{1/2})^{-r_0+1}$$

$$+ \int \left((8k^{1/2})^{-r_0+1} + \|x\|^{r_0-1} \right) |\psi|(dx)$$

$$+ \int \left((8k^{1/2})^{-r_0+1} + \|x\|^{r_0-1} \right) |\psi' - \psi|(dx) \Big]$$

$$\leqslant c_{10} \bar{\Delta}_{n,s} n^{-(s-2)/2}, \tag{15.28}$$

by Lemmas 14.3, 14.6 and inequalities (14.12), (14.81). Hence [see (15.24), (15.28)]

$$\left| \int f d(Q_n - \psi) \right| \leqslant c_{11} M_r(f) \bar{\Delta}_{n,s} n^{-(s-2)/2}$$

$$+ \left| \int g_{a_n}(x)(1 + \|x\|^{r_0})(Q_n' - \psi')(dx) \right| \tag{15.29}$$

if it is noted that

$$M_{r_0}(f) = \sup_{x \in R^k} (1 + \|x\|^{r_0})^{-1} |f(x)|$$

$$\leqslant M_r(f) \sup_{x \in R^k} (1 + \|x\|^r)(1 + \|x\|^{r_0})^{-1} \leqslant 2M_r(f). \tag{15.30}$$

By Corollary 11.5 one has for every $\epsilon > 0$

$$\left| \int g_{a_n}(x)(1 + \|x\|^{r_0})(Q_n' - \psi')(dx) \right|$$

$$\leqslant 2M_{r_0}(f) \|(Q_n' - \psi')_{r_0} * K_\epsilon\| + 2\omega_g^*(2\epsilon : (\psi'^+)_{r_0}), \tag{15.31}$$

where

$$(Q_n' - \psi')_{r_0}(dx) = (1 + \|x\|^{r_0})(Q_n' - \psi')(dx),$$

$$(\psi'^+)_{r_0}(dx) = (1 + \|x\|^{r_0})\psi'^+(dx). \tag{15.32}$$

Now choose

$$\epsilon = 16c_9\rho_3 n^{-1/2}. \tag{15.33}$$

By Lemma 11.6,

$$\|(Q_n' - \psi')_{r_0} * K_\epsilon\| \leqslant c_{12} \max \int |D^\alpha (\hat{Q}_n' - \hat{\psi}') \cdot D^\beta \hat{K}_\epsilon|, \tag{15.34}$$

where the maximum is over all pairs of nonnegative integral vectots α, β satisfying

$$0 \leqslant |\alpha + \beta| \leqslant k + r_0 + 1, \tag{15.35}$$

and the integration is with respect to Lebesgue measure on R^k. Since $\hat{K}_\epsilon(t) = 0$ for $\|t\| > c_9/\epsilon = n^{1/2}/(16\rho_3)$, and

$$|(D^\beta \hat{K}_\epsilon)(t)| \leqslant \int \epsilon^{|\beta|} |x^\beta| K(dx) \leqslant c_{13}, \tag{15.36}$$

one has

$$\int |D^\alpha (\hat{Q}_n' - \hat{\psi}') \cdot D^\beta \hat{K}_\epsilon|$$

$$\leqslant c_{13} \int_{\{\|t\| \leqslant n^{1/2}/(16\rho_3)\}} |D^\alpha (\hat{Q}_n' - \hat{\psi}')(t)| \, dt. \tag{15.37}$$

Note that [see (14.60) with $s = 3$] ϵ is bounded above by a constant because of (15.11) [whatever the constant c_1 one may choose in (15.11)]. This fact was used in (15.36).

Now

$$\int_{\{\|t\| \leqslant n^{1/2}/(16\rho_3)\}} |D^\alpha (\hat{Q}_n' - \hat{\psi}')(t)| \, dt$$

$$\leqslant \int_{A_n} \left| D^\alpha \left[\hat{Q}_n'(t) - \sum_{m=0}^{k+s-1} n^{-m/2} \tilde{P}_m(it : \{\chi_\nu'\}) \exp\{-\tfrac{1}{2}\langle t, Dt\rangle\} \right] \right| dt$$

$$+ \int \left| D^\alpha \left[\sum_{m=s-1}^{k+s-1} n^{-m/2} \tilde{P}_m(it : \{\chi_\nu'\}) \exp\{-\tfrac{1}{2}\langle t, Dt\rangle\} \right] \right| dt$$

$$+ \int_{A_n'} |(D^\alpha \hat{Q}_n')(t)| \, dt + \int_{A_n'} |(D^\alpha \hat{\psi}')(t)| \, dt = I_1 + I_2 + I_3 + I_4, \tag{15.38}$$

where, writing Λ for the largest eigenvalue of D,

$$A_n=\left\{\|t\|\leqslant c_{14}\left(\frac{n^{(k+s)/2}}{\eta_{k+s+2}}\right)^{1/(k+s+2)}\Lambda^{-1/2}\right\},$$

$$A_n'=\left\{\|t\|\leqslant\frac{n^{1/2}}{16\rho_3}\right\}\backslash A_n,$$

$$\eta_m\equiv n^{-1}\sum_{j=1}^{n}E\|B\mathbf{Z}_j\|^m\qquad(m>0),\tag{15.39}$$

B being the symmetric positive-definite matrix satisfying $B^2=D^{-1}$. By Lemma 14.1(v) (with $s'=m$) and inequality (14.21) (Corollary 14.2), one has

$$\|B\|\leqslant\left(\tfrac{4}{3}\right)^{1/2},\qquad \Lambda\leqslant\tfrac{5}{4},$$

$$\eta_m\leqslant 2^m\left(\tfrac{4}{3}\right)^{m/2}n^{(m-s)/2}\Delta_{n,s}^*,$$

$$n^{-(m-2)/2}\eta_m\leqslant 2^m\left(\tfrac{4}{3}\right)^{m/2}n^{-(s-2)/2}\Delta_{n,s}^*\qquad(m\geqslant s+1).\tag{15.40}$$

Hence, putting $m=k+s+2$, we get

$$A_n\supset\left\{\|t\|\leqslant c_{15}\left(\frac{n^{(s-2)/2}}{\Delta_{n,s}^*}\right)^{1/(k+s+2)}\right\}.\tag{15.41}$$

If we choose c_{14} to be the constant $c_{22}(k+s+2,k)$ appearing in Theorem 9.11, then we get [using (15.40)]

$$I_1\leqslant c_{16}\eta_{k+s+2}n^{-(k+s)/2}\leqslant c_{17}n^{-(s-2)/2}\Delta_{n,s}^*.\tag{15.42}$$

For this we have used Theorem 9.11 and the remark after Theorem 9.12. Also, by Lemma 9.5 (with $r=m$, $s=m+2$) and Corollary 14.2, one has [still using (15.40)]

$$\int\left|D^\alpha\left[n^{-m/2}\tilde{P}_m(it:\{\chi_\nu'\})\exp\left\{-\tfrac{1}{2}\langle t,Dt\rangle\right\}\right]\right|dt$$
$$\leqslant c_{18}n^{-(s-2)/2}\Delta_{n,s}^*\qquad(s-1\leqslant m\leqslant k+s-1),\tag{15.43}$$

whereas (14.10) yields

$$n^{-m/2}\rho_{m+2}'\leqslant 2^{m+2}\rho_2=2^{m+2}k\qquad(1\leqslant m\leqslant s-2).\tag{15.44}$$

The inequality (15.43) immediately leads to

$$I_2 \leqslant c_{19} n^{-(s-2)/2} \Delta^*_{n,s}. \tag{15.45}$$

Next use Corollary 14.4 to get [see (15.39), (15.41)]

$$\begin{aligned} I_3 &\equiv \int_{A'_n} |D^\alpha \hat{Q}'_n| \\ &\leqslant c_{20} \int_{\left\{ \|t\| > c_{15} \left(\frac{n^{(s-2)/2}}{\Delta^*_{n,s}} \right)^{1/(k+s+2)} \right\}} (1 + \|t\|^{|\alpha|}) \exp\left\{ -\tfrac{5}{24} \|t\|^2 \right\} dt \\ &\leqslant c_{21} n^{-(s-2)/2} \Delta^*_{n,s} \int \|t\|^{k+s+2} (1 + \|t\|^{|\alpha|}) \exp\left\{ -\tfrac{5}{24} \|t\|^2 \right\} dt \\ &= c_{22} n^{-(s-2)/2} \Delta^*_{n,s}. \end{aligned} \tag{15.46}$$

A similar calculation using (15.44) gives

$$I_4 \equiv \int_{A'_n} |D^\alpha \hat{\psi}'| \leqslant c_{23} n^{-(s-2)/2} \Delta^*_{n,s}. \tag{15.47}$$

The estimation of the right side of (15.34) is complete, and we have

$$\|(Q'_n - \psi')_{r_0} * K_\epsilon\| \leqslant c_{24} n^{-(s-2)/2} \Delta^*_{n,s}. \tag{15.48}$$

Finally, by Lemma 14.6,

$$\begin{aligned} \left|\omega^*_g(2\epsilon : (\psi'^+)_{r_0}) - \omega^*_g(2\epsilon : (\psi^+)_{r_0})\right| &\leqslant \sup_{y \in R^k} \int \omega_{g_y}(x : 2\epsilon) \left|(\psi'^+)_{r_0} - (\psi^+)_{r_0}\right|(dx) \\ &\leqslant 2 M_{r_0}(f) \|(\psi'^+)_{r_0} - (\psi^+)_{r_0}\| \\ &\leqslant c_{25} M_r(f) \bar{\Delta}_{n,s} n^{-(s-2)/2}. \end{aligned} \tag{15.49}$$

Now use (15.48), (15.49) to get an estimate for the right side of (15.31) and then use this estimate in (15.29) to complete the proof, noting that $M_{r_0}(f) \leqslant 2M_r(f)$, $\bar{\Delta}_{n,s} \leqslant \Delta^*_{n,s}$. Q.E.D.

Remark. In view of relation (7.22),

$$\int f\, d(Q_n - \psi) = \int (f - c)\, d(Q_n - \psi)$$

for all $c \in R^1$. Hence $M_r(f)$ in (15.13) (and in all subsequent theorems involving it) may be replaced by

$$M^*_r(f) = \inf_{c \in R^1} M_r(f - c). \tag{15.50}$$

COROLLARY 15.2. *If* $\rho_3<\infty$, *then for all bounded Borel-measurable functions* f *one has*

$$\left|\int f d(Q_n-\Phi)\right| \leqslant c_{26}\omega_f(R^k)\rho_3 n^{-1/2}+2\omega_f^*\left(c_4\rho_3 n^{-1/2}:\Phi\right), \quad (15.51)$$

where c_{26} *depends only on* k.

Proof. The corollary follows from the above theorem if one takes $r=r_0=0$, $s=3$, and notes that in the present case the condition (15.11) is unnecessary. For if

$$\rho_3 \geqslant \bar{\Delta}_{n,3}(\tfrac{2}{3})>c_1 n^{1/2},$$

then

$$\left|\int f d(Q_n-\Phi)\right| \leqslant 2\omega_f(R^k) \leqslant \frac{2}{c_1}\omega_f(R^k)\rho_3 n^{-1/2}.$$

Now (15.13) yields [using (15.8)]

$$\left|\int f d\left(Q_n-\Phi-n^{-1/2}P_1(-\Phi:\{\chi_\nu\})\right)\right|$$

$$\leqslant \omega_f(R^k)n^{-1/2}\Delta_{n,3}^*+2\omega_f^*\left(c_4\rho_3 n^{-1/2}:\Phi+|n^{-1/2}P_1(-\Phi:\{\chi_\nu\})|\right)$$

$$\leqslant \omega_f(R^k)n^{-1/2}\rho_3+2\omega_f(R^k)n^{-1/2}\|P_1(-\Phi:\{\chi_\nu\})\|+2\omega_f^*\left(c_4\rho_3 n^{-1/2}:\Phi\right).$$

(15.52)

Lemma 13.1 does the rest. Q.E.D.

COROLLARY 15.3. *If* $\rho_3<\infty$, *then for all Borel sets* A *one has*

$$|Q_n(A)-\Phi(A)| \leqslant c_{26}\rho_3 n^{-1/2}+2\sup_{y\in R^k}\Phi((\partial A)^\eta+y), \quad (15.53)$$

where

$$\eta=c_4\rho_3 n^{-1/2}. \quad (15.54)$$

Proof. This follows immediately from the preceding corollary if one takes $f=I_A$ (the indicator function of A) and recalls that [see (2.40)–(2.43)]

$$\omega_{I_A}(x:\eta)=I_{A^\eta\setminus A^{-\eta}}(x)=I_{(\partial A)^\eta}(x) \qquad (x\in R^k,\quad A\subset R^k,\quad \eta>0),$$

$$\omega_{I_A}^*(\eta:\Phi)=\sup_{y\in R^k}\int\omega_{I_A}(x+y:\eta)\Phi(dx)=\sup_{y\in R^k}\Phi((\partial A)^\eta+y). \quad (15.55)$$

Q.E.D.

Remark. Suppose A is a Borel set such that $\Phi(A)=0$. With $r=r_0=0$, (15.13) then yields

$$|Q_n(A)-\Phi(A)|\leqslant c_2 n^{-(s-2)/2}\Delta^*_{n,s}+2\sup_{y\in R^k}\psi^+((\partial A)^\eta+y), \tag{15.56}$$

where $\eta=c_4\rho_3 n^{-1/2}$. If, in addition, A is a set having a small diameter, one may use inequalities (11.25) of Corollary 11.5, rather than inequalities (11.26) as used in (15.31), to get, instead of the term $M_0(I_A)\|(Q_n'-\psi')*K_\epsilon\|$ [in (15.31)], the expression

$$\gamma^*(I_{A^\epsilon}:\epsilon)\leqslant\lambda_k(A^\epsilon)\sup_{y\in R^k}|h_n(y)|,$$

where λ_k denotes the Lebesgue measure on R^k, and h_n is the density of $(Q_n'-\psi')*K_\epsilon$. With $\alpha=0$ in (15.38), (15.42), (15.45)–(15.47), one obtains

$$|h_n(y)|\leqslant c_{24}n^{-(s-2)/2}\Delta^*_{n,s}\qquad (y\in R^k),$$

and therefore

$$|Q_n(A)-\Phi(A)|\leqslant c_{27}n^{-(s-2)/2}\Delta^*_{n,s}\lambda_k(A^\epsilon)$$
$$+2\sup_{y\in R^k}\psi^+((\partial A)^\eta+y), \tag{15.57}$$

with $\epsilon=c_9\rho_3 n^{-1/2}$, $\eta=c_4\rho_3 n^{-1/2}$. We make use of (15.56) and (15.57) in Section 17 (Theorems 17.4, 17.5).

Theorem 15.1 provides sharp bounds for a wide class of functions f. This is seen, for example, by comparing it with the more specialized asymptotic expansions in Chapters 4 and 5, or with the general expansions for trigonometric polynomials f as provided by Theorems 9.10–9.12. Section 17 contains several applications. However note that the right side of (15.50) (Corollary 15.2) does not go to zero (as $n\to\infty$) for *every* bounded Φ-continuous f when Q_n is the distribution of $n^{-1/2}(\mathbf{X}_1+\cdots+\mathbf{X}_n)$ and $\{\mathbf{X}_n:n\geqslant 1\}$ is a sequence of independent and identically distributed random vectors with zero means, covariance matrices I, and finite third absolute moments ρ_3. The reason for this is that the bound is in terms of ω_f^* [and $\omega_f(R^k)$] rather than $\bar\omega_f$. Recall that [see Lemma 1.2(iii), Theorem 1.3(iii)] that $\int f\,dQ_n\to\int f\,d\Phi$ for all bounded Φ-continuous f and that a bounded f is Φ-continuous if and only if

$$\lim_{\epsilon\downarrow 0}\bar\omega_f(\epsilon:\Phi)=0\qquad\left[\bar\omega_f(\epsilon:\Phi)=\int\omega_f(x:\epsilon)\Phi(dx)\right].$$

On the other hand, it is not difficult to construct bounded Φ-continuous functions f such that $\omega_f^*(\epsilon:\Phi)$ does not go to zero with ϵ. For example, let $f=I_A$, the indicator function of the following Borel subset A of R^1.

$$A=\bigcup_{r=2}^{\infty}\ \bigcup_{i=1}^{[(r-1)/2]}\left\{x\in R^1:r+\frac{2i}{r}<x\leqslant r+\frac{(2i+1)}{r}\right\},$$

with $[(r-1)/2]$ denoting the integer part of $(r-1)/2$. It is easy to see that

$$\omega^*_{I_A}(\epsilon:\Phi)=\sup_{y\in R^k}\Phi((\partial A)^{\epsilon}+y)=1,$$

$$\overline{\omega}_{I_A}(\epsilon:\Phi)=\Phi((\partial A)^{\epsilon})\leqslant(2\pi)^{-1/2}\left[\sum_{r=1}^{\infty} r\exp\left\{-\frac{r^2}{2}\right\}\right]\epsilon \qquad (\epsilon>0).$$

It would be ideal if one could replace ω_f^* in the bound (15.51) by $\overline{\omega}_f$. Unfortunately we are unable to do this. The situation is partially salvaged by the next theorem, which provides an effective bound for *every* bounded Φ-continuous f. However a price is paid for greater generality. For "nice" functions and sets there is a loss in precision. On the other hand, apart from its greater generality, this theorem is more useful than Theorem 15.1 for estimating tail probabilities (see Section 17).

THEOREM 15.4. *There exist positive constants* c_i' $(i=1,2,3)$ *depending only on* k *and* s *such that if*

$$\overline{\Delta}_{n,s}\left(\tfrac{2}{3}\right)\leqslant c_1' n^{(s-2)/2} \tag{15.58}$$

holds for some integer $s\geqslant 3$, *then for every Borel-measurable function* f *satisfying*

$$M_r(f)<\infty \tag{15.59}$$

for some integer r, $0\leqslant r\leqslant s$, *one has*

$$\left|\int f\,d(Q_n-\psi)\right|\leqslant c_2' M_r(f)(1+\eta^{k+r+1})n^{-(s-2)/2}\Delta^*_{n,s}+\overline{\omega}_f(\eta:\psi^+), \tag{15.60}$$

where

$$\psi=\sum_{m=0}^{s-2} n^{-(s-2)/2}P_m(-\Phi:\{\chi_\nu\}),$$

and

$$\eta=c_3'\rho_3([\log n]+1)n^{-1/2}, \tag{15.61}$$

where $[\log n]$ *is the integer part of* $\log n$. *The quantities* $\overline{\Delta}_{n,s}(\tfrac{2}{3})$, $\Delta^*_{n,s}$ *are defined by* (15.6) *and* (15.7), *respectively*.

Proof. We continue to use the same notation as in the proof of Theorem 15.1. Define ψ', ψ'' as in (15.15). As in (15.24), one has

$$\left|\int f d(Q_n-\psi)\right| \leqslant c_{28} M_r(f)\bar{\Delta}_{n,s} n^{-(s-2)/2} + \left|\int f_{a_n} d(Q_n'-\psi')\right|. \tag{15.62}$$

Choose a probability measure K' on R^k satisfying

$$\begin{aligned} &K'(\{x:\|x\|<1\})=1,\\ &|D^\alpha \hat{K}'(t)| \leqslant c_{29}\exp\{-\|t\|^{1/2}\} \qquad (t\in R^k,\quad |\alpha|\leqslant k+s+1). \end{aligned} \tag{15.63}$$

Such a choice is possible by Corollary 10.4. For $\epsilon>0$ and a positive integer p define

$$\begin{aligned} K_\epsilon'(B) &= K'(\epsilon^{-1}B) \qquad (B\in\mathfrak{B}^k),\\ K_{\epsilon,p} &= K_\epsilon'^{*p}, \end{aligned} \tag{15.64}$$

and note that

$$K_{\epsilon,p}(\{x:\|x\|<p\epsilon\})=1, \qquad \hat{K}_{\epsilon,p}(t)=(\hat{K}(\epsilon t))^p \qquad (t\in R^k). \tag{15.65}$$

By Corollary 11.2, with $\mu=Q_n'$, $\nu=\psi'$, $K_\epsilon=K_{\epsilon,p}$, and $f=f_{a_n}$, one has

$$\left|\int f_{a_n} d(Q_n'-\psi')\right| \leqslant \gamma(f_{a_n}:p\epsilon)+\bar{\omega}_{f_{a_n}}(2p\epsilon:(\psi')^+), \tag{15.66}$$

where, by definition of γ [see (11.8)] and of $M_r(f)$ [see (15.4)]

$$\begin{aligned} \gamma(f_{a_n}:p\epsilon) &\leqslant M_r(f)(1+\|x\|+\|a_n\|+p\epsilon)^r|(Q_n'-\psi')*K_{\epsilon,p}|(dx)\\ &\leqslant M_r(f)\int(c_{30}+2^r\|x\|^r+p^r\epsilon^r)|(Q_n'-\psi')*K_{\epsilon,p}|(dx)\\ &\leqslant c_{31}M_r(f)\Big[\max_{|\alpha+\beta|\leqslant k+r+1}\|D^\alpha(\hat{Q}_n'-\hat{\psi}')\cdot D^\beta\hat{K}_{\epsilon,p}\|_1\\ &\qquad + p^r\epsilon^r\max_{|\alpha+\beta|\leqslant k+1}\|D^\alpha(\hat{Q}_n'-\hat{\psi}')\cdot D^\beta\hat{K}_{\epsilon,p}\|_1\Big]. \end{aligned} \tag{15.67}$$

If $r=0$, then the first inequality in (15.67) holds because $\psi'(R^k)=\Phi'(R^k)=1$ [see (7.22) and (11.13)]. The last inequality follows from Lemma 11.6.

In view of (15.63), one has

$$\begin{aligned} &\|D^\alpha(\hat{Q}_n'-\hat{\psi}')\cdot D^\beta\hat{K}_{\epsilon,p}\|_1\\ &\qquad\leqslant c_{32}(p\epsilon)^{|\beta|}\int|D^\alpha(\hat{Q}_n'-\hat{\psi}')(t)|\exp\{-p\|\epsilon t\|^{1/2}\}\,dt. \end{aligned} \tag{15.68}$$

Now choose

$$\epsilon = c_{33}\rho_3 n^{-1/2}, \qquad p = [\log n] + 1, \tag{15.69}$$

and write

$$\int |D^{\alpha}(\hat{Q}_n' - \hat{\psi}')(t)| \exp\{-p\|\epsilon t\|^{1/2}\}\,dt \leqslant I_1 + I_2' + I_3 + I_4'' + I_5'', \tag{15.70}$$

where, as in (15.38)–(15.47),

$$I_1 \leqslant c_{34} n^{-(s-2)/2}\Delta_{n,s}^*,$$

$$I_2' \equiv \sum_{m=s-1}^{k+s-1} \int \left| D^{\alpha}\left[n^{-m/2}\tilde{P}_m(it : \{\chi_\nu'\}) \exp\{-\tfrac{1}{2}\langle t, Dt\rangle\}\right]\right| dt$$

$$\leqslant c_{35} n^{-(s-2)/2}\Delta_{n,s}^*,$$

$$I_3 \leqslant c_{36} n^{-(s-2)/2}\Delta_{n,s}^*,$$

$$I_4'' \equiv \int_{R^k \setminus A_n} |(D^{\alpha}\hat{\psi}')(t)|\,dt \leqslant c_{37} n^{-(s-2)/2}\Delta_{n,s}^*,$$

$$I_5'' \equiv \int_{\{\|t\| > n^{1/2}/(16\rho_3)\}} |(D^{\alpha}\hat{Q}_n')(t)| \exp\{-p\|\epsilon t\|^{1/2}\}\,dt$$

$$\leqslant c_{38} \int_{\{\|t\| > n^{1/2}/(16\rho_3)\}} \exp\{-p\|\epsilon t\|^{1/2}\}\,dt. \tag{15.71}$$

The last inequality follows from Corollary 14.4 if one notes that

$$|D^{\alpha}\hat{Q}_n'(t)| \leqslant E\|n^{-1/2}(\mathbf{Z}_1 + \cdots + \mathbf{Z}_n)\|^{|\alpha|}$$

$$\leqslant \left(E\|n^{-1/2}(\mathbf{Z}_1 + \cdots + \mathbf{Z}_n)\|^{m'}\right)^{|\alpha|/m'}, \tag{15.72}$$

for every even integer $m' \geqslant |\alpha|$, and that for an even integer m', $E\|n^{-1/2}(\mathbf{Z}_1 + \cdots + \mathbf{Z}_n)\|^{m'})$ may be expressed in terms of ordinary (as opposed to absolute) moments that are estimated by putting $t=0$ in

Corollary 14.4. Also,

$$\int_{\{\|t\|>n^{1/2}/(16\rho_3)\}} \exp\{-p\|\epsilon t\|^{1/2}\}\,dt$$

$$= \epsilon^{-k}\int_{\{\|t\|>c_{33}/16\}} \exp\{-p\|t\|^{1/2}\}\,dt$$

$$= c_{39}\epsilon^{-k}\int_{c_{33}/16}^{\infty} u^{k-1}\exp\{-pu^{1/2}\}\,du$$

$$= 2c_{39}(\epsilon p^2)^{-k}\int_{p(c_{33}/16)^{1/2}}^{\infty} v^{2k-1}\exp\{-v\}\,dv$$

$$\leqslant c_{40}(\epsilon p^2)^{-k}\int_{p(c_{33}/16)^{1/2}}^{\infty} \exp\left\{\frac{-v}{2}\right\}dv. \tag{15.73}$$

Now choose

$$c_{33} = 64\left(k+\frac{s-2}{2}\right)^2 \tag{15.74}$$

to get

$$I_5'' \leqslant c_{41}(n^{1/2}\log^2 n)^{-k} n^{-(s-2)/2} \leqslant c_{42} n^{-(s-2)/2}\Delta_{n,s}^*. \tag{15.75}$$

The last inequality in (15.75) derives from [see (14.7), (14.10), and (15.58)]

$$n^{-1/2}c_{43} \leqslant n^{-1/2}\left[n^{-1}\sum_{j=1}^{n}\int\|\mathbf{Y}_j\|^2\right]^{(s+1)/2}$$

$$\leqslant n^{-1/2}\left[n^{-1}\sum_{j=1}^{n}\int\|\mathbf{Y}_j\|^{s+1}\right] \leqslant \Delta_{n,s}^*. \tag{15.76}$$

Hence

$$\|D^{\alpha}(\hat{Q}_n' - \hat{\psi}')\cdot D^{\beta}\hat{K}_{\epsilon,p}\|_1 \leqslant c_{44}(p\epsilon)^{|\beta|} n^{-(s-2)/2}\Delta_{n,s}^*, \tag{15.77}$$

which leads to

$$\gamma(f_{a_n} : p\epsilon) \leqslant c_{45} M_r(f)\left(1+(p\epsilon)^{k+r+1}\right) n^{-(s-2)/2}\Delta_{n,s}^*. \tag{15.78}$$

Finally, by Lemma 14.6,

$$\bar{\omega}_{f_{a_n}}(2p\epsilon:(\psi')^+)=\bar{\omega}_f(2p\epsilon:\psi^+)+\int\omega_{f_{a_n}}(x:2p\epsilon)((\psi')^+-\psi^+)(dx)$$
$$+\int(\omega_{f_{a_n}}(x:2p\epsilon)-\omega_f(x:2p\epsilon))\psi^+(dx)$$
$$\leqslant\bar{\omega}_f(2p\epsilon:\psi^+)+M_r(f)\int(1+(\|x\|+\|a_n\|$$
$$+2p\epsilon)^r)|\psi'-\psi|(dx)$$
$$+\left|\int\omega_f(x:2p\epsilon)((\psi'')^+-\psi)(dx)\right|$$
$$\leqslant\bar{\omega}_f(2p\epsilon:\psi^+)+c_{46}M_r(f)(1+(p\epsilon)^r)n^{-(s-2)/2}\Delta^*_{n,s}.$$
(15.79)

In the last step we also used the inequality

$$\omega_f(x:2p\epsilon)\leqslant 2M_r(f)(1+(\|x\|+2p\epsilon)^r)\qquad(r\geqslant 0,\quad x\in R^k).\quad(15.80)$$

The proof is now complete by (15.78), (15.79), (15.66), and (15.62) if one writes $\eta=2p\epsilon$. Q.E.D.

COROLLARY 15.5 *Under the hypothesis of Theorem* 15.4, *one has*

$$\left|\int fd(Q_n-\Phi)\right|\leqslant c_{47}(s,k)M_r(f)(1+\eta^{k+r+1})(\rho_3n^{-1/2}+\rho_sn^{-(s-2)/2})$$
$$+\bar{\omega}_f(\eta:\Phi),\qquad(15.81)$$

where, as in (15.61),

$$\eta=c_3'\rho_3([\log n]+1)n^{-1/2}.$$

Proof. From the expression (14.74) one has

$$|P_m(-\phi:\{\chi_\nu\})(x)|\leqslant c_{48}(s,k)\rho_{m+2}(1+\|x\|^{3m})\phi(x)$$
$$(x\in R^k,\quad 1\leqslant m\leqslant s-2).\qquad(15.82)$$

For, if $|\nu_i| \geqslant 3$, $\Sigma_{i=1}^{p}(|\nu_i|-2)=m$, then (recalling that $\rho_2 = k$)

$$|\chi_{\nu_1} \cdots \chi_{\nu_p}| \leqslant c_{49}(m,k)\rho_{|\nu_1|} \cdots \rho_{|\nu_p|}$$

$$\leqslant c_{50}(m,k)\rho_{m+2}^{\sum_{i=1}^{p}(|\nu_i|-2)/m} = c_{50}(m,k)\rho_{m+2}$$

$$(1 \leqslant m \leqslant s-2). \qquad (15.83)$$

Now

$$n^{-m/2}\rho_{m+2} = n^{-1}\sum_{j=1}^{n} n^{-m/2}\left(\int_{\{\|\mathbf{X}_j\| \leqslant n^{1/2}\}} + \int_{\{\|\mathbf{X}_j\| > n^{1/2}\}}\right)\|\mathbf{X}_j\|^{m+2}$$

$$\leqslant n^{-1}\sum_{j=1}^{n}\left(n^{-m/2+(m-1)/2}\int \|\mathbf{X}_j\|^3 + n^{-m/2-(s-m-2)/2}\int \|\mathbf{X}_j\|^s\right)$$

$$= \rho_3 n^{-1/2} + \rho_s n^{-(s-2)/2} \qquad (1 \leqslant m \leqslant s-2). \qquad (15.84)$$

Therefore

$$\|n^{-m/2}P_m(-\Phi : \{\chi_\nu\})\| \leqslant c_{51}(s,k)\left(\rho_3 n^{-1/2} + \rho_s n^{-(s-2)/2}\right)$$

$$(1 \leqslant m \leqslant s-2). \qquad (15.85)$$

Also, using (15.80), (15.82), and (15.84), we get

$$\bar{\omega}_f\left(\eta : |n^{-m/2}P_m(-\Phi : \{\chi_\nu\})|\right)$$

$$\leqslant c_{52}(s,k)(1+\eta^r)M_r(f)\left(\rho_3 n^{-1/2} + \rho_s n^{-(s-2)/2}\right)$$

$$(1 \leqslant m \leqslant s-2). \qquad (15.86)$$

Now (15.81) follows from (15.60) if one uses (15.85) and (15.86), as well as the first inequality in (15.8). Q.E.D.

Remark. The stipulation (15.58) [which is the same as (15.11)] is redundant when one has $r=0$ in Theorem 15.4 or Corollary 15.5. The proof of this assertion in the Corollary 15.5 case is essentially contained in the proof of Corollary 15.2. In the Theorem 15.4 case one needs the additional fact [see (7.22)]

$$P_m(-\Phi : \{\chi_\nu\})(R^k) = 0 \qquad (1 \leqslant m \leqslant s-2). \qquad (15.87)$$

16. NORMALIZATION

The main theorems in the last section have been stated for independent random vectors $\mathbf{X}_1,\dots,\mathbf{X}_n$ satisfying (*standard normalization*)

$$E\mathbf{X}_j=0 \qquad (1\leqslant j\leqslant n), \qquad n^{-1}\sum_{j=1}^{n}\operatorname{Cov}(\mathbf{X}_j)=I \tag{16.1}$$

where I is the $k\times k$ identity matrix. This was done for the sake of simplicity. It is a relatively minor matter to extend all results of the last section to independent random vectors $\mathbf{X}_1,\dots,\mathbf{X}_n$ satisfying

$$E\mathbf{X}_j=0 \qquad (1\leqslant j\leqslant n), \qquad n^{-1}\sum_{j=1}^{n}\operatorname{Cov}(\mathbf{X}_j)=V, \tag{16.2}$$

where V is an arbitrary symmetric, positive-definite matrix. One could also take the mean vectors in (16.2) to be arbitrary. But since this is taken care of merely by changing the integrand f to a translate of it, we assume (16.2) throughout this section.

Let Q_n denote the distribution of $n^{-1/2}(\mathbf{X}_1+\cdots+\mathbf{X}_n)$. If f is integrable with respect to Q_n and $\Phi_{0,V}$, then by changing variables $x\to Tx$ one has

$$\int f\,d(Q_n-\Phi_{0,V})=\int f\circ T^{-1}d(G_n-\Phi), \tag{16.3}$$

where T is the symmetric positive-definite matrix satisfying

$$T^2=V^{-1}, \tag{16.4}$$

and G_n is the distribution of $n^{-1/2}(T\mathbf{X}_1+\cdots+T\mathbf{X}_n)$. Note that the random vectors $T\mathbf{X}_1,\dots,T\mathbf{X}_n$ satisfy standard normalization, that is,

$$E(T\mathbf{X}_j)=0 \qquad (1\leqslant j\leqslant n), \qquad n^{-1}\sum_{j=1}^{n}\operatorname{Cov}(T\mathbf{X}_j)=I. \tag{16.5}$$

Hence one may use the results of the last section to estimate the right side of (16.3) in terms of the moments

$$\tau_r\equiv n^{-1}\sum_{j=1}^{n}E\|T\mathbf{X}_j\|^r \qquad (r>0), \tag{16.6}$$

and $M_r(f\circ T^{-1})$ and $\omega^*_{g_T}$ or $\bar{\omega}_{f\circ T^{-1}}$, where

$$g_T(x)=\begin{cases}(1+\|x\|^{r_0})^{-1}(f\circ T^{-1})(x) & \text{if } r>0,\\ (f\circ T^{-1})(x) & \text{if } r=0.\end{cases} \tag{16.7}$$

Since T is easily computable, we may leave things at this stage. If one would like the bound to involve moments of $\mathbf{X}_j$'s and not those of $T\mathbf{X}_j$'s, then the simple inequality

$$\tau_r \leqslant \|T\|^r \rho_r = \lambda^{-r/2} \rho_r \qquad (r>0) \tag{16.8}$$

may be used. Here λ is the *smallest eigenvalue* of V. Note that

$$\|T\| = \lambda^{-1/2}, \qquad \|T^{-1}\| = \Lambda^{1/2}, \tag{16.9}$$

where Λ is the *largest eigenvalue* of V. We now rewrite (or extend) the results of Section 15 in a series of theorems and corollaries, assuming that the n independent random vectors $\mathbf{X}_1, \dots, \mathbf{X}_n$ satisfy (16.2).

THEOREM 16.1. *There exist positive constants* c_i $(i=1,2,3,4)$ *depending only on* k *and* s *such that if*

$$n^{-1} \sum_{j=1}^{n} \int_{\{\|T\mathbf{X}_j\| > (2/3) n^{1/2}\}} \|T\mathbf{X}_j\|^s \leqslant c_1 n^{(s-2)/2} \tag{16.10}$$

holds for some integer $s \geqslant 3$, *then for every real-valued, Borel-measurable function* f *satisfying*

$$M_r(f) < \infty \tag{16.11}$$

for some integer r, $0 \leqslant r \leqslant s$, *one has*

$$\begin{aligned} \left| \int f\, d(Q_n - \Phi_{0,V}) \right| &\leqslant M_r(f \circ T^{-1}) \left(c_2 \tau_3 n^{-1/2} + c_3 \tau_s n^{-(s-2)/2} \right) \\ &\quad + 2\omega^*_{g_T}\left(c_4 \tau_3 n^{-1/2} : \Phi_{r_0} \right) \\ &\leqslant \max\{1, \Lambda^{r/2}\} M_r(f) \left(c_2 \lambda^{-3/2} \rho_3 n^{-1/2} \right. \\ &\qquad\qquad \left. + c_3 \lambda^{-s/2} \rho_s n^{-(s-2)/2} \right) \\ &\quad + 2\omega^*_{g_T}\left(c_4 \lambda^{-3/2} \rho_3 n^{-1/2} : \Phi_{r_0} \right). \end{aligned} \tag{16.12}$$

Proof. The first inequality in (16.12) follows from (16.3) and Theorem 15.1, as well as the relations (15.82), (15.84), and (15.86). Also, one has

$$\begin{aligned} M_r(f \circ T^{-1}) &= \sup_{x \in R^k} (1 + \|x\|^r)^{-1} \left| f(T^{-1}x) \right| \\ &\leqslant \left(\sup_{x \in R^k} (1 + \|T^{-1}x\|^r)^{-1} \left| f(T^{-1}x) \right| \right) \left(\sup_{x \in R^k} (1 + \|T^{-1}x\|^r)(1 + \|x\|^r)^{-1} \right) \\ &= M_r(f) \left(\sup_{x \in R^k} (1 + \|T^{-1}x\|^r)(1 + \|x\|^r)^{-1} \right) \\ &\leqslant M_r(f) \max\{1, \|T^{-1}\|^r\} = M_r(f) \max\{1, \Lambda^{r/2}\}. \end{aligned} \tag{16.13}$$

The second inequality in (16.12) follows from (16.9) and (16.13). Q.E.D.

The constants c's *below depend only on* s *and* k.

COROLLARY 16.2. *If* $\rho_3 < \infty$, *then for all bounded, Borel-measurable functions* f *one has*

$$\left|\int f d(Q_n - \Phi_{0,V})\right| \leqslant c_5 \omega_f(R^k)\tau_3 n^{-1/2}$$

$$+ 2\omega^*_{f\circ T^{-1}}(c_4 \tau_3 n^{-1/2} : \Phi)$$

$$\leqslant c_5 \omega_f(R^k)\tau_3 n^{-1/2} + 2\omega^*_f(c_4 \Lambda^{1/2}\tau_3 n^{-1/2} : \Phi_{0,V})$$

$$\leqslant c_5 \omega_f(R^k)\lambda^{-3/2}\rho_3 n^{-1/2}$$

$$+ 2\omega^*_f(c_4 \Lambda^{1/2}\lambda^{-3/2}\rho_3 n^{-1/2} : \Phi_{0,V}). \tag{16.14}$$

Proof. The first inequality follows from Theorem 16.1 on taking $r = r_0 = 0$, $s = 3$, and recalling

$$M_0(f\circ T^{-1}) = \omega_{f\circ T^{-1}}(R^k) = \omega_f(R^k). \tag{16.15}$$

To get the second inequality in (16.14) note that

$$\omega_{f\circ T^{-1}}(x:\epsilon) = \sup\{|f(T^{-1}y) - f(T^{-1}z)| : y, z \in B(x:\epsilon)\}$$

$$\leqslant \sup\{|f(y') - f(z')| : y', z' \in B(T^{-1}x : \|T^{-1}\|\epsilon)\}$$

$$= \omega_f(T^{-1}x : \|T^{-1}\|\epsilon), \tag{16.16}$$

so that

$$\omega^*_{f\circ T^{-1}}(\epsilon:\Phi) \leqslant \sup_{y\in R^k} \int \omega_f(T^{-1}(x+y) : \|T^{-1}\|\epsilon)\Phi(dx)$$

$$= \sup_{y'\in R^k} \int \omega_f(T^{-1}x + y' : \|T^{-1}\|\epsilon)\Phi(dx)$$

$$= \sup_{y'\in R^k} \int \omega_f(z + y' : \|T^{-1}\|\epsilon)\Phi_{0,V}(dz)$$

$$= \omega^*_f(\|T^{-1}\|\epsilon : \Phi_{0,V}) = \omega^*_f(\Lambda^{1/2}\epsilon : \Phi_{0,V}). \tag{16.17}$$

The last inequality in (16.14) now follows from (16.8). Finally, note that the condition (16.10) may be replaced by $\rho_3<\infty$ as in the proof of Corollary 15.2. Q.E.D.

COROLLARY 16.3. *If* $\rho_3<\infty$, *then for all Borel sets* A *one has*

$$|Q_n(A)-\Phi_{0,V}(A)|\leqslant c_5\tau_3 n^{-1/2}+2\sup_{y\in R^k}\Phi_{0,V}\big((\partial A)^{\eta'}+y\big)$$

$$\leqslant c_5\lambda^{-3/2}\rho_3 n^{-1/2}+2\sup_{y\in R^k}\Phi_{0,V}\big((\partial A)^{\eta'}+y\big), \quad (16.18)$$

where

$$\eta=c_4\tau_3 n^{-1/2}\leqslant c_4\lambda^{-3/2}\rho_3 n^{-1/2}, \qquad \eta'=\|T^{-1}\|\eta=\Lambda^{1/2}\eta. \quad (16.19)$$

Proof. The proof follows from Corollary 16.2 and the relations (15.55). Also note that $I_A\circ T^{-1}=I_{TA}$. Q.E.D.

Before stating the analog of Theorem 15.4, we note that if we start with the finite signed measure $P_m(-\Phi_{0,V}:\{\chi_\nu\})$, where χ_ν is the average of the νth cumulants of $\mathbf{X}_j$'s and make the transformation $x\to Tx$ on R^k, then the induced signed measure is $P_m(-\Phi:\{\tilde{\chi}_\nu\})$, where $\tilde{\chi}_\nu$ is the average of the νth cumulants of $T\mathbf{X}_j$'s, $1\leqslant j\leqslant n$. To see this, observe that the Fourier–Stieltjes transform of $P_m(-\Phi_{0,V}:\{\chi_\nu\})$ is

$$\hat{P}_m(-\Phi_{0,V}:\{\chi_\nu\})(t)=\tilde{P}_m(it:\{\chi_\nu\})\exp\{-\tfrac{1}{2}\langle t,Vt\rangle\} \qquad (t\in R^k), \quad (16.20)$$

and that of the induced signed measure [see Theorem 5.1(v)] is

$$t\to\tilde{P}_m(iTt,\{\chi_\nu\})\exp\{-\tfrac{1}{2}\|t\|^2\} \qquad (t\in R^k). \quad (16.21)$$

Now look at the expression (7.3) to conclude

$$\tilde{P}_m(iTt,\{\chi_\nu\})=\tilde{P}_m(it:\{\tilde{\chi}_\nu\}) \qquad (t\in R^k). \quad (16.22)$$

THEOREM 16.4. *There exist positive constants* c_1, c_i' $(i=2,3)$ *depending only on* k *and* s *such that under the hypothesis of Theorem* 16.1 *one has*

$$\left|\int f\,d\left(Q_n-\sum_{m=0}^{s-2}n^{-m/2}P_m(-\Phi_{0,V}:\{\chi_\nu\})\right)\right|$$

$$\leqslant c_2'\max\{1,\Delta^{r/2}\}M_r(f)(1+\eta^{k+r+1})\lambda^{-s/2}\rho_s n^{-(s-2)/2}$$

$$+\bar{\omega}_f\left(\Lambda^{1/2}\eta:\left(\sum_{m=0}^{s-2}n^{-m/2}P_m(-\Phi_{0,V}:\{\chi_\nu\})\right)^+\right), \quad (16.23)$$

where

$$\eta = c_3'\lambda^{-3/2}\rho_3([\log n]+1)n^{-1/2}. \tag{16.24}$$

Proof. This is an immediate consequence of Theorem 15.4 and the relation

$$\int fd\left(Q_n - \sum_{m=0}^{s-2} n^{-m/2}P_m(-\Phi_{0,V}:\{\chi_\nu\})\right)$$
$$= \int (f\circ T^{-1})d\left(G_n - \sum_{m=0}^{s-2} n^{-m/2}P_m(-\Phi:\{\tilde{\chi}_\nu\})\right), \tag{16.25}$$

which follows from the discussion preceding the statement of the present theorem. One of course also uses (16.8), (16.9), (16.13), and (16.16), as in the proof of Theorem 16.1. Q.E.D.

COROLLARY 16.5. *Under the hypothesis of Theorem* 16.4 *one has*

$$\left|\int fd(Q_n - \Phi_{0,V})\right| \le c_6 \max\{1, \Lambda^{r/2}\} M_r(f)(1+\eta^{k+r+1})$$
$$\times\left(\lambda^{-3/2}\rho_3 n^{-1/2} + \lambda^{-s/2}\rho_s n^{-(s-2)/2}\right)$$
$$+\bar{\omega}_f(\Lambda^{1/2}\eta:\Phi_{0,V}), \tag{16.26}$$

where η *is defined by* (16.24).

Proof. This follows from Corollary 15.5 if one uses (16.3), (16.8), (16.9), (16.13), and (16.16). Q.E.D.

Remark. We have already pointed out (see Corollaries 16.2, 16.3) that the stipulation (16.10) is redundant in Theorem 16.1 in the case $r=0$. This is also true for Theorem 16.4 and Corollary 16.5, and the proof is also essentially the same. Note that (16.10) may be replaced by the slightly more stringent but simpler condition

$$\tau_s \le c_1 n^{(s-2)/2} \tag{16.27}$$

or by

$$\rho_s \le c_1 \lambda^{s/2} n^{(s-2)/2}. \tag{16.28}$$

17. SOME APPLICATIONS

The present section is devoted to some important applications of the main theorems of Sections 15 and 16. We continue to use the same notation as in Section 16. Thus $\mathbf{X}_1,\ldots,\mathbf{X}_n$ are n independent random vectors with values in R^k satisfying

$$EX_j=0 \qquad (1\leqslant j\leqslant n), \qquad n^{-1}\sum_{j=1}^{n} \mathrm{Cov}(\mathbf{X}_j)=V, \tag{17.1}$$

where V is a symmetric, positive-definite matrix. Also, T is the symmetric positive-definite matrix satisfying (16.4), and we write

$$\rho_r=n^{-1}\sum_{j=1}^{n} E\|\mathbf{X}_j\|^r, \qquad \tau_r=n^{-1}\sum_{j=1}^{n} E\|T\mathbf{X}_j\|^r \qquad (r>0). \tag{17.2}$$

Recall that λ, Λ are the *smallest and largest eigenvalues* of V, respectively, and Q_n is the distribution of $n^{-1/2}(\mathbf{X}_1+\cdots+\mathbf{X}_n)$.

We define the class $\mathcal{Q}_\alpha^*(d:\mu)$ as *the class of all Borel subsets* A *of* R^k *satisfying*

$$\sup_{y\in R^k} \mu\big((\partial A)^\epsilon+y\big)\leqslant d\epsilon^\alpha \qquad (\epsilon>0), \tag{17.3}$$

for a given pair of positive numbers α, d, *and a measure* μ.

THEOREM 17.1. *There exist positive constants* c_1 *and* c_2 *depending only on* k *such that if* $\rho_3<\infty$, *then*

$$\begin{aligned}\sup_{A\in\mathcal{Q}_\alpha^*(d:\Phi_{0,V})} &|Q_n(A)-\Phi_{0,V}(A)| \\ &\leqslant c_1\tau_3 n^{-1/2}+2d\big(c_2\Lambda^{1/2}\tau_3 n^{-1/2}\big)^\alpha \\ &\leqslant c_1\lambda^{-3/2}\rho_3 n^{-1/2}+2d\big(c_2\Lambda^{1/2}\lambda^{-3/2}\rho_3 n^{-1/2}\big)^\alpha. \end{aligned}\tag{17.4}$$

Proof. This follows immediately from Corollary 16.3. Q.E.D.

COROLLARY 17.2. *There exists a constant* c_3 *depending only on* k *such that if* $\rho_3<\infty$, *then*

$$\sup_{C\in\mathcal{C}} |Q_n(C)-\Phi_{0,V}(C)|\leqslant c_3\tau_3 n^{-1/2}\leqslant c_3\lambda^{-3/2}\rho_3 n^{-1/2}, \tag{17.5}$$

where $\mathcal{C}$ *is the class of all Borel-measurable convex subsets of* R^k.

Proof. Let G_n be the distribution of $n^{-1/2}(T\mathbf{X}_1+\cdots+T\mathbf{X}_n)$. Applying Theorem 17.1 to the random vectors $T\mathbf{X}_1,\dots,T\mathbf{X}_n$ and using Corollary 3.2 [or inequality (13.42)], one has

$$\begin{aligned}\sup_{C\in\mathcal{C}} |Q_n(C)-\Phi_{0,V}(C)|&=\sup_{C\in\mathcal{C}} |G_n(C)-\Phi(C)| \\ &\leqslant c_1\tau_3 n^{-1/2}+2d\big(c_2\tau_3 n^{-1/2}\big), \end{aligned}\tag{17.6}$$

where

$$d=2^{5/2}\frac{\Gamma((k+1)/2)}{\Gamma(k/2)}. \tag{17.7}$$

Note that $\mathcal{C}$ is invariant under translation, as well as nonsingular, linear transformations. Q.E.D.

COROLLARY 17.3. *Let* k=2. *Denote by* $\mathfrak{D}(l)$ *the class of all Borel subsets of* R^2 *each having a boundary contained in some rectifiable curve of length not exceeding* l. *There are absolute constants* c_4, c_5 *such that if* $\rho_3<\infty$, *then*

$$\sup_{D\in\mathfrak{D}(l)}|Q_n(D)-\Phi_{0,V}(D)|\leqslant c_4\lambda^{-3/2}\rho_3 n^{-1/2}$$
$$+c_5(\lambda^{-1}l+\lambda^{-1/2})\Lambda^{1/2}\lambda^{-3/2}\rho_3 n^{-1/2}. \tag{17.8}$$

Proof. From the estimate (2.60) we get

$$\Phi_{0,V}((\partial D)^{\epsilon})\leqslant(2\pi)^{-1}(\operatorname{Det}V)^{-1/2}(4\pi l+8\pi\epsilon^2)$$
$$\leqslant\lambda^{-1}(2l\epsilon+4\epsilon^2)\leqslant(2\lambda^{-1}l+2\lambda^{-1/2})\qquad\left(\epsilon\leqslant\frac{\lambda^{1/2}}{2}\right), \tag{17.9}$$

since $\operatorname{Det}V\geqslant\lambda^2$. It is enough to consider $\epsilon\leqslant\lambda^{1/2}/2$; otherwise $\lambda^{-1}(2l\epsilon+4\epsilon^2)$ exceeds one. Now apply the second inequality in (17.4), observing that $\mathfrak{D}(l)$ is translation invariant. Q.E.D.

There are Borel subsets A of R^k for which

$$\sup_{y\in R^k}\Phi((\partial A)^{\epsilon}+y)\leqslant c_6\epsilon^{\alpha}\qquad(\epsilon>0) \tag{17.10}$$

for some positive constants c_6 and α, $\alpha>1$. Examples of such sets are affine subspaces of dimensions $k'<k-1$ (and their subsets and complements) and many other manifolds of dimension $k'<k-1$, for which $\alpha=k-k'$. Below we assume that $V=I$ merely to avoid notational complexity.

THEOREM 17.4. *Let* V=I. *Assume that* $\rho_s<\infty$ *for some integer* $s\geqslant3$. *If* A *is a Borel set satisfying* (17.10) *for some* $\alpha>1$, *then*

$$|Q_n(A)-\Phi(A)|\leqslant c_7 n^{-(s-2)/2}\Delta^*_{n,s}$$
$$+c_8(\rho_3 n^{-1/2})^{\alpha}\left(1+\sum_{m=1}^{s-2}n^{-m/2}\rho_{m+2}\log^{3m/2}n\right), \tag{17.11}$$

where c_7 *depends only on* s *and* k, *and* c_8 *depends only on* s, k, α, *and* c_6. *In particular, if* $\mathbf{X}_1,\dots,\mathbf{X}_n$ *are identically distributed, then*

$$|Q_n(A)-\Phi(A)|=O(n^{-\alpha/2}) \tag{17.12}$$

provided that $s \geqslant \alpha+2$.

Proof. First assume that $\Phi(A)=0$ or $\Phi(A)=1$. In this case $P_m(-\Phi:\{\chi_\nu\})(A)=0$ for all m, $1\leqslant m\leqslant s-2$, because of (7.22). Hence (15.56) holds, and it remains to show that

$$n^{-m/2}\int_{(\partial A)^\eta+y}|P_m(-\phi:\{\chi_\nu\})(x)|\,dx$$
$$\leqslant c_8'\left(\rho_3 n^{-1/2}\right)^\alpha\left(1+n^{-m/2}\rho_{m+2}\log^{3m/2}n\right) \qquad (1\leqslant m\leqslant s-2).$$

This follows from (15.82) and the estimate

$$\int_{(\partial A)^\eta+y}\|x\|^{3m}\phi(x)\,dx\leqslant\int_{(\partial A)^\eta+y}(2\alpha\log n)^{3m/2}\phi(x)\,dx$$
$$+\int_{\{\|x\|\geqslant(2\alpha\log n)^{1/2}\}}\|x\|^{3m}\phi(x)\,dx$$
$$\leqslant c_6(2\alpha\log n)^{3m/2}\left(\rho_3 n^{-1/2}\right)^\alpha+n^{-\alpha/2}.$$

We now show that $\Phi(A)=0$ or 1. For this first observe that for all y, $z\in R^k$ $(y\neq z)$, one has

$$A+y\backslash A+z\subset Cl\left(A^{\|y-z\|}\right)+y,$$

so that, by (17.10),

$$|\Phi(A+y)-\Phi(A+z)|\leqslant c_6\|y-z\|^\alpha.$$

In other words, the function $f(y)\equiv\Phi(A+y)$ is Lipschitzian of order $\alpha>1$. It follows that f is differentiable and has a zero differential on all of R^k. Hence f is a constant on R^k; that is,

$$\Phi(A+y)=\Phi(A) \qquad (y\in R^k). \tag{17.13}$$

Define the measure μ on R^k by

$$\mu(B)=\int_B I_A(x)\,dx=\lambda_k(B\cap A) \qquad (B\in\mathfrak{B}^k),$$

where I_A is the indicator function of A and λ_k is Lebesgue measure on R^k. We show that if (17.13) holds, then μ is translation invariant. This would imply that $\mu = c\lambda_k$ for some c, $0 \leqslant c < \infty$, which is possible only if $I_A = 0$ (almost everywhere) or $I_A = 1$ (almost everywhere), and the proof of the theorem is complete. If B is a bounded Borel set, then for every $\epsilon > 0$ there exist $c_i \in R^1, y_i \in R^k$, $1 \leqslant i \leqslant m$, say, such that $\|I_B - \sum_{i=1}^m c_i \phi(x+y_i)\|_1 < \epsilon/2$, where ϕ is the standard normal density on R^k. This follows because linear combinations of translates of any function whose fourier transform does not vanish anywhere on R^k are dense in $L^1(R^k)$.† One then has, using (17.13) in the last step,

$$
\begin{aligned}
|\mu(B-y) - \mu(B)| &= \left| \int_A I_{B-y}(x)\,dx - \int_A I_B(x)\,dx \right| \\
&= \left| \int_{A+y} I_B(x)\,dx - \int_A I_B(x)\,dx \right| \\
&\leqslant \left| \int_{A+y} I_B(x)\,dx - \sum_{i=1}^m \int_{A+y} c_i\phi(x+y_i)\,dx \right| \\
&\quad + \left| \sum_{i=1}^m \int_{A+y} c_i\phi(x+y_i)\,dx - \sum_{i=1}^m \int_A c_i\phi(x+y_i)\,dx \right| \\
&\quad + \left| \sum_{i=1}^m \int_A c_i\phi(x+y_i)\,dx - \int_A I_B(x)\,dx \right| \\
&< \frac{\epsilon}{2} + 0 + \frac{\epsilon}{2} = \epsilon
\end{aligned}
$$

for all $y \in R^k$. Hence $\mu(B-y) = \mu(B)$ for all $y \in R^k$. Q.E.D.

For the special case (of Theorem 17.4) when A is an affine subspace of dimension $k' < k-1$, the hypothesis $\rho_s < \infty$ may be relaxed. Indeed, one has

THEOREM 17.5. *There exists a positive number* c_{10} *depending only on* k *such that if the hypothesis of Theorem* 17.4 *holds with* s=3, *then for every affine subspace* A (*of* R^k) *of dimension* $k' \leqslant k-1$, *one has*

$$
Q_n(A) \leqslant c_{10}\left(\rho_3 n^{-1/2}\right)^{(k-k')}. \tag{17.14}
$$

†See Wiener [1], Theorem 9, p. 98.

Proof. An affine subspace A of dimension $k'<k$ has a representation

$$A=\left\{x\in R^k:\langle t_i,x\rangle=c_i,\ 1\leqslant i\leqslant k-k'\right\},$$

where $t_1,\dots,t_{k-k'}$ are mutually orthogonal vectors in R^k each with norm one, and $c_i\in R^1$, $1\leqslant i\leqslant k-k'$. Let T denote the $(k-k')\times k$ matrix whose ith row vector is t_i $(1\leqslant i\leqslant k-k')$. Then $TT'=\bar{I}$, the $(k-k')\times(k-k')$ identity matrix. Define $(k-k')$-dimensional random vectors $\mathbf{U}_j$ by

$$\mathbf{U}_j=T\mathbf{X}_j \qquad (1\leqslant j\leqslant n)$$

and note that

$$E\mathbf{U}_j=0 \qquad (1\leqslant j\leqslant n), \qquad n^{-1}\sum_{j=1}^{n}\operatorname{Cov}\mathbf{U}_j=\bar{I},$$

$$n^{-1}\sum_{j=1}^{n}E\|\mathbf{U}_j\|^3\leqslant n^{-1}\sum_{j=1}^{n}E\|\mathbf{X}_j\|^3=\rho_3.$$

Let $\bar{Q}_n$ denote the distribution of $n^{-1/2}\Sigma_{j=1}^{n}\mathbf{U}_j$ and let $\bar{\Phi}$ denote the standard normal distribution on $R^{k-k'}$. Now let c be the vector in $R^{k-k'}$ whose ith coordinate is c_i $(1\leqslant i\leqslant k-k')$. Then

$$Q_n(A)=\bar{Q}_n(\{c\}), \qquad \Phi(A)=\bar{\Phi}(\{c\})=0,$$

and using (15.57) to $\bar{Q}_n$, $\bar{\Phi}$, and $\{c\}$ (in place of Q_n, Φ, and A, respectively) for $s=3$,

$$\begin{aligned}Q_n(A)&=|Q_n(A)-\Phi(A)|=|\bar{Q}_n(\{c\})-\bar{\Phi}(\{c\})|\\&\leqslant c_{10}n^{-1/2}\rho_3\left(\rho_3n^{-1/2}\right)^{k-k'}+2\sup_{y\in R^{k-k'}}\bar{\psi}^{+}\left((\partial\{c+y\})^{\eta}\right)\\&\leqslant c_{10}\left(n^{-1/2}\rho_3\right)^{k+1-k'}+c'_{10}\left(\rho_3n^{-1/2}\right)^{k-k'},\end{aligned}$$

where $\bar{\psi}=\bar{\Phi}+n^{-1/2}P_1(-\bar{\Phi}:\{\bar{\chi}_\nu\})$, $\bar{\chi}_\nu$ denoting the average of the νth cumulants of $\mathbf{U}_j$'s. Q.E.D.

Remark. It is fairly straightforward to extend the assertion (17.12) to a sequence $\{\mathbf{X}_n:n\geqslant 1\}$ of independent random vectors for which

$$\liminf_n \lambda_n>0, \qquad \sup_n n^{-1}\sum_{j=1}^{n}E\|\mathbf{X}_j\|^s<\infty. \tag{17.15}$$

Here λ_n is the smallest eigenvalue of $V_n\equiv n^{-1}\Sigma_{j=1}^{n}\operatorname{Cov}(\mathbf{X}_j)$. Note that with $T_n=V_n^{-1/2}$ one has

$$n^{-1}\sum_{j=1}^{n}E\|T_n\mathbf{X}_j\|^s\leqslant\lambda_n^{-s/2}\left(n^{-1}\sum_{j=1}^{n}E\|\mathbf{X}_j\|^s\right), \tag{17.16}$$

so that applying (17.11) to random vectors $T_n\mathbf{X}_j$, $1 \leqslant j \leqslant n$, one arrives at (17.12), provided that $s \geqslant \alpha + 2$ and (17.15) holds. Similarly, in Corollaries 17.2 and 17.3 the error bounds are $O(n^{-1/2})$ if (17.15) holds with $s = 3$. In the same manner, orders of magnitude of errors for a sequence $\{\mathbf{X}_n : n \geqslant 1\}$ satisfying (17.15) may be obtained from the remaining theorems and corollaries of this section.

For an application of a different nature, consider a Borel set A and define the function f by

$$f(x) = (1 + d^s(0, \partial A))I_{A'}(x) \qquad (x \in R^k), \tag{17.17}$$

where

$$A' = \begin{cases} A & \text{if } 0 \in R^k \setminus A, \\ R^k \setminus A & \text{if } 0 \in A, \end{cases} \tag{17.18}$$

and $d(0, \partial A)$ is the *euclidean distance of* ∂A *from the origin* O. Note that

$$M_s(f) \leqslant 1 \qquad (s > 0). \tag{17.19}$$

Defining g by (15.14) with $r = s$, one has, for $\|z\| < 2\epsilon < c_{11}(k,s) < d(0, \partial A)$,

$$\begin{aligned}
&|g(x+y+z) - g(x+y)| \\
&\quad \leqslant (1 + \|x+y\|^{s_0})^{-1}|f(x+y+z) - f(x+y)| \\
&\qquad + |f(x+y+z)||(1 + \|x+y\|^{s_0})^{-1} - (1 + \|x+y+z\|^{s_0})^{-1}| \\
&\quad \leqslant (1 + \|x+\|^{s_0})^{-1}|f(x+y+z) - f(x+y)| + c_{12}(k,s)\epsilon \\
&\quad \leqslant (1 + \|x+y\|^{s_0})^{-1}(1 + d^s(0, \partial A))I_{(\partial A)^{2\epsilon}}(x+y) + c_{12}(k,s)\epsilon \\
&\quad \leqslant \left(1 + (d(0, \partial A) - 2\epsilon)^{s_0}\right)^{-1}(1 + d^s(0, \partial A))I_{(\partial A)^{2\epsilon}}(x+y) + c_{12}(k,s)\epsilon \\
&\quad \leqslant c_{13}(k,s)I_{(\partial A)^{2\epsilon}}(x+y) + c_{12}(k,s)\epsilon,
\end{aligned} \tag{17.20}$$

where

$$c_{13}(k,s) = \sup_{b \geqslant 0} \left(1 + (b + c_{11}(k,s))^s\right)(1 + b^{s_0})^{-1}. \tag{17.21}$$

Thus in this case

$$\omega_g^*(2\epsilon : \Phi_{s_0}) \leqslant c_{13}(k,s) \sup_{y \in R^k} \Phi_{s_0}\left((\partial A)^{2\epsilon} + y\right) + c_{14}(k,s)\epsilon, \tag{17.22}$$

where

$$c_{14}(k,s) = c_{12}(k,s)\Phi_{s_0}(R^k). \tag{17.23}$$

Specializing to convex sets, we can prove

THEOREM 17.6. *There exist positive constants* c_{15}, c_{16}, c_{17} *depending only on* k *and* s *such that if* V = I *and*

$$\rho_s < c_{15} n^{(s-2)/2} \tag{17.24}$$

Removed by Michel (Sankhya 1978 40 388-392 (Rare copy))

for some integer s ≥ 3, *then*

$$\sup_{C \in \mathcal{C}} (1 + d^s(0, \partial C)) |Q_n(C) - \Phi(C)| \leq c_{16} \rho_3 n^{-1/2} + c_{17} \rho_s n^{-(s-2)/2}. \tag{17.25}$$

Proof. Let C be a Borel-measurable convex set. Replacing A by C in (17.17) and using (17.22) in Theorem 15.1 (with $r = s$), one has

$$\begin{aligned}
(1 + d^s(0, \partial C))&|Q_n(C) - \Phi(C)| \\
&= \left| \int f d(Q_n - \Phi) \right| \\
&\leq \left| \int f d(Q_n - \psi) \right| + \sum_{m=1}^{s-2} n^{-m/2} \left| \int f dP_m(-\Phi : \{\chi_\nu\}) \right| \\
&\leq c_{18} \rho_s n^{-(s-2)/2} + c_{19} \omega_g^*(2\epsilon : (\psi^+)_{s_0}) \\
&\quad + \sum_{m=1}^{s-2} n^{-m/2} \left| \int f dP_m(-\Phi : \{\chi_\nu\}) \right|,
\end{aligned} \tag{17.26}$$

where $\epsilon = c_{10} \rho_3 n^{-1/2}$. Now, by (15.82) and (15.84),

$$\begin{aligned}
\int (1 + \|x\|^{s_0}) |P_m(-\phi : \{\chi_\nu\})(x)| dx &\leq c_{21} \rho_{m+2}, \\
n^{-m/2} \rho_{m+2} &\leq n^{-1/2} \rho_3 + n^{-(s-2)/2} \rho_s.
\end{aligned} \tag{17.27}$$

Hence (17.26) reduces to

$$\begin{aligned}
(1 + d^s(0, \partial C)) |Q_n(C) - \Phi(C)| &\leq c_{22} (\rho_3 n^{-1/2} + \rho_s n^{-(s-2)/2}) \\
&\quad + c_{19} \omega_g^*(2\epsilon : \Phi_{s_0}).
\end{aligned} \tag{17.28}$$

After (17.22) and Corollary 3.2 are used in (17.28) the proof is complete. Q.E.D.

Theorem 17.6 leads to the so-called *global* and *mean central limit theorems in* R^1.

COROLLARY 17.7. *Under the hypothesis of Theorem* 17.6 *with* $k=1$ *one has*

$$\sup_{x\in R^1}(1+|x|^s)|F_n(x)-\Phi(x)|\leqslant c_{23}\rho_3 n^{-1/2}+c_{24}\rho_s n^{-(s-2)/2} \tag{17.29}$$

where $F_n(\cdot)$ *is the distribution function of* $n^{-1/2}(\mathbf{X}_1+\cdots+\mathbf{X}_n)$ *and* $\Phi(\cdot)$ *is the standard normal distribution function. It follows that*

$$\|F_n-\Phi\|_p\equiv\left(\int_{R^1}|F_n(x)-\Phi(x)|^p\,dx\right)^{1/p}$$

$$\leqslant c_{25}(p,s)\left(c_{23}\rho_3 n^{-1/2}+c_{24}\rho_s n^{-(s-2)/2}\right), \tag{17.30}$$

for all $p>1/s$.

Proof. The inequality (17.29) follows from Theorem 17.6 on taking $k=1$, $C=(-\infty,x]$, $x\in R^1$. Inequality (17.30) is immediate from this. Q.E.D.

Remark. The validity of (17.30) for $p>\frac{1}{2}$ may be proved without the assumption (17.24). Let $\rho_3<\infty$. Then

$$|F_n(x)-\Phi(x)|\leqslant(1+|x|^3)^{-1}c_{26}\rho_3 n^{-1/2}$$

$$\leqslant 2(1+x^2)^{-1}c_{26}\rho_3 n^{-1/2}\qquad (x\in R^1) \tag{17.31}$$

if (17.24) holds, that is, if

$$\rho_3<c_{27}n^{1/2} \tag{17.32}$$

for a suitable positive number c_{27}. However

$$x^2|F_n(x)-\Phi(x)|\leqslant x^2(F_n(x)+\Phi(x))$$

$$\leqslant\int_{\{|y|>|x|\}}y^2Q_n(dy)+\int_{\{|y|>|x|\}}y^2\Phi(dy)\leqslant 2\qquad (x<0),$$

$$x^2|F_n(x)-\Phi(x)|=x^2|(1-F_n(x))-(1-\Phi(x))|$$

$$\leqslant\int_{\{|y|>x\}}y^2Q_n(dy)+\int_{\{|y|>x\}}y^2\Phi(dy)\leqslant 2\qquad (x\geqslant 0),$$

$$|F_n(x)-\Phi(x)|\leqslant 1\qquad (x\in R^1), \tag{17.33}$$

so that the inequality

$$(1+x^2)|F_n(x)-\Phi(x)|\leqslant 3\qquad (x\in R^1) \tag{17.34}$$

holds whenever $\rho_2=1$. Now

$$|F_n(x)-\Phi(x)| \leqslant c_{28}(1+x^2)^{-1}\rho_3 n^{-1/2} \qquad (x\in R^1) \tag{17.35}$$

holds whenever $\rho_3<\infty$ if $c_{28}=\max\{2c_{26}, 3/c_{27}\}$. This is true since (17.35) holds because of (17.31), provided that (17.32) holds. If (17.32) is violated, then (17.35) holds by virtue of (17.34).

The final application of Theorem 15.1 is to Lipschitzian functions.

THEOREM 17.8. *Let* (17.1) *hold with* V = I. *There exist positive constants* c_{29}, c_{30}, c_{31} *depending only on* r, k, *and* s *such that if*

$$\rho_s < c_{29} n^{(s-2)/2} \tag{17.36}$$

for some integer $s \geqslant 3$, *then for all* f *satisfying*

$$M_r(f)<\infty, \qquad |f(x)-f(y)| \leqslant d\|x-y\|^{\alpha} \qquad (x,y\in R^k) \tag{17.37}$$

for some integer r, $0\leqslant r\leqslant s$, *and a pair of positive numbers* d, α $(0<\alpha\leqslant 1)$, *one has the inequality*

$$\left|\int f\,d(Q_n-\Phi)\right| \leqslant c_{29} M_r(f)\left(\rho_3 n^{-1/2}+\rho_s n^{-(s-2)/2}\right) + c_{30} d\left(c_{31}\rho_3 n^{-1/2}\right)^{\alpha}. \tag{17.38}$$

If r = 0, (17.36) *may be replaced by* $\rho_3<\infty$.

Proof. By Theorem 16.1 (with $V=I=T$) one has

$$\left|\int f\,d(Q_n-\Phi)\right| \leqslant c_{32} M_r(f)\left(\rho_3 n^{-1/2}+\rho_s n^{-(s-2)/2}\right) + 2\omega_g^*\left(c_{33}\rho_3 n^{-1/2} : \Phi_{r_0}\right). \tag{17.39}$$

Now if $r=0$, then $r_0=0$, $g=f$, and $\omega_f^*(\epsilon:\Phi)\leqslant d(2\epsilon)^{\alpha}$ and we are done by Corollary 15.2. If $r>0$, then

$$\begin{aligned}
|g(x)-g(y)| &\\
&= |(1+\|x\|^r)^{-1} f(x)-(1+\|y\|^r)^{-1} f(y)|\\
&\leqslant (1+\|x\|^r)^{-1}|f(x)-f(y)|+|f(y)\left[(1+\|x\|^r)^{-1}-(1+\|y\|^r)^{-1}\right]|\\
&\leqslant |f(x)-f(y)|+|f(y)(1+\|y\|^r)^{-1}|(1+\|x\|^r)^{-1}|\,\|x\|^r-\|y\|^r|\\
&\leqslant |f(x)-f(y)|+M_r(f) r\|x-y\|(1+\|x\|^r)^{-1}\max\left\{\|x\|^{r-1},\|y\|^{r-1}\right\}\\
&\leqslant d(2\epsilon)^{\alpha}+rM_r(f)(1+\|x\|^{r-1})(\|x\|+2\epsilon)^{r-1}2\epsilon\\
&\leqslant d(2\epsilon)^{\alpha}+2^r rM_r(f)(1+\|x\|^r)^{r-1}\left(\|x\|^{r-1}+(2\epsilon)^{r-1}\right)\epsilon\\
&\leqslant d(2\epsilon)^{\alpha}+2^r rM_r(f)\left[1+(2\epsilon)^{r-1}\right]\epsilon \qquad (\|x-y\|\leqslant 2\epsilon).
\end{aligned} \tag{17.40}$$

Letting $\epsilon = c_{33}\rho_3 n^{-1/2}$, one has $\epsilon \leqslant c_{34}$ by (14.60). Hence (17.38) follows from (17.39) and (17.40). Q.E.D.

We now turn to some applications of Theorem 15.4. Specialize Theorem 15.4 to the case $r=0$, $s=3$, $f=I_A$, where A is a Borel subset of R^k. If $V=I$ and $\rho_3 < \infty$, then one can show that

$$Q_n(A) - \Phi(A) \leqslant c_{35}(1+\eta^{k+1})n^{-1/2}\Delta^*_{n,3} + \Phi(A^\eta \backslash A), \tag{17.41}$$

where, as in (15.61),

$$\eta = c_3'\rho_3([\log n]+1)n^{-1/2}. \tag{17.42}$$

The one-sided inequality (17.41) follows if one uses Corollary 11.3 instead of Corollary 11.2 [see (15.66)] in the proof of Theorem 15.4. The rest of the proof need not be altered.

Let

$$\delta_n = \max\{\eta, c_{35}(1+\eta^{k+1})n^{-1/2}\Delta^*_{n,3}\}. \tag{17.43}$$

THEOREM 17.9. *Let* V=I, $\rho_3 < \infty$. *Then the Prokhorov distance* $\mathrm{d_p}$ *and the bounded Lipschitzian distance* $\mathrm{d_{BL}}$ *between* $\mathrm{Q_n}$ *and* Φ *are estimated by*

$$d_p(Q_n, \Phi) \leqslant \delta_n, \qquad d_{BL}(Q_n, \Phi) \leqslant c_{36}\rho_3 n^{-1/2}, \tag{17.44}$$

where c_{36} *depends only on* k.

Proof. The assertion concerning d_{BL} follows immediately from Theorem 17.8 with $r=0$ [see (2.51) for the definition of this distance]. As to d_p, note that for any two probability measures G_1, G_2 on R^k one has

$$\begin{aligned} d_p(G_1,G_2) &\equiv \inf\{\epsilon>0: G_1(A) \leqslant G_2(A^\epsilon)+\epsilon \text{ and} \\ &\qquad G_2(A) \leqslant G_1(A^\epsilon)+\epsilon \text{ for all } A \in \mathcal{B}^k\} \\ &= \inf\{\epsilon>0: G_1(F) \leqslant G_2(F^\epsilon)+\epsilon, \\ &\qquad G_2(F) \leqslant G_1(F^\epsilon)+\epsilon \text{ for all closed } F\} \end{aligned} \tag{17.45}$$

since $A^\epsilon = (\text{closure of } A)^\epsilon$ for all $A \subset R^k$ and all $\epsilon > 0$. Also,

$$d_p(G_1,G_2) = \inf\{\epsilon>0: G_1(F) \leqslant G_2(F^\epsilon)+\epsilon \text{ for all closed } F\} \tag{17.46}$$

since, given $G_1(F) \leqslant G_2(F^\epsilon)+\epsilon$ for all closed F and some $\epsilon > 0$, one obtains

$$\begin{aligned} 1 - G_1(F^\epsilon) &= G_1(R^k \backslash F^\epsilon) \leqslant G_2\big((R^k\backslash F^\epsilon)^\epsilon\big)+\epsilon \\ &\leqslant G_2(R^k\backslash F)+\epsilon = 1 - G_2(F)+\epsilon \end{aligned} \tag{17.47}$$

for all closed sets F, using

$$R^k \backslash F \supset (R^k \backslash F^\epsilon)^\epsilon. \tag{17.48}$$

Now (17.41) and (17.43) yield

$$Q_n(A) \leqslant \Phi(A) + \delta_n + \Phi(A^{\delta_n} \backslash A) = \Phi(A^{\delta_n}) + \delta_n \qquad (A \in \mathcal{B}^k). \tag{17.49}$$

Together with (17.46) this gives the first inequality in (17.44). Q.E.D.

Recall that $\Delta^*_{n,3} \leqslant \rho_3$ [see (15.8)], so that one may replace $\Delta^*_{n,3}$ in the expression (17.43) for δ_n by ρ_3. Also observe that the more commonly used Kolmogorov distance [see (2.72)] is bounded above by the distance

$$d_0(G_1, G_2) \equiv \sup_{C \in \mathcal{C}} |G_1(C) - G_2(C)| \tag{17.50}$$

where $\mathcal{C}$ is the class of all Borel-measurable convex subsets of R^k, and G_1, G_2 are probability measures on R^k. By Corollary 17.2 one has

$$d_0(Q_n, \Phi) \leqslant c_{37} \rho_3 n^{-1/2}, \tag{17.51}$$

under the hypothesis of Theorem 17.9.

For two positive numbers d, α, where $0 < \alpha \leqslant 1$, let $\mathcal{Q}_\alpha(d : \Phi_{0,V})$ denote *the class of all Borel subsets* A *of* R^k *satisfying*

$$(\Phi_{0,V}(\partial A)^\epsilon) \leqslant d\epsilon^\alpha \qquad (\epsilon > 0). \tag{17.52}$$

THEOREM 17.10. *If* $\rho_3 < \infty$, *then there exist constants* c_{38}, c_{39} *depending only on* k *such that*

$$\sup_{A \in \mathcal{Q}_\alpha(d : \Phi_{0,V})} |Q_n(A) - \Phi_{0,V}(A)|$$

$$\leqslant c_{38}(1 + \eta^{k+1}) \lambda^{-3/2} \rho_3 n^{-1/2} + d(\Lambda^{1/2}\eta)^\alpha, \tag{17.53}$$

where

$$\eta = c_{39} \lambda^{-3/2} \rho_3([\log n] + 1) n^{-1/2}, \tag{17.54}$$

and λ, Λ *are the smallest and largest eigenvalues, respectively, of the average covariance matrix* V *of* $\mathbf{X}_1, \ldots, \mathbf{X}_n$.

Proof. This is an immediate consequence of Corollary 16.5 with $s = 3$, $r = 0$, $f = I_A$. Q.E.D.

The class $\mathcal{Q}_\alpha(d:\Phi_{0,V})$ may be shown to be larger than $\mathcal{Q}_\alpha^*(d:\Phi_{0,V})$ defined by (17.3), for any pair of positive numbers d, α, where $0<\alpha\leqslant 1$ (see, e.g., the discussion preceding the statement of Theorem 15.4). The latter is the largest translation invariant subclass of the former.

A final application of Theorem 15.4 provides an estimate for tail probabilities of Q_n.

THEOREM 17.11. *Let* $\{\mathbf{X}_n : n\geqslant 1\}$ *be a sequence of independent random vectors with values in* R^k *and having zero means and finite* sth *absolute moments for some integer* $s\geqslant 3$. *Let*

$$V_n = n^{-1}\sum_{j=1}^{n}\operatorname{Cov}(\mathbf{X}_j),\qquad \lambda_n=\text{smallest eigenvalue of } V_n,$$

$$\Lambda_n=\text{largest eigenvalue of } V_n,\qquad \bar{\rho}_{r,n}=n^{-1}\sum_{j=1}^{n}E\|\mathbf{X}_j\|^r,$$

$$\tilde{\Delta}^*_{n,s}=\inf_{0\leqslant\epsilon\leqslant 1}\left[\epsilon n^{-1}\sum_{j=1}^{n}\lambda_n^{-s/2}\int_{\{\|\mathbf{X}_j\|\leqslant\Lambda_n^{1/2}\epsilon n^{1/2}\}}\|\mathbf{X}_j\|^s\right.$$
$$\left.+n^{-1}\sum_{j=1}^{n}\lambda_n^{-s/2}\int_{\{\|\mathbf{X}_j\|>\lambda_n^{1/2}\epsilon n^{1/2}\}}\|\mathbf{X}_j\|^s\right].\tag{17.55}$$

Assume that

$$\liminf_{n\to\infty}\lambda_n>0,\qquad \lim_{n\to\infty}n^{-1/2}\bar{\rho}_{3,n}\log^{1/2}n=0,\qquad \lim_{n\to\infty}n^{-(s-2)/2}\bar{\rho}_{s,n}=0.\tag{17.56}$$

Then one has

$$\sup_{a\geqslant((s-2+\delta)\log n)^{1/2}}\left\{a^s\operatorname{Prob}\left(\|n^{-1/2}(\mathbf{X}_1+\cdots+\mathbf{X}_n)\|\geqslant\Lambda_n^{1/2}a\right)\right\}$$
$$\leqslant c_{40}(s,k)\left(1+\Lambda_n^{s/2}\right)\left(1+\left(n^{-1/2}\lambda_n^{-3/2}\log n\right)^{k+s+1}\right)n^{-(s-2)/2}\tilde{\Delta}^*_{n,s}$$
$$+\Theta_n(\delta)n^{-(s-2)/2},\tag{17.57}$$

where $\Theta_n(\delta)\to 0$ *as* $n\to\infty$ *for each* $\delta>0$.

Proof. Without loss of generality, assume $\lambda_n>0$ for all n. Let Q_n denote the distribution of $n^{-1/2}(T_n\mathbf{X}_1+\cdots+T_n\mathbf{X}_n)$, where T_n is the symmetric, positive-definite matrix satisfying

$$T_n^2=V_n^{-1}\qquad(n\geqslant 1).\tag{17.58}$$

Write

$$\bar{\tau}_{m+2,n}=n^{-1}\sum_{j=1}^{n}E\|T_n\mathbf{X}_j\|^{m+2}\qquad(0\leqslant m\leqslant s-2),$$

$$\tilde{\chi}_{\nu,n}=n^{-1}\sum_{j=1}^{n}(\nu\text{th cumulant of }T_n\mathbf{X}_j)\qquad(|\nu|\leqslant s).\tag{17.59}$$

Define the function f by

$$f(x)=\begin{cases}0 & \text{if}\quad \|x\|<a\\ a^s & \text{if}\quad \|x\|\geqslant a,\end{cases}\tag{17.60}$$

and use Theorem 15.4 with this f to get

$$\begin{aligned}a^s&\left|\left(Q_n-\sum_{m=0}^{s-2}n^{-m/2}P_m(-\Phi:\{\tilde{\chi}_{\nu,n}\})\right)(\{x:\|x\|\geqslant a\})\right|\\ &\leqslant c_{41}(s,k)M_s(f)\left[1+\left(n^{-1/2}\bar{\tau}_{3,n}\log n\right)^{k+s+1}\right]\Delta^*_{n,s}n^{-(s-2)/2}\\ &\quad+a^s\sum_{m=0}^{s-2}n^{-m/2}|P_m(-\Phi:\{\tilde{\chi}_{\nu,n}\})|\\ &\quad\left(\{x:\|x\|\geqslant a-c_{42}n^{-1/2}\bar{\tau}_{3,n}\log n\}\right)\end{aligned}\tag{17.61}$$

where

$$\begin{aligned}\Delta^*_{n,s}\equiv\inf_{0\leqslant\epsilon\leqslant 1}&\left[\epsilon n^{-1}\sum_{j=1}^{n}\int_{\{\|T_n\mathbf{X}_j\|\leqslant\epsilon n^{1/2}\}}\|T_n\mathbf{X}_j\|^s\right.\\ &\left.+n^{-1}\sum_{j=1}^{n}\int_{\{\|T_n\mathbf{X}_j\|>\epsilon n^{1/2}\}}\|T_n\mathbf{X}_j\|^s\right]\leqslant\tilde{\Delta}^*_{n,s},\end{aligned}\tag{17.62}$$

by an easy computation using

$$\Lambda_n^{-1/2}\|\mathbf{X}_j\|\leqslant\|T_n\mathbf{X}_j\|\leqslant\lambda_n^{-1/2}\|\mathbf{X}_j\|\qquad(1\leqslant j\leqslant n),$$

$$\bar{\tau}_{m+2,n}\leqslant\lambda_n^{-(m+2)/2}\bar{\rho}_{m+2,n}\qquad(0\leqslant m\leqslant s-2).\tag{17.63}$$

The assumption (17.56) implies [in view of (17.63), (15.84)]

$$n^{-m/2}\bar{\tau}_{m+2,n}\to 0 \qquad \text{as } n\to\infty \qquad (1\leqslant m\leqslant s-2), \tag{17.64}$$

and

$$a-c_{42}n^{-1/2}\bar{\tau}_{3,n}\log n\geqslant\frac{a}{2},$$

$$a-c_{42}n^{-1/2}\bar{\tau}_{3,n}\log n\geqslant\left(\left(s-2+\frac{\delta}{2}\right)\log n\right)^{1/2} \tag{17.65}$$

for all sufficiently large n if $a\geqslant((s-2+\delta)\log n)^{1/2}$. Hence

$$a^s n^{-m/2}|P_m(-\Phi:\{\tilde{\chi}_{\nu,n}\})|\left(\left\{x:\|x\|\geqslant\frac{a}{2}\right\}\right)$$

$$\leqslant c_{43}n^{-m/2}\bar{\tau}_{m+2,n}\int_{\{\|x\|^2\geqslant(s-2+\delta/2)\log n\}}\|x\|^{3m+s}\exp\left\{-\frac{\|x\|^2}{2}\right\}dx$$

$$=\Theta_n(\delta)n^{-(s-2)/2} \qquad (0\leqslant m\leqslant s-2). \tag{17.66}$$

Also note that

$$M_s(f)=\frac{a^s}{1+a^s}<1. \tag{17.67}$$

The estimates (17.62), (17.66) are now used in (17.61) to yield

$$a^sQ_n(\{x:\|x\|\geqslant a\})$$

$$\leqslant c_{41}(s,k)\left[1+\left(n^{-1/2}\lambda_n^{-3/2}\bar{\rho}_{3,n}\log n\right)^{k+s+1}\right]\tilde{\Delta}^*_{n,s}n^{-(s-2)/2}$$

$$+\Theta_n(\delta)n^{-(s-2)/2}. \tag{17.68}$$

Finally observe that

$$Q_n(\{x:\|x\|\geqslant a\})=\text{Prob}\left(\|n^{-1/2}(T_n\mathbf{X}_1+\cdots+T_n\mathbf{X}_n)\|\geqslant a\right)$$

$$\geqslant\text{Prob}\left(\|n^{-1/2}(\mathbf{X}_1+\cdots+\mathbf{X}_n)\|\geqslant a\Lambda_n^{1/2}\right), \tag{17.69}$$

since $\|T_nx\|\geqslant\Lambda_n^{-1/2}\|x\|$. Q.E.D.

COROLLARY 17.12. *Let* $\{\mathrm{X}_n:n\geqslant 1\}$ *be a sequence of independent and identically distributed random vectors having a common mean zero and common covariance matrix* V. *If* $\rho_s\equiv E\|\mathbf{X}_1\|^s$ *is finite for some integer* $s\geqslant 3$, *then*

$$P\left(\|\mathbf{X}_1+\cdots+\mathbf{X}_n\|>a_n\Lambda^{1/2}n^{1/2}\right)=\delta_n n^{-(s-2)/2}a_n^{-s}, \tag{17.70}$$

where $\delta_n \to 0$ *as* $n \to \infty$ *uniformly for every sequence* $\{a_n : n \geq 1\}$ *of numbers satisfying*

$$a_n \geq (s-2+\delta)^{1/2} \log^{1/2} n \tag{17.71}$$

for any fixed $\delta > 0$, *and* Λ *is the largest eigenvalue of* V.

Proof. Note that in this case

$$\tilde{\Delta}^*_{n,s} \leq n^{-1/4} \lambda^{-s/2} \rho_s + \lambda^{-s/2} \int_{\{\|\mathbf{X}_1\| > \lambda^{1/2} n^{1/4}\}} \|\mathbf{X}_1\|^s \to 0 \tag{17.72}$$

as $n \to \infty$. Here λ, Λ are the smallest and largest eigenvalues of V, respectively. Q.E.D.

COROLLARY 17.13. *Let* $\{\mathbf{X}_n : n \geq 1\}$ *be a sequence of independent random vectors having zero means and finite* sth *absolute moments for some integer* $s \geq 3$. *Assume that*

$$\liminf_{n \to \infty} \lambda_n > 0, \qquad \sup_{n \to \infty} \bar{\rho}_{s,n} < \infty, \tag{17.73}$$

where the notation is the same as in Theorem 17.11. *Then*

$$P\left(\|\mathbf{X}_1 + \cdots + \mathbf{X}_n\| > a_n \Lambda_n^{1/2} n^{1/2}\right) = \delta'_n n^{-(s-2)/2} a_n^{-s}, \tag{17.74}$$

where $\{\delta'_n : n \geq 1\}$ *remains uniformly bounded for every sequence of numbers* $\{a_n : n \geq 1\}$ *satisfying* (17.71) *for any fixed* $\delta > 0$.

Proof. In view of (17.73) the sequence $\{\Lambda_n : n \geq 1\}$ is bounded since, writing $V_n = ((v_{ij}))$, one has

$$\Lambda_n = \sup_{\|x\|=1} \langle x, V_n x \rangle = \sup_{\|x\|=1} \sum_{i,j=1}^{k} v_{ij} x_i x_j$$

$$\leq \sum_{i,j=1}^{k} (v_{ii} v_{jj})^{1/2} |x_i x_j| = \left(\sum_{i=1}^{k} |x_i| v_{ii}^{1/2} \right)^2 \leq \sum_{i=1}^{k} v_{ii} = \bar{\rho}_{2,n},$$

$$\bar{\rho}_{m,n} \leq (\bar{\rho}_{s,n})^{m/s} \qquad (1 \leq m \leq s). \tag{17.75}$$

Also [putting $\epsilon = 0$ in the expression within square brackets in (17.55)],

$$\tilde{\Delta}^*_{n,s} \leq \lambda_n^{-s/2} n^{-1} \sum_{j=1}^{n} E(\|\mathbf{X}_j\|^s) = \lambda_n^{-s/2} \bar{\rho}_{s,n}, \tag{17.76}$$

so that $\{\tilde{\Delta}^*_{n,s}:n\geqslant 1\}$ is a bounded sequence. The relation (17.74) now follows from (17.57). Q.E.D.

Note that the sequence $\{\delta'_n:n\geqslant 1\}$ in Corollary 17.13 may be shown to go to zero in the same manner as $\{\delta_n:n\geqslant 1\}$ in Corollary 17.12 if

$$\lim_{n\to\infty} n^{-1}\sum_{j=1}^{n}\int_{\{\|\mathbf{X}_j\|>\Lambda_n^{1/2}n^{1/4}\}}\|\mathbf{X}_j\|^s=0. \tag{17.77}$$

18. RATES OF CONVERGENCE UNDER FINITENESS OF SECOND MOMENTS

Most of the main results in Sections 15, 16, and 17 have appropriate analogs when only the second moments are assumed finite. Here we prove an analog of Theorem 15.1 and derive some corollaries.

As before, $\mathbf{X}_1,\ldots,\mathbf{X}_n$ are independent random vectors with values in R^k, and

$$E\mathbf{X}_j=0 \qquad (1\leqslant j\leqslant n), \qquad n^{-1}\sum_{j=1}^{n}\operatorname{Cov}(\mathbf{X}_j)=V, \tag{18.1}$$

where V is a symmetric, positive-definite matrix, and we write

$$\rho_r=n^{-1}\sum_{j=1}^{n}E\|\mathbf{X}_j\|^r \qquad (r>0), \tag{18.2}$$

and denote by T the symmetric, positive-definite matrix satisfying

$$T^2=V^{-1}. \tag{18.3}$$

We also define

$$\bar{\Delta}_{n,s}(\epsilon)=n^{-1}\sum_{j=1}^{n}\int_{\{\|\mathbf{X}_j\|>\epsilon n^{1/2}\}}\|\mathbf{X}_j\|^s \qquad (\epsilon\geqslant 0,\quad s>0),$$

$$\bar{\Delta}_{n,s}=\bar{\Delta}_{n,s}(1),$$

$$\delta^*_{n,s}=\inf_{0\leqslant\epsilon\leqslant 1}\left[\epsilon^{(3-s)}n^{-1}\sum_{j=1}^{n}\int_{\{\|\mathbf{X}_j\|\leqslant\epsilon n^{1/2}\}}\|\mathbf{X}_j\|^s+\bar{\Delta}_{n,s}(\epsilon)\right]. \tag{18.4}$$

Recalling the definition (15.7) of $\Delta^*_{n,s}$, we see that

$$\delta^*_{n,2}=\Delta^*_{n,2}, \qquad \bar{\Delta}_{n,2}(1)\leqslant\delta^*_{n,2}\leqslant\rho_2. \tag{18.5}$$

Also, as before, for a function f on R^k we define

$$M_r(f)=\sup_{x\in R^k}(1+\|x\|^r)^{-1}|f(x)| \qquad (r>0),$$

$$M_0(f)=\sup_{x,y\in R^k}|f(x)-f(y)|=\omega_f(R^k). \tag{18.6}$$

Finally, let Q_n denote the distribution of $n^{-1/2}(\mathbf{X}_1+\cdots+\mathbf{X}_n)$ and let $\Phi_{a,V}$ denote the normal distribution on R^k with mean a and covariance matrix V. We write $\Phi=\Phi_{0,I}$, where I is the identity matrix.

THEOREM 18.1. *Let* $V=I$ *and* $\rho_s<\infty$ *for some* s, $2\leqslant s\leqslant 3$. *There exist positive constants* c_1, c_2, c_3 *depending only on* k *and* s *such that for every Borel-measurable function* f *on* R^k *satisfying*

$$M_r(f)<\infty \tag{18.7}$$

for an integer r, $0\leqslant r\leqslant s$, *one has*

$$\left|\int f\,d(Q_n-\Phi)\right|\leqslant c_1M_r(f)n^{-(s-2)/2}\delta^*_{n,s}+c_2\omega^*_g\left(c_3n^{-(s-2)/2}\delta^*_{n,s}:\Phi_{r_0}\right), \tag{18.8}$$

where $r_0=r$ *if* r *is even*, $r_0=r+1$ *if* r *is odd*,

$$\begin{aligned} g(x)&=(1+\|x\|^{r_0})^{-1}f(x) &&\text{if } r>0,\\ g(x)&=f(x) &&\text{if } r=0, \end{aligned} \tag{18.9}$$

and Φ_{r_0} *is the measure*

$$\Phi_{r_0}(dx)=(1+\|x\|^{r_0})\Phi(dx). \tag{18.10}$$

Proof. Let the truncated random vectors $\mathbf{Y}_j$, $\mathbf{Z}_j$ $(1\leqslant j\leqslant n)$ be defined as in (14.2). Let Q_n'', Q_n' be the distributions of $n^{-1/2}(\mathbf{Y}_1+\cdots+\mathbf{Y}_n)$, $n^{-1/2}(\mathbf{Z}_1+\cdots+\mathbf{Z}_n)$, respectively. Let

$$a_n=n^{-1/2}\sum_{j=1}^{n}E\mathbf{Y}_j, \qquad D=n^{-1}\sum_{j=1}^{n}\operatorname{Cov}(\mathbf{Y}_j), \tag{18.11}$$

and write $\Phi''=\Phi_{-a_n,I}$, $\Phi'=\Phi_{0,D}$. As in the proof of Theorem 15.1,

$$\left|\int f\,d(Q_n-\Phi)\right| \leqslant \left|\int f\,d(Q_n-Q_n'')\right| + \left|\int f_{a_n}\,d(\Phi''-\Phi)\right|$$
$$+\left|\int f_{a_n}\,d(\Phi-\Phi')\right| + \left|\int f_{a_n}\,d(Q_n'-\Phi')\right|$$
$$\leqslant c_4 M_r(f)\bar{\Delta}_{n,s}n^{-(s-2)/2} + \left|\int f_{a_n}\,d(Q_n'-\Phi')\right|. \qquad (18.12)$$

Note that

$$\int \|x\|^r Q_n(dx) \leqslant \rho_2^{r/2} = k^{r/2}, \qquad \int \|x\|^r \Phi(dx) \leqslant k^{r/2} \qquad (0 \leqslant r \leqslant 2), \qquad (18.13)$$

so that (18.8) may be shown to be trivially ture if (15.11) is violated, and one may therefore assume, without loss of generality, that (15.11) holds. Next apply Theorem 16.1 to random vectors $\mathbf{Z}_1,\dots,\mathbf{Z}_n$ [letting $s=3$ in (16.12)] to get

$$\left|\int f_{a_n}\,d(Q_n'-\Phi')\right| \leqslant c_5 M_r(f_{a_n})(1+\Lambda^{r/2})\lambda^{-3/2}\rho_3' n^{-1/2}$$
$$+2\omega_{g'}^*\left(c_6\lambda^{-3/2}\rho_3' n^{-1/2} : \Phi_{r_0}\right), \qquad (18.14)$$

where λ, Λ are the smallest and largest eigenvalues, respectively, of D, and

$$\rho_3' = n^{-1}\sum_{j=1}^{n} E\|\mathbf{Z}_j\|^3,$$

$$g'(x) = \begin{cases} (1+\|x\|^{r_0})^{-1}\left(f_{a_n}\circ B^{-1}\right)(x) & \text{if } r>0,\\ \left(f_{a_n}\circ B^{-1}\right)(x) & \text{if } r=0. \end{cases} \qquad (18.15)$$

Here B is the symmetric, positive-definite matrix satisfying $B^2=D^{-1}$. By (14.10) one has

$$\rho_3' n^{-1/2} \leqslant 2^3 n^{-3/2}\sum_{j=1}^{n} E\|\mathbf{Y}_j\|^3 \leqslant 2^3\delta_{n,s}^* n^{-(s-2)/2}. \qquad (18.16)$$

Now by writing

$$g''(x) = \left(1+\|B^{-1}x+a_n\|^{r_0}\right)^{-1}\left(f_{a_n}\circ B^{-1}\right)(x) \qquad \text{for } r>0,$$

it is simple to check that [see (14.12), (14.19), (14.81), and (18.5)]

$$\begin{aligned}|g''(x)-g'(x)| &\leq M_{r_0}(f)|\,\|B^{-1}x+a_n\|^{r_0}-\|x\|^{r_0}|(1+\|x\|^{r_0})^{-1}\\ &\leq r_0M_{r_0}(f)(\|B^{-1}-I\|\|x\|+\|a_n\|)\\ &\quad\times(\|x\|+\|B^{-1}-I\|\|x\|+\|a_n\|)^{r_0-1}(1+\|x\|^{r_0})^{-1}\\ &\leq c_7M_r(f)\delta^*_{n,s}n^{-(s-2)/2}.\end{aligned}$$

As a consequence

$$\omega^*_{g'}(\epsilon:\Phi_{r_0})\leq\omega^*_{g''}(\epsilon:\Phi_{r_0})+c_8M_r(f)\delta^*_{n,s}n^{-(s-2)/2}. \tag{18.17}$$

Changing variables $x\to B^{-1}x+a_n$, and using Lemma 14.6,

$$\begin{aligned}\omega^*_{g''}(\epsilon:\Phi_{r_0}) &= \sup_{y\in R^k}\int\omega_{g''}(x+y:\epsilon)\Phi_{r_0}(dx)\\ &\leq \sup_{y\in R^k}\int\omega_g(B^{-1}(x+y)+a_n:\|B^{-1}\|\epsilon)\Phi_{r_0}(dx)\\ &\leq \sup_{y\in R^k}\int\omega_g(z+y:c_9\epsilon)[1+\|B(z-a_n)\|^{r_0}]\Phi_{a_n,D}(dz)\\ &\leq c_{10}\omega^*_g(c_{11}\epsilon:\Phi_{r_0})c_{12}M_r(f)\delta^*_{n,s}n^{-(s-2)/2}.\end{aligned} \tag{18.18}$$

The proof is now complete by inequalities (18.12), (18.14), (18.16), (18.17), and (18.18). Q.E.D.

COROLLARY 18.2. *For each* $n\geq 1$ *let* $\mathbf{X}_1^{(n)},\dots,\mathbf{X}_{k_n}^{(n)}$ *be independent random vectors with values in* R^k, *having zero means and an average positive-definite covariance matrix* $V_n=k_n^{-1}\sum_{j=1}^{k_n}\operatorname{Cov}(\mathbf{X}_j^{(n)})$. *Let* G_n *denote the distribution of* $k_n^{-1/2}T_n(\mathbf{X}_1^{(n)}+\cdots+\mathbf{X}_{k_n}^{(n)})$, *where* T_n *is the symmetric, positive-definite matrix satisfying* $T_n^2=V_n^{-1}$, $n\geq 1$. *Suppose* $k_n\to\infty$ *as* $n\to\infty$. *If*

$$\Theta_n(\epsilon)\equiv k_n^{-1}\sum_{j=1}^{k_n}\int_{\{\|T_n\mathbf{X}_j^{(n)}\|>\epsilon k_n^{1/2}\}}\|T_n\mathbf{X}_j^{(n)}\|^2\to 0\qquad\text{as } n\to\infty, \tag{18.19}$$

for every $\epsilon>0$, *then* $\{G_n:n\geq 1\}$ *converges weakly to the standard normal distribution* Φ.

Proof. Apply Theorem 18.1 (with $r=r_0=0$) to the random vectors $T_n\mathbf{X}_1^{(n)},\dots,T_n\mathbf{X}_{k_n}^{(n)}$. For every Lipschitzian function f bounded by one (or for the indicator function f of an arbitrary Borel-measurable convex set) one has

$$\left|\int f(G_n-\Phi)\right| \leqslant (c_1+c_{13}d)\delta_n^* \tag{18.20}$$

where

$$\delta_n^* = \inf_{0\leqslant\epsilon\leqslant 1}\left\{\epsilon k_n^{-1}\sum_{j=1}^{k_n}\int_{\{\|T_n\mathbf{X}_j^{(n)}\|>\epsilon k_n^{1/2}\}}\|T_n\mathbf{X}_j^{(n)}\|^2+\Theta_n(\epsilon)\right\}, \tag{18.21}$$

and d depends on f. Note that

$$\delta_n^* \leqslant \epsilon k_n^{-1}\sum_{j=1}^{k_n}\int \|T_n\mathbf{X}_j^{(n)}\|^2+\Theta_n(\epsilon)=k\epsilon+\Theta_n(\epsilon) \qquad (0\leqslant\epsilon\leqslant 1).$$

Given $\eta>0$, choose $\epsilon=\eta/2k$ and let $n_0(\eta)$ be an integer such that for $n\geqslant n_0(\eta)$ one has $\Theta_n(\eta/2k)\leqslant\eta/2$. Then $\delta_n^*\leqslant\eta$ for all $n\geqslant n_0(\eta)$. Q.E.D.

The above corollary is an extension of *Lindeberg's central limit theorem to the multidimensional case.*

COROLLARY 18.3. *If in Corollary* 18.2 *one replaces* (18.19) *by*

$$E\|\mathbf{X}_j^{(n)}\|^s<\infty \qquad (1\leqslant j\leqslant k_n,\quad n\geqslant 1) \tag{18.22}$$

for some s, $2\leqslant s\leqslant 3$, *then*

$$\begin{aligned}\sup_{C\in\mathcal{C}}|G_n(C)-\Phi(C)| &\leqslant c_{14}k_n^{-(s-2)/2}\inf_{0\leqslant\epsilon\leqslant 1}\left[\epsilon^{3-s}k_n^{-1}\sum_{j=1}^{k_n}\int_{\{\|T_n\mathbf{X}_j^{(n)}\|>\epsilon k_n^{1/2}\}}\|T_n\mathbf{X}_j^{(n)}\|^s\right.\\ &\qquad\left.+k_n^{-1}\sum_{j=1}^{k_n}\int_{\{\|T_n\mathbf{X}_j^{(n)}\|>\epsilon k_n^{1/2}\}}\|T_n\mathbf{X}_j^{(n)}\|^s\right]\\ &\leqslant c_{14}k_n^{-(s-2)/2}\left[k_n^{-1}\sum_{j=1}^{k_n}E\|T_n\mathbf{X}_j^{(n)}\|^s\right],\end{aligned} \tag{18.23}$$

where $\mathcal{C}$ *is the class of all Borel-measurable convex subsets of* R^k, *and* c_{14} *depends only on* k.

Proof. The first inequality in (18.23) follows from Theorem 18.1 (with $r=0=r_0$) applied to random vectors $T_n\mathbf{X}_j^{(n)}$, $1\leqslant j\leqslant k_n$, and from Corollary 3.2 (with $s=0$). The second inequality is obtained from the first by letting $\epsilon=0$ in the expression within square brackets. Q.E.D.

The above corollary contains a multidimensional extension of *Liapounov's central limit theorem*: $\{G_n : n\geqslant 1\}$ converges weakly to Φ if

$$\lim_{n\to\infty} k_n^{-(s-2)/2}\left(k_n^{-1}\sum_{j=1}^{k_n} E\,\|T_n\mathbf{X}_j\|^s\right)=0, \qquad (18.24)$$

for some s, $2<s\leqslant 3$. The first inequality in (18.23), however, is sharper. For example, if $\{\mathbf{X}_n : n\geqslant 1\}$ is a sequence of independent and identically distributed random vectors with common mean zero, common positive-definite covariance matrix V, and finite sth absolute moments for some s, $2\leqslant s<3$, then one has [letting $k_n=n$, $\epsilon=n^{-1/4}$ in (18.23)]

$$\sup_{C\in\mathcal{C}} \left|\operatorname{Prob}\left(n^{-1/2}(\mathbf{X}_1+\cdots+\mathbf{X}_n)\in C\right)-\Phi_{0,V}(C)\right|$$
$$=o\left(n^{-(s-2)/2}\right) \qquad (n\to\infty). \qquad (18.25)$$

One may in the same manner obtain analogs of Theorem 17.6 and the mean central limit theorem Corollary 17.7. For example, if $k=1$, then there exists a constant c_{15} depending only on p such that under the hypothesis of Theorem 18.1 (with $k=1$)

$$\|F_n-\Phi\|_p\leqslant c_{15}\delta_{n,s}^{*}n^{-(s-2)/2} \qquad \left(p>\tfrac{1}{2}\right), \qquad (18.26)$$

where $F_n(\cdot)$ is the distribution function of $n^{-1/2}(\mathbf{X}_1+\cdots+\mathbf{X}_n)$, and $\Phi(\cdot)$ is the standard normal distribution function on R^1. If $\{\mathbf{X}_n : n\geqslant 1\}$ is an independent and identically distributed sequence of random variables, then the right side is $o(n^{-(s-2)/2})$ as $n\to\infty$.

NOTES

The first central limit theorem was proved for i.i.d. Bernoulli random variables by DeMoivre [1]; Laplace [1] elucidated and refined it, and also gave a statement (as well as some reasoning for the validity) of a rather general central limit theorem. Chebyshev [1] proved (with a complement due to Markov [1]) the first general central limit theorem by his famous method of moments; however, Chebyshev's moment conditions were very severe. Then came Liapounov's pioneering investigations [1, 2] in which he introduced the characteristic function in probability theory and used it to prove convergence to the normal distribution under the extremely mild hypothesis (18.24) (for $k=1$). Finally Lindeberg [1] proved Corollary 18.2 (for $k=1$). In the i.i.d. case this reduces to the so-called *classical central limit theorem*: if $\{\mathbf{X}_n : n \geq 1\}$ is a sequence of i.i.d. random variables each with mean zero and variance one, then the distribution of $n^{-1/2}(\mathbf{X}_1 + \cdots + \mathbf{X}_n)$ converges weakly to the standard normal distribution Φ. This classical central limit theorem was also proved by Lévy [1] (p. 233). Feller [1] proved that the Lindeberg condition (18.19) is also necessary in order that (i) the distribution of $k_n^{-1/2} s_n^{-1}(\mathbf{X}_1^{(n)} + \cdots + \mathbf{X}_{k_n}^{(n)})$ converge weakly to Φ and (ii) $m_n^2 / k_n s_n^2 \to 0$ as $n \to \infty$; here $k=1$, and we write s_n^2 for V_n, s_n^{-1} for T_n, and $m_n^2 = \max\{\operatorname{var}(\mathbf{X}_j^{(n)}) : 1 \leq j \leq k_n\}$. Many authors have obtained multidimensional extensions of the central limit theorem, for example, Bernstein [1], Khinchin [1, 2]; the Lindeberg–Feller theorem was extended to R^k by Takano [1].

Section 11. Lemma 11.1–Corollary 11.5 are due to Bhattacharya [1–5]. These easily extend to metric groups, and Bhattacharya [6] used them to derive rates of convergence of the n-fold convolution of a probability measure on a compact group to the normalized Haar measure as $n \to \infty$. Lemma 11.6 is perhaps well known to analysts.

Section 12. The first result on the speed of convergence is due to Liapounov [2], who proved

$$\sup_{x \in R^1} |F_n(x) - \Phi(x)| \leq c\rho_3 n^{-1/2} \log n \qquad (n \geq 2), \tag{N.1}$$

where F_n is the distribution function of the normalized sum of n independent random variables with zero means, average variance one (normalization), and average third absolute moment ρ_3. Under an additional hypothesis [Cramér's condition (20.1)] Cramér [1, 3] (Chapter VII) was able to remove the factor $\log n$ from (N.1). The best result was obtained independently by Berry [1] and Esseen [1], who eliminated the logarithmic factor in (N.1) under Liapounov's hypothesis. The constants appearing in Theorem 12.4 (the Berry–Esseen theorem) are not the best known. After initial work by Zolotarev [1, 2], the constant 0.7975 [appearing in (12.49)] was obtained by van Beek [1]. Lemmas 12.1 and 12.2 are refinements due to Zolotarev [1] of some inequalities of Berry [1] and Esseen [1]. The multidimensional extension of the Berry–Esseen theorem is due to Bergström [1], who used an ingenious induction argument. Bergström's approach does not require the Fourier analytic machinery and has been extended in recent years by Bergström [2], Sazonov [1, 2], and Paulauskas [1].

Sections 13–18. Apart from a special (and deep) result of Esseen [1] [see Notes, Chapter 5] the first result going beyond distribution functions was obtained by Rao [1, 2], who proved that

$$\sup_{C \in \mathcal{C}} |Q_n(C) - \Phi(C)| \leq c(k)\rho_4^{3/2} n^{-1/2} (\log n)^{(k-1)/(2(k+1))} \qquad (n \geq 2), \tag{N.2}$$

where $\mathcal{C}$ is the class of all Borel-measurable convex subsets of R^k, and Q_n is the distribution of the normalized sum of n i.i.d. random vectors. von Bahr [3] and Bhattacharya [1, 2] independently extended it to much more general classes of sets [e.g., the class $\mathcal{Q}_\alpha^*(d:\Phi)$ in (17.3)] and at the same time made it precise by eliminating the logarithmic factor on the right side of (N.2). The moment condition in Bhattacharya [1, 2] was $\rho_{3+\delta} < \infty$ for some $\delta > 0$, whereas von Bahr [3] essentially assumed that the random vectors are i.i.d. and that $\rho_3 < \infty$, $\rho_{k+1} < \infty$. For the class $\mathcal{C}$, Sazonov [1] finally relaxed the moment condition to $\rho_3 < \infty$, proving Corollary 17.2 in the i.i.d. case (Bergström [3] later proved this independently of

Sazonov), while Rotar' [1] relaxed it for the general non-i.i.d. case. For more general classes of sets this relaxation of the moment condition is due to Bhattacharya [7]. Paulauskas [1] also has a result that goes somewhat beyond the class $\mathcal{C}$.

The results of Section 13 are due to Bhattacharya [3], although the explicit computation of constants given here is new.

The first effective use of truncation in the present context is due to Bikjalis [4]; Lemma 14.1 and Corollary 14.2 are essentially due to him. Lemma 14.3 is due to Rotar' [1]. Lemmas 14.6 and 14.8 were obtained by Bhattacharya [7]; a result analogous to the inequality (14.107) was obtained earlier by Bikjalis [4]. Analogs of Lemma 14.7 were obtained earlier by Doob [1], pp. 225–228, for a stationary Markov chain, by Brillinger [1] for the i.i.d. case, and by von Bahr [1] for the case considered by us; but we are unable to deduce the present explicit form needed by us from their results.

Theorems 15.1, 15.4, and Corollary 15.2 are due to Bhattacharya [7], as is the present form of Corollary 15.3; earlier, a version of Corollary 15.3 was independently proved by von Bahr [3] and Bhattacharya [1, 2].

Theorems 17.1, 17.4, 17.8–17.10, and Corollary 17.3 are due to Bhattacharya [4, 5, 7]. Corollaries 17.5 and 17.12 were proved by von Bahr [2, 3] in the i.i.d. case; the corresponding results (Theorems 17.4, 17.11, and Corollary 17.13) in the non-i.i.d. case are new. The first global, or mean central limit theorems are due to Esseen [1, 3], and Agnew [1]. The fairly precise result Corollary 17.7 was proved for $s=3$ by Nagaev [1] in the i.i.d. case (a slightly weaker result was proved earlier by Esseen [1]) and later by Bikjalis [2] in the non-i.i.d. case; afterwards, the much more powerful Theorem 17.6 was proved by Rotar' [1] for $s=3$. Rotar' [1] also stated a result which implies Theorem 17.6 for all $s \geqslant 3$; however we are unable to verify it.

Theorem 18.3 is new, as is perhaps Corollary 18.3; however Osipov and Petrov [1] and Feller [2] contain fairly general inequalities for the difference between the distribution functions F_n and Φ in the non-i.i.d. case in one dimension. More precise results than (18.25), (18.26) are known in one dimension. Ibragimov [1] has proved the following result. *Suppose that* $\{\mathbf{X}_n : n \geqslant 1\}$ *is a sequence of i.i.d. random variables each with mean zero and variance one*; *let* $0<\delta<1$, $1<p\leqslant\infty$; *then* $\|F_n-\Phi\|_p = O(n^{-\delta/2})$ *if and only if*

$$\int_{\{|\mathbf{X}_1|>x\}} \mathbf{X}_1^2 = O(x^{-\delta}) \qquad (x\to\infty); \tag{N.3}$$

also,

$$\|F_n-\Phi\|_p = O(n^{-1/2})$$

if and only if (N.3) *holds with* $\delta=1$ *and*

$$\int_{\{|\mathbf{X}_1|\leqslant x\}} \mathbf{X}_1^3 = O(1) \qquad (x\to\infty); \tag{N.4}$$

here $\|F_n-\Phi\|_\infty$ *denotes the Kolmogorov distance* [see (2.72)]. Under the same hypothesis Heyde [1] has shown that *if* $0<\delta<1$, *then*

$$\sum_{n=1}^{\infty} n^{-1+\delta/2}\|F_n-\Phi\|_\infty < \infty \tag{N.5}$$

if and only if $E|\mathbf{X}_1|^{2+\delta}<\infty$; *also* (N.5) *holds with* $\delta=0$ *if and only if*

$$E\left(\mathbf{X}_1^2 \log(1+|\mathbf{X}_1|)\right) < \infty.$$

An extension of Ibragimov's results to R^k has been recently obtained by Bikjalis [6].

CHAPTER 4

Asymptotic Expansions-Nonlattice Distributions

We have seen in Chapter 2 that the characteristic function of Q_n (distribution of nth normalized partial sum of a sequence $\{\mathbf{X}_n: n \geqslant 1\}$ of independent and identically distributed random vectors) admits an *asymptotic expansion* in powers of $n^{-1/2}$ in the sense that the remainder is of a smaller order of magnitude than the last term in the expansion. If the sth absolute moment of $\mathbf{X}_1$ is finite for some integer $s \geqslant 3$, then there exist functions $P_r(-\phi: \{\chi_\nu\})$, $0 \leqslant r \leqslant s-2$, which are polynomial multiples of the standard normal density ϕ, with $P_0 = \phi$, such that

$$\int \exp\{i\langle t,x\rangle\} Q_n(dx) = \sum_{r=0}^{s-2} n^{-r/2} \int \exp\{i\langle t,x\rangle\} P_r(-\phi: \{\chi_\nu\})(x)\,dx$$
$$+ o(n^{-(s-2)/2}) \qquad (t \in R^k,\ n \to \infty). \tag{1}$$

In fact, we know from Theorem 9.12 that one may replace the function $x \to \exp\{i\langle t,x\rangle\}$ by any polynomial multiple of it of degree s or less. Unfortunately such an expansion of $\int f\,dQ_n$ does not hold for a large enough class of functions (e.g., indicator functions of even those sets that have "smooth" boundaries) unless some further assumption is made on the nature of the distribution of $\mathbf{X}_1$. In Section 19 we assume that the distribution of $\mathbf{X}_1$ has a bounded density (or at least an r-fold convolution of it has one, for some r), and show that in this case there is an expansion of the density of Q_n as well as of $\int f\,dQ_n$ for every bounded measurable f. In Section 20, the assumption of Section 1 is relaxed to read "the distribution of $\mathbf{X}_1$ satisfies Cramér's condition (20.1)," and an expansion of $\int f\,dQ_n$ is obtained for a very large class of functions f, although not for every bounded measurable f. In particular, one has an expansion of $Q_n(A)$ for every Borel set A satisfying

$$\Phi((\partial A)^{\varepsilon}) = o((-\log\varepsilon)^{-(s-2)/2}) \qquad \varepsilon\downarrow 0, \tag{2}$$

the remainder term in the expansion being $o(n^{-(s-2)/2})$ uniformly over every class $\mathcal{Q}$ of sets A satisfying (2) uniformly.

19. LOCAL LIMIT THEOREMS AND ASYMPTOTIC EXPANSIONS FOR DENSITIES

Suppose that $\{\mathbf{X}_n: n \geqslant 1\}$ is a sequence of independent random vectors such that the distributions $\{Q_n: n \geqslant 1\}$ of their normalized partial sums have densities $\{q_n: n \geqslant 1\}$, at least for large n, satisfying

$$\lim_{n\to\infty} q_n(x) = \psi(x) \qquad (x \in R^k), \tag{19.1}$$

where ψ is the density of the limiting distribution Ψ (e.g., the standard normal distribution on R^k). Then this assertion (19.1) is called a *local limit theorem* (for densities of $\{Q_n: n \geqslant 1\}$). We shall always take the *continuous version* of the density q_n, if there exists one. Recall that by Scheffé's theorem (Lemma 2.1), (19.1) implies

$$\lim_{n\to\infty} \|Q_n - \Psi\| = 0. \tag{19.2}$$

We shall usually consider *uniform local limit theorems*, that is, assertions of the form

$$\lim_{n\to\infty} \sup_{x\in R^k} |q_n(x) - \psi(x)| = 0. \tag{19.3}$$

Since assertions involving densities are made in this section under hypotheses guaranteeing existence of continuous versions of q_n for all large n, there is no scope for ambiguity in interpreting (19.3) or similar statements.

THEOREM 19.1 *Let* $\{\mathbf{X}_n: n \geqslant 1\}$ *be a sequence of i.i.d. random vectors with values in* R^k *and*

$$E\mathbf{X}_1 = 0, \qquad \operatorname{Cov}(\mathbf{X}_1) = V, \tag{19.4}$$

where V *is a symmetric positive-definite matrix. Let* Q_n *denote the distribution of* $n^{-1/2}(\mathbf{X}_1 + \cdots + \mathbf{X}_n)$ $(n = 1, 2, \cdots)$. *The following statements are all equivalent.*

(i) $\hat{Q}_1 \in L^p(R^k)$ *for some* $p \geqslant 1$.

(ii) *For every sufficiently large* n, Q_n *has a density* q_n *and*

$$\lim_{n\to\infty} \sup_{x\in R^k} |q_n(x)-\phi_{0,V}(x)|=0. \tag{19.5}$$

(iii) *There exists an integer* m *such that* Q_1^{*m} *(or, equivalently,* Q_m*) has a bounded (almost everywhere) density.*

Proof. We first show that (i)⇒(ii). If $|\hat{Q}_1|^p$ is integrable, then so is $|\hat{Q}_n|$ for every $n \geqslant p$. This implies [by the Fourier inversion theorem 4.1(iv)] that, for $n \geqslant p$, Q_n has a bounded continuous density q_n and

$$\sup_{x\in R^k} |q_n(x)-\phi_{0,V}(x)| \leqslant (2\pi)^{-k}\int |\hat{q}_n(t)-\hat{\phi}_{0,V}(t)|\,dt \qquad (n \geqslant p). \tag{19.6}$$

By the classical central limit theorem (see, e.g., Corollary 18.2 specialized to the i.i.d. case)

$$\lim_{n\to\infty} \int_{\{\|t\| \leqslant a\}} |\hat{q}_n(t)-\hat{\phi}_{0,V}(t)|\,dt=0 \tag{19.7}$$

for each positive a. Also, since

$$\hat{Q}_1(t)=1-\tfrac{1}{2}\langle t,Vt\rangle+o(\|t\|^2) \qquad (t\to 0), \tag{19.8}$$

there exists a positive number b such that

$$|\hat{Q}_1(t)| \leqslant 1-\tfrac{1}{4}\langle t,Vt\rangle \leqslant \exp\{-\tfrac{1}{4}\langle t,Vt\rangle\} \qquad (\|t\| \leqslant b). \tag{19.9}$$

This implies

$$|\hat{Q}_n(t)| = \left|\hat{Q}_1\left(\frac{t}{n^{1/2}}\right)\right|^n \leqslant \exp\{-\tfrac{1}{4}\langle t,Vt\rangle\} \qquad (\|t\| \leqslant bn^{1/2}). \tag{19.10}$$

Let

$$\delta \equiv \sup\{|\hat{Q}_1(t)| : \|t\| > b\}. \tag{19.11}$$

Then $\delta < 1$, since $\delta = 1$ implies $\sup\{|\hat{Q}_1(t)|^m : \|t\| > b\} = 1$ for all $m \geqslant 1$; but the Riemann–Lebesgue lemma [Theorem 4.1(iii)] applies for $m \geqslant p$, so that there must exist $t_0 \in R^k$ such that $|\hat{Q}_1(t_0)| = 1$, which means that $\mathbf{X}_1$ assigns all its mass to a countable set of parallel hyperplanes (see Section 21); this would imply singularity of Q_m with respect to Lebesgue measure for all $m \geqslant 1$, contradicting the fact that Q_m is absolutely continuous for all $m \geqslant p$.

Next, for $\|t\|>bn^{1/2}$

$$|\hat{q}_n(t)| \leqslant \left|\hat{Q}_1\left(\frac{t}{n^{1/2}}\right)\right|^p \cdot \left(\sup\left\{\left|\hat{Q}_1\left(\frac{t}{n^{1/2}}\right)\right| : \|t\|>bn^{1/2}\right\}\right)^{n-p}$$

$$= \delta^{n-p}\left|\hat{Q}_1\left(\frac{t}{n^{1/2}}\right)\right|^p \qquad (n \geqslant p). \tag{19.12}$$

Now

$$\begin{aligned} \overline{\lim_{n\to\infty}} \int |\hat{q}_n(t) - \hat{\phi}_{0,V}(t)|\,dt &\leqslant \overline{\lim_{n\to\infty}} \int_{\{\|t\|\leqslant a\}} |\hat{q}_n(t) - \hat{\phi}_{0,V}(t)|\,dt \\ &\quad + 2\int_{\{\|t\|>a\}} \exp\{-\tfrac{1}{4}\langle t, Vt\rangle\}\,dt \\ &\quad + \overline{\lim_{n\to\infty}}\, \delta^{n-p} \int_{\{\|t\|>bn^{1/2}\}} \left|\hat{Q}_1\left(\frac{t}{n^{1/2}}\right)\right|^p dt \\ &= 2\int_{\{\|t\|>a\}} \exp\{-\tfrac{1}{4}\langle t, Vt\rangle\}\,dt \end{aligned} \tag{19.13}$$

for all $a>0$. Letting $a\to\infty$, one gets

$$\overline{\lim_{n\to\infty}} \sup_{x\in R^k} |q_n(x) - \phi_{0,V}(x)| \leqslant (2\pi)^{-k}\, \overline{\lim_{n\to\infty}} \int |\hat{q}_n(t) - \hat{\phi}_{0,V}(t)|\,dt = 0. \tag{19.14}$$

Next note that (19.5) clearly implies boundedness of each q_n for all sufficiently large n. Hence (ii)⇒(iii). To complete the proof we show that (iii)⇒(i). If q_m is bounded above by c, then $q_m^2 \leqslant cq_m$, so that $q_m \in L^2(R^k)$. This implies $\hat{q}_m \in L^2(R^k)$ [Theorem 4.1(vi)]; that is, $\hat{Q}_1 \in L^{2m}(R^k)$. Q.E.D.

The three statements (i), (ii), and (iii) are each equivalent to (iv): *There exist* $r>1$ *and an integer* m *such that* Q_1^{*m} *has a density belonging to* $L^r(R^k)$. It is clear that (iii)⇒(iv) with $r=2$. Conversely, by the so-called Hausdorff–Young Theorem,† if $f\in L^1(R^k)\cap L^r(R^k)$ for some $r\in(1,2)$, then $\hat{f}\in L^{r'}(R^k)$, where $r'=r/(r-1)$, and

$$\|\hat{f}\|_{r'} \leqslant (2\pi)^{k/r'}\|f\|_{r}. \tag{19.15}$$

---

†See Katznelson [1], p. 142.

Hence if (iv) holds with $r\in(1,2)$, then $\hat{Q}_1^m\in L^{r'}(R^k)$; that is, $\hat{Q}_1\in L^{mr'}(R^k)$. If (iv) holds for some $r\geqslant 2$, then $q_m\in L^2(R^k)$ (since $q_m^2\leqslant q_m+q_m^r$) and, therefore, $\hat{q}_m\in L^2(R^k)$, that is, $\hat{Q}_1\in L^{2m}(R^k)$.

As an example of an absolutely continuous probability measure on R^1 having compact support (and mean zero) and density q_1 such that q_1^{*m} (hence q_m) is unbounded for every m, define†

$$q_1=h*\tilde{h},\qquad h(x)=\begin{cases}(\log 2)\left(x\log^2 x\right)^{-1} & \text{if } x\in(0,\frac{1}{2})\\ 0 & \text{if } x\not\in(0,\frac{1}{2})\end{cases}\tag{19.16}$$

$$\tilde{h}(x)=h(-x)\qquad (x\in R^1).$$

The following result, which provides asymptotic expansions for densities, is more important from our point of view than Theorem 19.1.

THEOREM 19.2 *Let* $\{\mathbf{X}_n\colon n\geqslant 1\}$ *be a sequence of i.i.d. random vectors with values in* R^k, *having a (common) mean zero and a positive-definite covariance matrix* V. *Assume* $\rho_s\equiv E\|\mathbf{X}_1\|^s<\infty$ *for some integer* $s\geqslant 3$ *and that the characteristic function* $\hat{Q}_1$ *of* $\mathbf{X}_1$ *belongs to* $L^p(R^k)$ *for some* $p\geqslant 1$. *A bounded continuous density* q_n *of the distribution* Q_n *of* $n^{-1/2}(\mathbf{X}_1+\cdots+\mathbf{X}_n)$ *exists for every* $n\geqslant p$, *and one has the asymptotic expansion*

$$\sup_{x\in R^k}(1+\|x\|^s)\left|q_n(x)-\sum_{j=0}^{s-2}n^{-j/2}P_j(-\phi_{0,V}\colon\{\chi_\nu\})(x)\right|$$
$$=o(n^{-(s-2)/2})\qquad (n\to\infty),\tag{19.17}$$

where χ_ν *denotes the* ν*th cumulant of* X_1 $(3\leqslant|\nu|\leqslant s)$.

Proof. Without loss of generality, assume p to be an integer (else, take $[p]+1$ for p). For $n\geqslant p+s$, $D^\alpha\hat{Q}_n$ is integrable for $0\leqslant|\alpha|\leqslant s$. Writing, for $n\geqslant p+s$, $|\alpha|\leqslant s$,

$$h_n(x)=x^\alpha\left(q_n(x)-\sum_{j=0}^{s-2}n^{-j/2}P_j(-\phi_{0,V}\colon\{\chi_\nu\})(x)\right)\qquad(x\in R^k),$$

$$\hat{h}_n(t)=(-i)^{|\alpha|}D^\alpha\left[\hat{Q}_n(t)-\sum_{j=0}^{s-2}n^{-j/2}\tilde{P}_j(it\colon\{\chi_\nu\})\exp\{-\tfrac{1}{2}\langle t,Vt\rangle\}\right]\quad(t\in R^k),\tag{19.18}$$

†See Gnedenko and Kolmogorov [1], pp. 223, 224 for a proof of the unboundedness of q_m for every m.

one has (by the Fourier inversion theorem)

$$h_n(x)=(2\pi)^{-k}\int \exp\{-i\langle t,x\rangle\}\hat{h}_n(t)\,dt \qquad (x\in R^k). \qquad (19.19)$$

Let B be the positive-definite symmetric matrix satisfying $B^2=V^{-1}$. Define

$$\eta_s = E\|B\mathbf{X}_1\|^s. \qquad (19.20)$$

By Theorem 9.12 (and the remark following it)

$$|\hat{h}_n(t)| \leqslant \delta(n)n^{-(s-2)/2}(\langle t,Vt\rangle^{(s-|\alpha|)/2} + \langle t,Vt\rangle^{(3(s-2)+|\alpha|)/2})\exp\{-\tfrac{1}{4}\langle t,Vt\rangle\} \qquad (19.21)$$

for all t satisfying

$$\|t\| \leqslant \Lambda^{-1/2}c_{20}(s,k)n^{1/2}\eta_s^{-1/(s-2)} = an^{1/2}, \qquad (19.22)$$

say, where Λ is the largest eigenvalue of V. In view of (19.19), (19.21), and (19.22), it is enough to show that

$$\int_{\{\|t\|>an^{1/2}\}} |D^\alpha \hat{Q}_n(t)|\,dt = o(n^{-(s-2)/2}) \qquad (n\to\infty),$$

$$\int_{\{\|t\|>an^{1/2}\}} \left|D^\alpha\left(\sum_{j=0}^{s-2} n^{-j/2}\tilde{P}_j(it:\{\chi_\nu\})\exp\{-\tfrac{1}{2}\langle t,Vt\rangle\}\right)\right|dt = o(n^{-(s-2)/2}) \qquad (n\to\infty). \qquad (19.23)$$

The second assertion in (19.23) is true because of the presence of the exponential term. The first follows easily from the estimate (obtained by application of Leibniz' formula for differentiation of a product of functions)

$$|D^\alpha \hat{Q}_n(t)| \leqslant c(s,k)\rho_{|\alpha|}n^{|\alpha|/2}\delta^{n-|\alpha|-p}\left|\hat{Q}_1\left(\frac{t}{n^{1/2}}\right)\right|^p \qquad (\|t\|>an^{1/2}), \qquad (19.24)$$

where $n \geqslant p+s$, $|\alpha| \leqslant s$, and from

$$\delta \equiv \sup\{|\hat{Q}_1(t)|: \|t\|>a\} < 1. \qquad (19.25)$$

Q.E.D.

Remark. It should be pointed out that Theorem 19.2 holds *even with* $s=2$. This is true because Theorem 9.12 holds with $s=2$. Therefore a sharper assertion than (19.5) holds, namely,

$$\lim_{n\to\infty} \sup_{x\in R^k} (1+\|x\|^2)|q_n(x)-\phi_{0,V}(x)|=0. \tag{19.26}$$

The next theorem deals with the non-i.i.d. case.

THEOREM 19.3 *Let* $\{\mathbf{X}_n\colon n\geqslant 1\}$ *be a sequence of independent random vectors with values in* R^k *having zero means and average positive-definite covariance matrices* V_n *for large* n. *Assume that*

$$\overline{\lim_{n\to\infty}}\, n^{-1}\sum_{j=1}^{n} E\|B_n\mathbf{X}_j\|^s<\infty \tag{19.27}$$

for some integer $s\geqslant 3$, *where* B_n *is the positive-definite symmetric matrix satisfying*

$$B_n^2=V_n^{-1}, \qquad V_n=n^{-1}\sum_{j=1}^{n}\operatorname{Cov}(\mathbf{X}_j), \tag{19.28}$$

defined for all sufficiently large n. *Also, assume that there exists a positive integer* p *such that the functions*

$$g_{m,n}(t)\equiv\prod_{j=m+1}^{m+p}|E(\exp\{i\langle t,B_n\mathbf{X}_j\rangle\})| \qquad (0\leqslant m\leqslant n-p, \quad n\geqslant p+1)$$

satisfy

$$\gamma\equiv\sup_{\substack{0\leqslant m\leqslant n-p,\\ n\geqslant p+1}}\int g_{m,n}(t)\,dt<\infty \tag{19.29}$$

and, for all positive numbers b,

$$\delta(b)\equiv\sup\{g_{m,n}(t)\colon \|t\|>b,\ 0\leqslant m\leqslant n-p,\ n\geqslant p+1\}<1. \tag{19.30}$$

Then the distribution Q_n *of* $n^{-1/2}B_n(\mathbf{X}_1+\cdots+\mathbf{X}_n)$ *has a density* q_n *for all sufficiently large* n, *and*

$$\sup_{x\in R^k}(1+\|x\|^s)\left|q_n(x)-\sum_{r=0}^{s-3}n^{-r/2}P_r(-\phi\colon\{\bar{\chi}_{\nu,n}\})(x)\right|$$

$$=O(n^{-(s-2)/2}) \qquad (n\to\infty), \tag{19.31}$$

where $\bar{\chi}_{\nu,n}$ *is the average of the* ν*th cumulants* $(3 \leqslant |\nu| \leqslant s)$ *of* $B_n X_j$ $(1 \leqslant j \leqslant n)$.

Proof. For a given nonnegative integral vector α, $|\alpha| \leqslant s$, write (for all large n)

$$h_n(x) = x^{\alpha}\left(q_n(x) - \sum_{r=0}^{s-3} n^{-r/2} P_r(-\phi : \{\bar{\chi}_{\nu,n}\})(x) \right) \qquad (x \in R^k),$$

$$\bar{\eta}_{r,n} = n^{-1} \sum_{j=1}^{n} E\|B_n \mathbf{X}_j\|^r \qquad (2 \leqslant r \leqslant s). \tag{19.32}$$

The statements below are meant to hold for all sufficiently large n. By the Fourier inversion theorem, one has

$$\sup_{x \in R^k} |h_n(x)| \leqslant (2\pi)^{-k} \int |\hat{h}_n(t)|\, dt. \tag{19.33}$$

By Theorem 9.11 and hypothesis (19.27), the inequality

$$|\hat{h}_n(t)| \leqslant b_1 n^{-(s-2)/2} (\|t\|^{s-|\alpha|} + \|t\|^{3(s-2)+|\alpha|}) \exp\left\{ -\frac{\|t\|^2}{4} \right\} \tag{19.34}$$

holds for all t in R^k satisfying

$$\|t\| \leqslant b_2 n^{(s-2)/2s}, \tag{19.35}$$

where b_1 and b_2 are two appropriate positive numbers independent of n. Also, mimicking the proof of Lemma 14.3 (however in this case $|\alpha| \leqslant s$ is a necessary condition) one has, under the present hypothesis,

$$|D^{\alpha} \hat{Q}_n(t)| \leqslant b_3 \exp\left\{ -\frac{\|t\|^2}{6} \right\} \tag{19.36}$$

for all t satisfying

$$\|t\| \leqslant b_4 n^{1/2}. \tag{19.37}$$

Again, b_3 and b_4, as well as all b's below, are positive numbers independent of n. Now, by (19.33),

$$\sup_{x \in R^k} |h_n(x)| \leqslant (2\pi)^{-k} (I_1 + I_2 + I_3), \tag{19.38}$$

where

$$I_1 \equiv \int_{\{\|t\| \leqslant b_2 n^{(s-2)/2s}\}} |\hat{h}_n(t)|\,dt \leqslant b_5 n^{-(s-2)/2},$$

$$I_2 \equiv \int_{\{b_2 n^{(s-2)/2s} < \|t\| \leqslant b_4 n^{1/2}\}} |\hat{h}_n(t)|\,dt,$$

$$I_3 \equiv \int_{\{\|t\| > b_4 n^{1/2}\}} |\hat{h}_n(t)|\,dt. \tag{19.39}$$

The estimate for I_1 follows from (19.34), (19.35). Write

$$f(t) \equiv D^{\alpha}\left[\sum_{r=0}^{s-3} n^{-r/2}\tilde{P}_r(it: \{\bar{\chi}_{\nu,n}\})\exp\left\{\frac{-\|t\|^2}{2}\right\}\right] \qquad (t \in R^k). \tag{19.40}$$

Applying Lemma 9.5 to the random vectors $B_n\mathbf{X}_1, \dots, B_n\mathbf{X}_n$, one obtains, using (19.27),

$$|f(t)| \leqslant b_6(1 + \|t\|^{3s})\exp\left\{\frac{-\|t\|^2}{2}\right\} \qquad (t \in R^k). \tag{19.41}$$

Thus by (19.36), (19.37), and (19.41), one gets

$$I_2 + I_3 \leqslant b_7 \int_{\{\|t\| > b_2 n^{(s-2)/2s}\}} (1 + \|t\|^{3s})\exp\left\{\frac{-\|t\|^2}{6}\right\} dt$$

$$+ \int_{\{\|t\| > b_4 n^{1/2}\}} |D^{\alpha}\hat{Q}_n(t)|\,dt. \tag{19.42}$$

As in (19.24), differentiate $D^{\alpha}\hat{Q}_n$ using Leibniz' formula so that $D^{\alpha}\hat{Q}_n$ is expressed as a sum of $n^{|\alpha|}$ terms, a typical term being

$$a(t) \equiv \left(\prod_{j \notin \{j_1, \dots, j_m\}} E\left(\exp\{i\langle tn^{-1/2}, B_n\mathbf{X}_j\rangle\}\right)\right)$$

$$\times \left(D^{\alpha_1}E\left(\exp\{i\langle tn^{-1/2}, B_n\mathbf{X}_{j_1}\rangle\}\right)\right)^{r_1} \cdots$$

$$\times \left(D^{\alpha_m}E\left(\exp\{i\langle tn^{-1/2}, B_n\mathbf{X}_{j_m}\rangle\}\right)\right)^{r_m}, \tag{19.43}$$

where $j_1, \dots, j_m$ are distinct indices in $\{1, \dots, n\}$, $r_1, \dots, r_m$ are positive integers, and $\alpha_1,\ldots,\alpha_m$ are nonnegative (nonzero) integral vectors satisfying $\sum_{i=1}^{m} r_i\alpha_i=\alpha$. Now

$$\begin{aligned}
&\left|\left(D^{\alpha_1}E\left(\exp\{i\langle tn^{-1/2},B_n\mathbf{X}_{j_1}\rangle\}\right)\right)^{r_1}\cdots\left(D^{\alpha_m}E\left(\exp\{i\langle tn^{-1/2},B_n\mathbf{X}_{j_m}\rangle\}\right)\right)^{r_m}\right| \\
&\qquad\leqslant n^{-|\alpha|/2}\left(E\|B_n\mathbf{X}_{j_1}\|^{|\alpha_1|}\right)^{r_1}\cdots\left(E\|B_n\mathbf{X}_{j_m}\|^{|\alpha_m|}\right)^{r_m} \\
&\qquad\leqslant n^{-|\alpha|/2}\left(E\|B_n\mathbf{X}_{j_1}\|^{|\alpha|}\right)^{r_1|\alpha_1|/|\alpha|}\cdots\left(E\|B_n\mathbf{X}_{j_m}\|^{|\alpha|}\right)^{r_m|\alpha_m|/|\alpha|} \\
&\qquad\leqslant n^{-|\alpha|/2}\sum_{i=1}^{m}\left(E\|B_n\mathbf{X}_{j_i}\|^{|\alpha|}\right)\leqslant n^{-|\alpha|/2}mn\bar{\eta}_{|\alpha|,n}. \qquad (19.44)
\end{aligned}$$

Also, since $m\leqslant|\alpha|$, there are at least $(n-|\alpha|)/(|\alpha|p)-1$ sets of p consecutive indices in $\{1,2,\ldots,n\}\backslash\{j_1,\ldots,j_m\}$. Hence

$$\begin{aligned}
\int_{\{\|t\|>b_4n^{1/4}\}}|a(t)|\,dt &\leqslant n^{-|\alpha|/2+1}m\bar{\eta}_{|\alpha|,n}(\delta(b_4))^{(n-|\alpha|)/(|\alpha|p)-2} \\
&\qquad\times\int_{R^k}g_{m',n}\left(\frac{t}{n^{1/2}}\right)dt \\
&\leqslant b_8n^{-|\alpha|/2+k/2+1}\bar{\eta}_{|\alpha|,n}(\delta(b_4))^{(n-|\alpha|)/(|\alpha|p)-2} \qquad (19.45)
\end{aligned}$$

for some m', $0\leqslant m'\leqslant n-p$ and, therefore,

$$\begin{aligned}
\int_{\{\|t\|>b_4n^{1/2}\}}|D^{\alpha}\hat{Q}_n(t)|\,dt &\leqslant b_8n^{(|\alpha|+k+2)/2}\bar{\eta}_{|\alpha|,n}(\delta(b_4))^{(n-|\alpha|)/(|\alpha|p)-2} \\
&=o(n^{-(s-2)/2}) \qquad (n\to\infty). \qquad (19.46)
\end{aligned}$$

The desired conclusion (19.31) now easily follows from (19.33), (19.39), (19.42), and (19.46) on taking $\alpha=0$, $\alpha=(s,0,\ldots,0),\ldots,$ $\alpha=(0,0,\ldots,0,s)$. Q.E.D.

The following variant of Theorem 19.3 is perhaps easier to use.

COROLLARY 19.4 *Let* $\{\mathbf{X}_n: n\geqslant 1\}$ *be a sequence of independent random vectors having zero means and finite* s*th absolute moments* $\{\rho_{s,n}\equiv E\|\mathbf{X}_n\|^s: n\geqslant 1\}$ *for some integer* $s\geqslant 3$. *Let* $V_n=n^{-1}\sum_{j=1}^{n}\mathrm{Cov}(\mathbf{X}_j)$ *and let* λ_n *denote the smallest eigenvalue of* V_n $(n\geqslant 1)$. *Write* $\rho_s=n^{-1}\sum_{j=1}^{n}E\|\mathbf{X}_j\|^s$. *Suppose that*

$$\varliminf_{n}\lambda_n>0, \qquad \sup_n\rho_s<\infty. \qquad (19.47)$$

Also assume that there exists an integer p *such that the functions* g_m $(m\geqslant 0)$

defined by

$$g_m(t) \equiv \prod_{j=m+1}^{m+p} |E(\exp\{i\langle t, \mathbf{X}_j\rangle\})| \qquad (m=0,1,2,\ldots) \qquad (19.48)$$

satisfy

$$\sup_{m\geq 0} \int g_m(t)\,dt < \infty, \quad \sup\{g_m(t) \colon \|t\| > b, m \geq 0\} < 1 \qquad (\text{all } b > 0). \qquad (19.49)$$

Then (19.31) *holds.*

Proof. Note that, with B_n defined by (19.28), one has

$$\overline{\lim_n}\, n^{-1} \sum_{j=1}^{n} E\|B_n \mathbf{X}_j\|^s \leq \left(\overline{\lim_n}\, \|B_n\|^s\right) \overline{\lim_n} \left(n^{-1} \sum_{j=1}^{n} E\|\mathbf{X}_j\|^s\right)$$

$$= \left(\underline{\lim_n}\, \lambda_n\right)^{-s/2} \overline{\lim_n} \left(n^{-1} \sum_{j=1}^{n} \rho_{s,j}\right) < \infty, \qquad (19.50)$$

by (19.47), since $\|B_n\| = \lambda_n^{-1/2}$. Also, letting $g_{m,n}$ be as in Theorem 19.3,

$$\sup_{\substack{0 \leq m \leq n-p,\\ n \geq p+1}} \int g_{m,n}(t)\,dt \leq \sup_{\substack{m \geq 0,\\ n \geq p+1}} (\operatorname{Det} V_n)^{1/2} \int g_m(t)\,dt. \qquad (19.51)$$

But, writing Λ_n for the largest eigenvalue of V_n,

$$\operatorname{Det}(V_n) \leq \Lambda_n^k, \qquad \Lambda_n \leq \rho_2 \leq \rho_s^{2/s}. \qquad (19.52)$$

Hence (19.29) holds. Finally,

$$g_{m,n}(t) = g_m(B_n t),$$

$$\sup_{\substack{0 \leq m \leq n-p,\\ n \geq p+1,\\ \|t\| > b}} g_{m,n}(t) \leq \sup_{\substack{m \geq 0\\ n \geq p+1\\ \|t\| > b\Lambda_n^{-1/2}}} g_m(t) < 1 \qquad (19.53)$$

for a sufficiently large p. Q.E.D.

If the integer s in Theorem 19.2 (never smaller than 3) is larger than k, then (19.17) immediately implies

$$\left\| Q_n - \sum_{j=0}^{s-2} n^{-j/2} P_j(-\Phi_{0,V} \colon \{\chi_\nu\}) \right\| = o(n^{-(s-2)/2}) \qquad (n \to \infty). \qquad (19.54)$$

Similarly, if $s \geqslant k+1$, then under the hypothesis of Theorem 19.3, one has

$$\left\| Q_n - \sum_{j=0}^{s-3} n^{-j/2} P_j(-\Phi: \{\bar{\chi}_{\nu,n}\}) \right\| = O(n^{-(s-2)/2}) \qquad (n \to \infty).$$

The following theorem deals with the general (i.i.d.) case.

THEOREM 19.5 *Let* $\{\mathbf{X}_n: n \geqslant 1\}$ *be a sequence of independent and identically distributed random vectors with mean zero, nonsingular covariance matrix* V *and a nonzero, absolutely continuous component. If* $\rho_s \equiv E\|\mathbf{X}_1\|^s < \infty$ *for some integer* $s \geqslant 3$, *then, writing* Q_n *for the distribution of* $n^{-1/2} \times (\mathbf{X}_1 + \cdots + \mathbf{X}_n)$, *one has*

$$\int (1+\|x\|^s) \left| Q_n - \sum_{r=0}^{s-2} n^{-r/2} P_r(-\Phi_{0,V}: \{\chi_\nu\}) \right| (dx)$$
$$= o(n^{-(s-2)/2}) \qquad (n \to \infty), \quad (19.55)$$

where χ_ν *is the* ν*th cumulant of* $\mathbf{X}_1$.

Proof. By changing variables from x to Tx, where $T' = T$ and $T^2 = V^{-1}$, one may immediately check that it is enough to prove the theorem for the case $V = I$. Hence we assume $V = I$. Define the truncated random vectors $\mathbf{Y}_{j,n}$, $\mathbf{Z}_{j,n}$ by

$$\mathbf{Y}_{j,n} = \begin{cases} \mathbf{X}_j & \text{if } \|\mathbf{X}_j\| \leqslant n^{1/2} \\ 0 & \text{if } \|\mathbf{X}_j\| > n^{1/2} \end{cases} \qquad \mathbf{Z}_{j,n} = \mathbf{Y}_{j,n} - E\mathbf{Y}_{j,n} \quad (1 \leqslant j \leqslant n, \quad n \geqslant 1).$$
(19.56)

These are all the same as $\mathbf{Y}_j$, $\mathbf{Z}_j$ defined by (14.2); the additional subscript n is introduced to emphasize that the truncations change with n. Let Q_n'', Q_n' denote the distributions of $n^{-1/2}\sum_{j=1}^n \mathbf{Y}_{j,n}$, $n^{-1/2}\sum_{j=1}^n \mathbf{Z}_{j,n}$, respectively. We write

$$\Psi_{n,s'} = \sum_{r=0}^{s'-2} n^{-r/2} P_r(-\Phi: \{\chi_\nu\}),$$

$$\Psi'_{n,s'} = \sum_{r=0}^{s'-2} n^{-r/2} P_r(-\Phi_{0,D_n}: \{\chi_{\nu,n}\}), \qquad (19.57)$$

where

$$D_n = \operatorname{Cov}(\mathbf{Y}_{1,n}) = \operatorname{Cov}(\mathbf{Z}_{1,n}),$$

$$\chi_{\nu,n} = \nu\text{th cumulant of } \mathbf{Z}_{1,n}. \tag{19.58}$$

Also, let $\Psi''_{n,s'}$ denote the signed measure whose density at x equals the density of $\Psi'_{n,s'}$ at $x - a_n$, where

$$a_n = n^{-1/2} \sum_{j=1}^{n} E\mathbf{Y}_{j,n} = n^{1/2} E\mathbf{Y}_{1,n}. \tag{19.59}$$

Now by Lemmas 14.6, 14.8,

$$\begin{aligned}\int (1+\|x\|^s)|Q_n - \Psi_{n,s}|(dx) &\leqslant \int (1+\|x\|^s)|Q_n - Q_n''|(dx)\\ &\quad + \int (1+\|x\|^s)|Q_n'' - \Psi_{n,s}''|(dx)\\ &\quad + \int (1+\|x\|^s)|\Psi_{n,s}'' - \Psi_{n,s}|(dx)\\ &= \int (1+\|x\|^s)|Q_n'' - \Psi_{n,s}''|(dx) + o(n^{-(s-2)/2})\\ &\qquad\qquad (n\to\infty).\end{aligned} \tag{19.60}$$

Also, by a change of variables $(x \to x - a_n)$,

$$\begin{aligned}\int (1+\|x\|^s)|Q_n'' - \Psi_{n,s}''|(dx) &= \int (1+\|x+a_n\|^s)|Q_n' - \Psi_{n,s}'|(dx)\\ &= \int (1+\|x\|^s)|Q_n' - \Psi_{n,s}'|(dx) + o(n^{-(s-2)/2})\\ &\qquad (n\to\infty),\end{aligned} \tag{19.61}$$

using (14.81) and the fact that $\Psi'_{n,s}$ has a bounded (uniformly in n) variation norm (since $\|\Psi'_{n,s} - \Psi_{n,s}\| \to 0$ as $n \to \infty$, by Lemma 14.6).

Let q_1 denote the density of the absolutely continuous component of Q_1. There exists a positive number c such that

$$1 > \theta \equiv \int_B q_1(x)\,dx > 0, \qquad B \subset \{x \in R^k : q_1(x) \leqslant c\}. \tag{19.62}$$

Write

$$B_n = \{x \in B : \|x\| \leqslant n^{1/2}\}, \qquad \theta_n = \int_{B_n} q_1(x)\,dx. \tag{19.63}$$

Then there exists n_0 such that

$$\theta \geqslant \theta_n \geqslant \frac{\theta}{2} \qquad (n \geqslant n_0). \tag{19.64}$$

The distribution $Q''_{1,n}$ of $\mathbf{Y}_{1,n}$ may then be expressed as

$$Q''_{1,n} = \theta_n G''_n + (1-\theta_n) H''_n, \tag{19.65}$$

where G''_n, H''_n are probability measures, and G''_n is absolutely continuous with density

$$\frac{1}{\theta_n} q_1(x) \cdot I_{B_n}(x) \qquad (x \in R^k). \tag{19.66}$$

Define the function p_n on R^k by

$$p_n(x) = \frac{1}{\theta_n} q_1(x + a_n) I_{B_n}(x + a_n) \qquad (x \in R^k). \tag{19.67}$$

Then the distribution $Q'_{1,n}$ of $\mathbf{Z}_{1,n}$ may be expressed as

$$Q'_{1,n} = \theta_n G_n + (1-\theta_n) H_n, \tag{19.68}$$

where G_n, H_n are probability measures on R^k, G_n being absolutely continuous with density p_n. Write G for the probability measure on R^k with density $q_1 \cdot I_B / \theta$. Then

$$\begin{aligned}\sup_{t \in R^k} |\hat{G}''_n(t) - \hat{G}(t)| &= \sup_{t \in R^k} \left| \frac{1}{\theta_n} \int_{B_n} \exp\{i\langle t, x\rangle\} q_1(x)\, dx \right. \\ &\qquad \left. - \frac{1}{\theta} \int_B \exp\{i\langle t, x\rangle\} q_1(x)\, dx \right| \\ &\leqslant \theta_n \left(\frac{1}{\theta_n} - \frac{1}{\theta} \right) + \frac{1}{\theta} \int_{B \setminus B_n} q_1(x)\, dx \to 0 \qquad (n \to \infty).\end{aligned} \tag{19.69}$$

Also observe that G_n has a bounded and, therefore, square integrable density p_n, so that $\hat{G}_n \in L^2(R^k)$. Clearly,

$$\hat{G}_n(t) = \hat{G}''_n(t) \exp\{i\langle t, -a_n\rangle\}, \qquad |\hat{G}_n(t)| = |\hat{G}''_n(t)| \qquad (t \in R^k). \tag{19.70}$$

Using the expression (19.68), one has

$$(Q'_{1,n})^{*n} = \sum_{j=0}^{n} \binom{n}{j} \theta_n^j (1-\theta_n)^{(n-j)} G_n^{*j} * H_n^{*(n-j)}, \tag{19.71}$$

where $G_n^{*0}=H_n^{*0}$ is the probability measure degenerate at zero. Write

$$\sum{}' = \sum_{\{j:0\leq j\leq n,\, j-n\theta_n \geq -n^{1/2}\log n\}},$$

$$\sum{}'' = \sum_{\{j;\, j-n\theta_n < -n^{1/2}\log n\}} = \sum_{j=0}^{n} - \sum{}'. \tag{19.72}$$

Applying Theorem 17.11 or Corollary 17.13 to a triangular array whose nth row consists of n independent centered Bernoulli random variables with parameter θ_n (note that such an extension of the quoted results is immediate in view of the fact that they deal with tail probabilities of $\{Q_n: n\geq 1\}$ and one only needs to be able to represent each Q_n as the distribution of the normalized sum of n independent random vectors whose average moments and average variance covariance matrix satisfy the given hypotheses), one has, using (19.64),

$$\sum{}'' \binom{n}{j}\theta_n^j(1-\theta_n)^{n-j} = o(n^{-m}) \qquad (n\to\infty), \tag{19.73}$$

for every positive integer m.

Define the measures $\overline{G}_n$, $\overline{H}_n$, $M_{n,j}$ by

$$\overline{G}_n(A) = G_n(n^{1/2}A), \qquad \overline{H}_n(A) = H_n(n^{1/2}A)$$

$$(A\in\mathcal{B}^k,\; n^{1/2}A = \{n^{1/2}x : x\in A\}),$$

$$M_{n,j} = \binom{n}{j}\theta_n^j(1-\theta_n)^{n-j}\overline{G}_n^{*j} * \overline{H}_n^{*(n-j)}. \tag{19.74}$$

Then write

$$Q_n' = \sum_{j=0}^{n} M_{n,j}. \tag{19.75}$$

One has

$$\int (1+\|x\|^s)|Q_n' - \Psi_{n,s}'|(dx)$$

$$= \int (1+\|x\|^s)\Big|\sum{}' M_{n,j} - \Psi_{n,s}'\Big|(dx) + o(n^{-(s-2)/2}), \tag{19.76}$$

since

$$\int (1+\|x\|^s)\left(\sum{}'' M_{n,j}\right)(dx) \leqslant \sum{}''\tbinom{n}{j}\theta_n^j(1-\theta_n)^{n-j}\left[1+\left(n+n^{1/2}\|E\mathbf{Y}_{1,n}\|\right)^s\right]$$
$$= o(n^{-(s-2)/2}) \qquad (n\to\infty). \tag{19.77}$$

Observe that the inequality

$$\|\mathbf{Z}_{1,n}\| \leqslant n^{1/2} + \|E\mathbf{Y}_{1,n}\| \tag{19.78}$$

implies

$$\overline{G}_n\left(\left\{\|x\| \leqslant 1 + n^{-1/2}\|E\mathbf{Y}_{1,n}\|\right\}\right) = 1$$
$$= \overline{H}_n\left(\left\{\|x\| \leqslant 1 + n^{-1/2}\|E\mathbf{Y}_{1,n}\|\right\}\right),$$

$$M_{n,j}\left(\left\{\|x\| > n + n^{1/2}\|E\mathbf{Y}_{1,n}\|\right\}\right) = 0,$$
$$M_{n,j}\left(\left\{\|x\| \leqslant n + n^{1/2}\|E\mathbf{Y}_{1,n}\|\right\}\right) = \tbinom{n}{j}\theta_n^j(1-\theta_n)^{n-j}. \tag{19.79}$$

By Lemma 11.6 we have

$$\int (1+\|x\|^s)\left|\sum{}' M_{n,j} - \Psi'_{n,s}\right|(dx) \leqslant b_9 \max_{|\beta| \leqslant k+s+1} \left\|D^\beta\left(\sum{}'\hat{M}_{n,j} - \hat{\Psi}'_{n,s}\right)\right\|_1. \tag{19.80}$$

Use Lemma 9.5, the fact that $\|D_n - I\|$ goes to zero as $n\to\infty$, and the relation [see Lemma 14.1(v)]

$$n^{-r/2}E\|\mathbf{Z}_{1,n}\|^{r+2} = o(n^{-(s-2)/2}) \qquad (r \geqslant s-1) \tag{19.81}$$

to get

$$\left\|D^\beta\left(\hat{\Psi}'_{n,k+s+1} - \hat{\Psi}'_{n,s}\right)\right\|_1 = o(n^{-(s-2)/2}) \qquad (n\to\infty). \tag{19.82}$$

Write (the constant c_{20} below is the same as in Theorem 9.10)

$$T_n = D_n^{-1/2}, \qquad \eta_r = E\|T_n\mathbf{Z}_{1,n}\|^r \leqslant \|T_n\|^r E\|\mathbf{Z}_{1,n}\|^r,$$
$$A_n = \left\{t\in R^k\colon \|t\| \leqslant \Lambda_n^{-1/2}c_{20}(k+s+2,k)n^{1/2}\eta_{k+s+1}^{-1/(k+s)}\right\},$$
$$A'_n = \left\{t\in R^k\colon \|t\| \leqslant \frac{n^{1/2}}{16\rho_3}\right\} \qquad (\rho_3 = E\|\mathbf{X}_1\|^3), \tag{19.83}$$

where Λ_n is the largest eigenvalue of D_n. Letting

$$\begin{aligned}
I_1 &= \int_{A_n} |D^\beta (\hat{Q}_n' - \hat{\Psi}_{n,k+s+1}')(t)|\,dt,\\
I_2 &= \int_{A_n'\setminus A_n} |D^\beta \hat{Q}_n'(t)|\,dt,\\
I_3 &= \int_{A_n'} \Big|D^\beta \sum{}'' \hat{M}_{n,j}(t)\Big|\,dt, \qquad (19.84)\\
I_4 &= \int_{R^k\setminus A_n'} \Big|D^\beta \sum{}' \hat{M}_{n,j}(t)\Big|\,dt,\\
I_5 &= \int_{R^k\setminus A_n} |D^\beta \hat{\Psi}_{n,k+s+1}'(t)|\,dt,
\end{aligned}$$

one has

$$\Big\| D^\beta \Big(\sum{}' \hat{M}_{n,j} - \hat{\Psi}_{n,k+s+1}'\Big)\Big\|_1 \leqslant I_1 + I_2 + I_3 + I_4 + I_5. \qquad (19.85)$$

By Theorem 9.10 and relation (19.81) (with $r = k+s$), one has

$$\begin{aligned}
I_1 \leqslant b_9 \Lambda_n^{|\beta|} \eta_{k+s+2} n^{-(k+s)/2} &\leqslant b_9 \| T_n \|^{k+s+2} E\, \|\mathbf{Z}_{1,n}\|^{k+s+2} n^{-(k+s)/2}\\
&= o(n^{-(s-2)/2}) \qquad (n \to \infty), \qquad (19.86)
\end{aligned}$$

since, by Corollary 14.2,

$$\begin{aligned}
\| T_n - I \| &= \| D_n^{-1/2} - I \| \to 0,\\
\| T_n^{-1} - I \| &= \| D_n^{1/2} - I \| \to 0 \qquad (n \to \infty). \qquad (19.87)
\end{aligned}$$

By Lemma 14.1(v) there exists a positive number b_{10} such that

$$\begin{aligned}
n^{1/2} \eta_{k+s+2}^{-1/(k+s)} &\geqslant n^{1/2} \big(\| T_n \|^{k+s+2} E\, \|\mathbf{Z}_{1,n}\|^{k+s+2} \big)^{-1/(k+s)}\\
&\geqslant n^{1/2} \big(\| T_n \|^{k+s+2} 2^{(k+s+2)} n^{(k+2)/2} \rho_s \big)^{-1/(k+s)}\\
&\geqslant b_{10} n^{(s-2)/(2(k+s))}
\end{aligned}$$

for all sufficiently large n. From this and (19.87) it follows that there exists a positive constant b_{11} such that

$$A_n \supset \left\{ t \in R^k: \|t\| \leqslant b_{11} n^{(s-2)/(2(k+s))} \right\},$$

$$A'_n \setminus A_n \subset \left\{ \|t\| > b_{11} n^{(s-2)/(2(k+s))},\ t \in A'_n \right\}. \tag{19.88}$$

By (19.88) and Lemma 14.3 one has

$$I_2 = o\left(n^{-(s-2)/2}\right) \qquad (n \to \infty). \tag{19.89}$$

Because of the presence of the exponential factor $\exp\{-\frac{1}{2}\langle t, D_n t\rangle\}$ in $\hat{\Psi}'_{n,k+s+1}$, (19.88) also implies

$$I_5 = o\left(n^{-(s-2)/2}\right) \qquad (n \to \infty). \tag{19.90}$$

It follows from (19.79) [and (19.74)] that

$$|D^{\beta} \hat{M}_{n,j}(t)| \leqslant \binom{n}{j} \theta_n^j (1-\theta_n)^{n-j} \left(n + n^{1/2} \|E\mathbf{Y}_{1,n}\|\right)^{|\beta|} \qquad (0 \leqslant j \leqslant n, \quad t \in R^k), \tag{19.91}$$

so that, in view of (19.73),

$$I_3 = o\left(n^{-(s-2)/2}\right). \tag{19.92}$$

It remains to estimate I_4. By virtue of (19.64) there exists an integer n_1 such that for all indices j entering in Σ'

$$j > n^{1/2} \qquad (n \geqslant n_1). \tag{19.93}$$

Therefore for all such j, using (19.79) and the Leibniz formula for differentiating a product of n functions,

$$|D^{\beta} \hat{M}_{n,j}(t)| \leqslant \binom{n}{j} \theta_n^j (1-\theta_n)^{n-j} \left(n^{|\beta|/2}\right) \left(n^{1/2} + \|E\mathbf{Y}_{1,n}\|\right)^{|\beta|} \left| \hat{G}_n\left(\frac{t}{n^{1/2}}\right) \right|^{n^{1/2} - |\beta|},$$

and one has

$$\begin{aligned} I_4 &\leqslant b_{12} n^{|\beta|} \int_{\{\|t\| > n^{1/2}/(16\rho_3)\}} \left| \hat{G}_n\left(\frac{t}{n^{1/2}}\right) \right|^{n^{1/2} - |\beta|} dt \\ &= b_{12} n^{|\beta| + k/2} \int_{\{\|t\| > 1/(16\rho_3)\}} |\hat{G}_n(t)|^{n^{1/2} - |\beta|} dt \\ &\leqslant b_{12} n^{|\beta| + k/2} \left(\delta_n^{n^{1/2} - |\beta| - 2}\right) \int |\hat{G}_n(t)|^2 dt, \end{aligned} \tag{19.94}$$

where b_{12} is a positive constant and

$$\delta_n \equiv \sup_{\{\|t\| > (16\rho_3)^{-1}\}} |\hat{G}_n(t)|. \tag{19.95}$$

By (19.69) and (19.70) (and remembering that G is absolutely continuous),

$$\overline{\lim_{n\to\infty}}\, \delta_n = \lim_{n\to\infty} \delta_n = \sup_{\|t\| > (16\rho_3)^{-1}} |\hat{G}(t)| < 1, \tag{19.96}$$

so that $\delta_n \leqslant \delta < 1$ for all large n (and some $\delta < 1$). Also, by the Plancherel theorem (Theorem 4.2),

$$\int |\hat{G}_n(t)|^2 dt = (2\pi)^k \int p_n^2(x)\, dx \leqslant (2\pi)^k \frac{1}{\theta_n^2} \int_B q_1^2(x)\, dx, \tag{19.97}$$

which is bounded away from infinity. Thus

$$I_4 = o(n^{-(s-2)/2}) \qquad (n \to \infty). \tag{19.98}$$

The above estimates of $I_1, \dots, I_5$ are now used in (19.85), and the resulting estimate is combined with (19.82) to yield [via (19.60), (19.76), (19.80)] the desired result. Q.E.D.

The following corollary is now immediate.

COROLLARY 19.6 *Under the hypothesis of Theorem* 19.4,

$$\|Q_n - \Psi_{n,s}\| = o(n^{-(s-2)/2}) \qquad (n \to \infty), \tag{19.99}$$

where $\|\ \ \|$ *denotes variation norm.*

It is possible to prove analogs of Theorem 19.5 (and the above Corollary) for a sequence $\{\mathbf{X}_n : n \geqslant 1\}$ of nonidentically distributed random vectors following the same method of proof as above. For example, it is not difficult to check that under the hypothesis of Theorem 19.3 or Corollary 19.4 one has

$$\int (1 + \|x\|^s) \left| Q_n - \sum_{r=0}^{s-3} n^{-r/2} P_r(-\Phi : \{\bar{\chi}_{\nu,n}\}) \right| (dx)$$
$$= O(n^{-(s-2)/2}) \qquad (n \to \infty), \tag{19.100}$$

where the notation is as in Theorem 19.3. Indeed, the proof is simpler because the sums of p consecutive $\mathbf{X}_n$'s have uniformly bounded densities.

Finally, we observe that the hypothesis of Theorem 19.5 (even when Q_1 is assumed to be absolutely continuous) is not sufficient to provide the uniform local expansion obtained in Theorem 19.2. The discussion following the proof of Theorem 19.1 [as well as the counterexample displayed by (19.16)] proves this point.

Remark. It is sometimes useful to relax in Theorem 19.5 the requirement that Q_1 has a nonzero absolutely continuous component and require instead that Q_1^{*m} has a nonzero absolutely continuous component for some positive integer m. The proof that (19.55) holds under this relaxation is essentially the same as above. Note that Theorem 19.2 is proved under a similar relaxation of the hypothesis that Q_1 has a bounded density.

20. ASYMPTOTIC EXPANSIONS UNDER CRAMÉR'S CONDITION

A probability measure Q on R^k satisfies *Cramér's condition* if

$$\overline{\lim_{\|t\|\to\infty}} |\hat{Q}(t)| < 1. \tag{20.1}$$

A probability measure Q having a nonzero, absolutely continuous component (with respect to Lebesgue measure) satisfies (20.1), as a result of the Riemann-Lebesgue lemma [Theorem 4.1(iii)]. In addition, there are many singular probability measures satisfying (20.1). Any Q having a nonzero singular component of this type also satisfies Cramér's condition. Note that (20.1) implies that Q is nondiscrete; if Q is purely discrete, then $\hat{Q}$ is *almost periodic*† and the lim sup in (20.1) equals one. Moreover, any Q satisfying (20.1) is *nondegenerate*, that is, does not assign all its mass to a hyperplane. In fact, (20.1) implies that Q is *strongly nonlattice* in the sense that there does not exist $t_0 \neq 0$ in R^k such that

$$|\hat{Q}(t_0)| = 1. \tag{20.2}$$

Notice (20.2) implies that $|\hat{Q}(nt_0)| = 1$ for all integers n, thus violating (20.1). We shall discuss (20.2) and lattice distributions in greater detail in the next chapter. We point out, however, that (20.2) means that the entire mass of Q is carried by a sequence of hyperplanes orthogonal to t_0. From the above discussion it follows that (20.1) is equivalent to

$$\sup_{\|t\|>b} |\hat{Q}(t)| < 1 \tag{20.3}$$

for all positive b (or, equivalently, some positive b).

---

†See Katznelson [1], Chapter VI.5.

THEOREM 20.1 *Let* $\{\mathbf{X}_n: n \geq 1\}$ *be an i.i.d. sequence of random vectors with values in* R^k *whose common distribution* Q_1 *satisfies Cramér's condition* (20.1). *Assume that* Q_1 *has mean zero and a finite* sth *absolute moment for some integer* $s \geq 3$. *Let* V *denote the covariance matrix of* Q_1 *and* χ_ν *its* νth *cumulant* $(3 \leq |\nu| \leq s)$. *Then for every real-valued, Borel-measurable function* f *on* R^k *satisfying*

$$M_{s'}(f) < \infty \tag{20.4}$$

for some s', $0 \leq s' \leq s$, *one has*

$$\left| \int f d\left(Q_n - \sum_{r=0}^{s-2} n^{-r/2} P_r(-\Phi_{0,V}: \{\chi_\nu\}) \right) \right|$$
$$\leq M_{s'}(f)\delta_1(n) + c(s,k)\bar{\omega}_f(2e^{-dn}: \Phi_{0,V}), \tag{20.5}$$

where Q_n *is the distribution of* $n^{-1/2}(\mathbf{X}_1 + \cdots + \mathbf{X}_n)$, d *is a suitable positive constant, and*

$$\delta_1(n) = o(n^{-(s-2)/2}), \qquad (n \to \infty). \tag{20.6}$$

Moreover, $c(s,k)$ *depends only on* s *and* k, *and the quantities* d, $\delta_1(n)$ *do not depend on* f.

Proof. Assume that $V = I$, without essential loss of generality. As in (19.56), introduce truncated random vectors $\mathbf{Y}_{j,n}$, $\mathbf{Z}_{j,n}$. Recall that

$$M_{s'}(f) = \begin{cases} \sup_{x \in R^k} (1 + \|x\|^{s'})^{-1} |f(x)|, & s' > 0, \\ \omega_f(R^k) = \sup_{x,y \in R^k} |f(x) - f(y)|, & s' = 0. \end{cases} \tag{20.7}$$

Let Q_n'' denote the distribution of $n^{-1/2}(\mathbf{Y}_{1,n} + \cdots + \mathbf{Y}_{n,n})$, and Q_n' that of $n^{-1/2}(\mathbf{Z}_{1,n} + \cdots + \mathbf{Z}_{n,n})$. By Lemma 14.8, for all large n one has

$$\left| \int f d(Q_n - Q_n'') \right| \leq M_{s'}(f) \int (1 + \|x\|^{s'}) |Q_n - Q_n''|(dx)$$
$$\leq M_{s'}(f) c'(s', s) n^{-(s-2)/2} \bar{\Delta}_{n,s}, \tag{20.8}$$

where

$$\bar{\Delta}_{n,s} = \int_{\{\|\mathbf{X}_1\| > n^{1/2}\}} \|\mathbf{X}_1\|^s = o(1) \qquad (n \to \infty). \tag{20.9}$$

Writing

$$a_n = n^{1/2} E\mathbf{Y}_{1,n}, \tag{20.10}$$

one has, by (14.81),

$$\|a_n\| \leqslant k^{1/2} n^{-(s-2)/2} \bar{\Delta}_{n,s} = o(n^{-(s-2)/2}) \qquad (n \to \infty). \tag{20.11}$$

Now

$$\int f\, dQ_n'' = \int f_{a_n}\, dQ_n',$$

$$\left| \int (f_{a_n} - f)\, d\left(\sum_{r=0}^{s-2} n^{-r/2} P_r(-\Phi : \{\chi_\nu\}) \right) \right|$$

$$= \left| \int f(x) \sum_{r=0}^{s-2} n^{-r/2} \big(P_r(-\phi : \{\chi_\nu\})(x) - P_r(-\phi : \{\chi_\nu\})(x - a_n) \big)\, dx \right|$$

$$\leqslant c_1(s', s, k) M_{s'}(f) n^{-(s-2)/2} \bar{\Delta}_{n,s}. \tag{20.12}$$

The last inequality follows from the second inequality in Lemma 14.6. Next use the first inequality in Lemma 14.6 to get

$$\left| \int f_{a_n}\, d\left(\sum_{r=0}^{s-2} n^{-r/2} P_r(-\Phi : \{\chi_\nu\}) - \sum_{r=0}^{s-2} n^{-r/2} P_r(-\Phi_{0,D_n} : \{\chi_{\nu,n}\}) \right) \right|$$

$$\leqslant c_2(s', s, k) M_{s'}(f) n^{-(s-2)/2} \bar{\Delta}_{n,s}, \tag{20.13}$$

where

$$D_n = \operatorname{Cov}(\mathbf{Z}_{1,n}), \qquad \chi_{\nu,n} = \nu\text{th cumulant of } \mathbf{Z}_{1,n}. \tag{20.14}$$

In view of (20.8), (20.11)–(20.13), it is enough to estimate

$$\int f_{a_n}\, d\left(Q_n' - \sum_{r=0}^{s-2} n^{-r/2} P_r(-\Phi_{0,D_n} : \{\chi_{\nu,n}\}) \right). \tag{20.15}$$

Write

$$H_n = Q_n' - \sum_{r=0}^{s+k-2} n^{-r/2} P_r(-\Phi_{0,D_n} : \{\chi_{\nu,n}\}). \tag{20.16}$$

We first estimate $\int f_{a_n}\,dH_n$. By Corollary 11.2,

$$\left|\int f_{a_n}\,dH_n\right| \leqslant M_{s'}(f)\int\left[1+(\|x\|+\varepsilon+\|a_n\|)^{s'}\right]|H_n * K_\varepsilon|(dx)$$
$$+\bar{\omega}_{f_{a_n}}\left(2\varepsilon:\left|\sum_{r=0}^{s+k-2} n^{-r/2}P_r(-\Phi_{0,D_n}:\{\chi_{\nu,n}\})\right|\right)$$
$$(\varepsilon>0), \quad (20.17)$$

where we choose the probability measure K_ε to satisfy

$$K_\varepsilon(\{x:\ \|x\|<\varepsilon\})=1, \tag{20.18}$$
$$|D^\alpha \hat{K}_\varepsilon(t)| \leqslant \varepsilon^{|\alpha|}c_3(s,k)\exp\{-(\varepsilon\|t\|)^{1/2}\} \qquad (t\in R^k,\ \ |\alpha|\leqslant s+k+1).$$

This is possible by Corollary 10.4 (with s replaced by $s+k+1$), on defining $K_\varepsilon(B)=K(\varepsilon^{-1}B)$ for all Borel sets B.

Since $\|x\| \leqslant \sum_{i=1}^k |x_i|$ and ε will be chosen to be smaller than one, Lemma 11.6 may be used to obtain

$$\int\left[1+(\|x\|+\|a_n\|+\varepsilon)^{s'}\right]|H_n * K_\varepsilon|(dx)$$
$$\leqslant c_4(s,s',k)\max_{0\leqslant|\beta|\leqslant k+s+1}\int|D^\beta(\hat{H}_n\hat{K}_\varepsilon)(t)|\,dt. \tag{20.19}$$

One can write

$$D^\beta(\hat{H}_n\hat{K}_\varepsilon)=\sum_{0\leqslant\alpha\leqslant\beta} c_5(\alpha,\beta)(D^{\beta-\alpha}\hat{H}_n)(D^\alpha\hat{K}_\varepsilon). \tag{20.20}$$

By Theorem 9.10,

$$\int_{\{\|t\|\leqslant A_n\}}|[D^{\beta-\alpha}\hat{H}_n(t)][D^\alpha\hat{K}_\varepsilon(t)]|\,dt \leqslant \int_{\{\|t\|\leqslant A_n\}}|D^{\beta-\alpha}\hat{H}_n(t)|\,dt$$
$$\leqslant c_6(s,k)n^{-(s+k-1)/2}E\|T_n\mathbf{Z}_{1,n}\|^{s+k+1}, \tag{20.21}$$

where T_n is the symmetric, positive-definite matrix defined, for all $n \geqslant n_0$, say, by $T_n^2=D_n^{-1}$, and, writing Λ_n for the largest eigenvalue of D_n,

$$A_n=\frac{c_7(s,k)n^{1/2}}{\left(E\|T_n\mathbf{Z}_{1,n}\|^{s+k+1}\right)^{1/(s+k-1)}\Lambda_n^{1/2}}. \tag{20.22}$$

By Corollary 14.2, $\{\|T_n\|: n \geqslant n_0\}$ is bounded, and [by Lemma 14.1(v)]

$$E\|\mathbf{Z}_{1,n}\|^{s+k+1} \leqslant 2^{s+k+1} E\|\mathbf{Y}_{1,n}\|^{s+k+1} = o(n^{(k+1)/2}) \qquad (n \to \infty),$$

$$A_n \geqslant \frac{c_8(s,k)n^{(1/2)(s-2)/(s+k-1)}}{\rho_s^{1/(s+k-1)}} \qquad (n \geqslant n_0), \tag{20.23}$$

where c_8 is positive. Use the first estimate of (20.23) in (20.21) to get

$$\int_{\{\|t\| \leqslant A_n\}} \left|\left[D^{\beta-\alpha}\hat{H}_n(t)\right]\left[D^{\alpha}\hat{K}_\varepsilon(t)\right]\right| dt = o(n^{-(s-2)/2}) \qquad (n \to \infty). \tag{20.24}$$

Write

$$c_n = \frac{n^{1/2}}{16\rho_3}. \tag{20.25}$$

By Lemma 14.3,

$$\begin{aligned}
&\int_{\{\|t\| > A_n\}} \left|\left[D^{\beta-\alpha}\hat{H}_n(t)\right]\left[D^{\alpha}\hat{K}_\varepsilon(t)\right]\right| dt \\
&\leqslant \int_{\{\|t\| > c_n\}} \left|\left[D^{\beta-\alpha}\hat{Q}_n'(t)\right]\left[D^{\alpha}\hat{K}_\varepsilon(t)\right]\right| dt \\
&\quad + \int_{\{\|t\| > A_n\}} c_9(s,k)(1 + \|t\|^{|\beta-\alpha|}) \exp\{-\tfrac{5}{24}\|t\|^2\} dt \\
&\quad + \int_{\{\|t\| > A_n\}} \left| D^{\beta-\alpha} \sum_{r=0}^{s+k-2} n^{-r/2}\tilde{P}_r(it : \{\chi_{\nu,n}\}) \exp\{-\tfrac{1}{2}\langle t, D_n t\rangle\}\right| dt \\
&= I_1 + I_2 + I_3,
\end{aligned} \tag{20.26}$$

say. By the second inequality in (20.23),

$$I_2 = o(n^{-(s-2)/2}) \qquad (n \to \infty). \tag{20.27}$$

The same is true of I_3 because of the presence of the exponential term. It remains to estimate I_1. This is where we use Cramér's condition (20.1).

Observe that

$$|D^{\beta-\alpha}\hat{Q}_n'(t)| \leqslant n^{|\beta-\alpha|}E\left\|\frac{\mathbf{Z}_{1,n}}{n^{1/2}}\right\|^{|\beta-\alpha|}\cdot|g_n(t)|^{n-|\beta-\alpha|}, \tag{20.28}$$

where

$$\begin{aligned} g_n(t) &= E\left(\exp\{i\langle n^{-1/2}t, \mathbf{Z}_{1,n}\rangle\}\right), \\ |g_n(t)| &= \left|E\left(\exp\{i\langle n^{-1/2}t, \mathbf{Y}_{1,n}\rangle\}\right)\right| \\ &\leqslant \left|E\left(\exp\{i\langle n^{-1/2}t, \mathbf{X}_1\rangle\}\right)\right| + 2P\left(\|\mathbf{X}_1\| > n^{1/2}\right), \end{aligned} \tag{20.29}$$

so that, by Cramér's condition [see (20.3)]

$$\sup_{\|t\|>c_n} |g_n(t)| < \theta < 1 \tag{20.30}$$

for all sufficiently large n. Here θ is a number independent of n. Hence by (20.28), (20.30), and (20.18), we get

$$\begin{aligned} I_1 &\leqslant c_{10}\varepsilon^{|\alpha|}n^{|\beta-\alpha|/2}\theta^{n-|\beta-\alpha|}\int_{\{\|t\|>n^{1/2}/16\rho_3\}} \exp\{-(\varepsilon\|t\|)^{1/2}\}\,dt \\ &\leqslant c_{10}n^{|\beta-\alpha|/2}\theta^{n-|\beta-\alpha|}\varepsilon^{|\alpha|-k}\int \exp\{-\|t\|^{1/2}\}\,dt \\ &\leqslant c_{11}n^{(s+k+1)/2}\theta^{n}\varepsilon^{-k} \end{aligned} \tag{20.31}$$

for all large n. Now choose

$$\varepsilon = e^{-dn}, \tag{20.32}$$

where d is any positive number satisfying

$$d < -\frac{1}{k}\log\theta, \tag{20.33}$$

so that

$$I_1 = o(n^{-(s-2)/2}) \qquad (n \to \infty). \tag{20.34}$$

Therefore we have shown

$$\begin{aligned} \left|\int f_{a_n}\,dH_n\right| \leqslant \bar{\omega}_{f_{a_n}}&\left(2e^{-dn} : \left|\sum_{r=0}^{s+k-2} n^{-r/2}P_r(-\Phi_{0,D_n} : \{\chi_{\nu,n}\})\right|\right) \\ &+ M_{s'}(f)o(n^{-(s-2)/2}) \qquad (n\to\infty). \end{aligned} \tag{20.35}$$

Now

$$\bar{\omega}_{f_{a_n}}\left(2e^{-dn}: \left|\sum_{r=0}^{s+k-2} n^{-r/2} P_r(-\Phi_{0,D_n}: \{\chi_{\nu,n}\})\right|\right)$$
$$\leqslant \sum_{r=0}^{s+k-2} \bar{\omega}_{f_{a_n}}\left(2e^{-dn}: n^{-r/2}|P_r(-\Phi_{0,D_n}: \{\chi_{\nu,n}\})|\right). \qquad (20.36)$$

For $0 \leqslant r \leqslant s-2$, using (14.74), (9.12) in the first step,

$$\bar{\omega}_{f_{a_n}}\left(2e^{-dn}: n^{-r/2}|P_r(-\Phi_{0,D_n}: \{\chi_{\nu,n}\})|\right)$$
$$\leqslant n^{-r/2}\int \omega_f(x: 2e^{-dn}) c_{12} E\|\mathbf{Z}_{1,n}\|^{r+2}(1+\|x\|^{3r})\cdot\phi_{a_n,D_n}(x)\,dx$$
$$\leqslant c_{13}\rho_s\left[\int_{\{\|x\|\leqslant n^{1/6}\}} \omega_f(x: 2e^{-dn})|\phi_{a_n,D_n}(x)-\phi(x)|\,dx\right.$$
$$\left.+\int_{\{\|x\|\leqslant n^{1/6}\}} \omega_f(x: 2e^{-dn})\phi(x)\,dx\right]$$
$$+c_{14}n^{-r/2}\rho_s\int_{\{\|x\|>n^{1/6}\}} \omega_f(x: 2e^{-dn})(1+\|x\|^{3r})\phi_{a_n,D_n}(x)\,dx$$
$$\leqslant c_{13}\rho_s\left[c_{15}M_{s'}(f)\int(1+\|x\|^{s'})|\phi_{a_n,D_n}(x)-\phi(x)|\,dx+\bar{\omega}_f(2e^{-dn}: \Phi)\right]$$
$$+c_{16}n^{-r/2}\rho_s\int_{\{\|x\|>n^{1/6}\}}(1+\|x\|^{3r+s'})\cdot\phi_{a_n,D_n}(x)\,dx$$
$$\leqslant M_{s'}(f)\cdot o(n^{-(s-2)/2})+c_{13}\rho_s\bar{\omega}_f(2e^{-dn}: \Phi). \qquad (20.37)$$

These inequalities are based on Lemma 14.6 and

$$\omega_f(x:\varepsilon)\leqslant 2M_{s'}(f)\left(1+(\|x\|+\varepsilon)^{s'}\right). \qquad (20.38)$$

For $s-1\leqslant r\leqslant s+k-2$,

$$\bar{\omega}_{f_{a_n}}\left(2e^{-dn}: n^{-r/2}|P_r(-\Phi_{0,D_n}: \{\chi_{\nu,n}\})|\right)$$
$$\leqslant c_{17}(r,s,k)n^{-r/2}E\|\mathbf{Z}_{1,n}\|^{r+2}M_{s'}(f)\int(1+\|x\|^{3r+s'})\cdot\phi_{a_n,D_n}(x)\,dx$$
$$= M_{s'}(f)\cdot o(^{n-(s-2)/2}) \qquad (n\to\infty), \qquad (20.39)$$

since, by Lemma 14.1(v) [or relation (19.81)],

$$n^{-r/2}E\|\mathbf{Z}_{1,n}\|^{r+2}=o(n^{-(s-2)/2}) \qquad (r\geqslant s-1). \tag{20.40}$$

Thus we have finally shown that

$$\left|\int f_{a_n}\,dH_n\right|\leqslant o(n^{-(s-2)/2})\cdot M_{s'}(f)+c_{18}(r,s,k)\bar{\omega}_f(2e^{-dn}\colon\Phi) \qquad (n\to\infty). \tag{20.41}$$

Since, as in (20.39),

$$\left|\int f_{a_n}\,d\left(\sum_{r=s-1}^{s+k-2} n^{-r/2}P_r(-\Phi_{0,D_n}\colon\{\chi_{\nu,n}\})\right)\right|$$

$$=M_{s'}(f)\cdot o(n^{-(s-2)/2}) \qquad (n\to\infty), \tag{20.42}$$

one has

$$\left|\int f_{a_n}\,d\left(Q_n'-\sum_{r=0}^{s-2} n^{-r/2}P_r(-\Phi_{0,D_n}\colon\{\chi_{\nu,n}\})\right)\right|$$

$$\leqslant M_{s'}(f)\cdot o(n^{-(s-2)/2})+c_{18}(r,s,k)\bar{\omega}_f(2e^{-dn}\colon\Phi). \tag{20.43}$$

The proof is now complete in view of (20.43), (20.13), (20.12), (20.9), and (20.8). Q.E.D.

The following corollaries are immediate.

COROLLARY 20.2 *Suppose that* f *is a bounded, Borel-measurable function. One has, under the hypothesis of Theorem* 20.1,

$$\left|\int f\,d\left(Q_n-\sum_{r=0}^{s-2} n^{-r/2}P_r(-\Phi_{0,V}\colon\{\chi_\nu\})\right)\right|=\omega_f(R^k)\delta_1(n)$$

$$+c(s,k)\bar{\omega}_f(2e^{-dn}\colon\Phi_{0,V}). \tag{20.44}$$

Let $\mathcal{F}$ be a class of real-valued, Borel-measurable functions on R^k satisfying

$$\sup_{f\in\mathcal{F}} M_s(f)<\infty,$$

$$\sup_{f\in\mathcal{F}} \bar{\omega}_f(\varepsilon\colon\Phi_{0,V})=o\left((-\log\varepsilon)^{-(s-2)/2}\right) \qquad (\varepsilon\downarrow 0). \tag{20.45}$$

COROLLARY 20.3 *Under the hypothesis of Theorem* 20.1,

$$\sup_{f\in\mathfrak{F}}\left|\int f\,d\left(Q_n-\sum_{r=0}^{s-2}n^{-r/2}P_r(-\Phi_{0,V}:\{\chi_\nu\})\right)\right|$$
$$=o(n^{-(s-2)/2})\qquad (n\to\infty). \tag{20.46}$$

Thus Theorem 20.1 provides asymptotic expansions (with an error term that is of smaller order of magnitude than the last term in the expansion) for an extremely large class of functions. In particular, for the class

$$\mathcal{Q}_\alpha(c:\Phi_{0,V})\equiv\left\{A\in\mathfrak{B}^k:\Phi_{0,V}\left((\partial A)^\varepsilon\right)\leqslant c\varepsilon^\alpha \text{ for all } \varepsilon>0\right\}\qquad (0<\alpha\leqslant 1), \tag{20.47}$$

one has

COROLLARY 20.4 *Under the hypothesis of Theorem* 20.1,

$$\sup_{A\in\mathcal{Q}_\alpha(c:\Phi_{0,V})}\left|Q_n(A)-\sum_{r=0}^{s-2}n^{-r/2}P_r(-\Phi_{0,V}:\{\chi_\nu\})(A)\right|=o(n^{-(s-2)/2})$$
$$(n\to\infty). \tag{20.48}$$

A simple consequence of (20.48) is

$$\sup_{C\in\mathcal{C}}\left|Q_n(C)-\sum_{r=0}^{s-2}n^{-r/2}P_r(-\Phi_{0,V}:\{\chi_\nu\})(C)\right|=o(n^{-(s-2)/2})\qquad (n\to\infty), \tag{20.49}$$

where $\mathcal{C}$ is the class of all Borel-measurable convex subsets of R^k. One may sharpen (20.49) somewhat as

COROLLARY 20.5 *Under the hypothesis of Theorem* 20.1,

$$\sup_{C\in\mathcal{C}}(1+d^s(0,\partial C))\left|Q_n(C)-\sum_{r=0}^{s-2}n^{-r/2}P_r(-\Phi_{0,V}:\{\chi_\nu\})(C)\right|$$
$$=o(n^{-(s-2)/2}), \tag{20.50}$$

where $\mathrm{d}(0,\partial \mathrm{C})$ *is the distance between the origin and the boundary* $\partial\mathrm{C}$ *of* C.

Proof. Take

$$f(x) = (1 + d^s(0, \partial C)) I_{C'}(x) \qquad (x \in R^k), \tag{20.51}$$

where $C' = R^k \backslash C$ or C according as $0 \in C$ or $0 \notin C$. One only needs to check

$$(1 + d^s(0, \partial C)) \Phi_{0,V}\big((\partial C)^{2\varepsilon}\big) \leqslant \int_{(\partial C)^{2\varepsilon}} \left[1 + (\|x\| + 2\varepsilon)^s\right] \Phi_{0,V}(dx)$$

$$= 0(\varepsilon) \qquad (\varepsilon \downarrow 0) \tag{20.52}$$

uniformly for $C \in \mathcal{C}$. The inequality in (20.52) is easy to check, and Corollary 3.2 does the rest. For $k = 1$, let F_n denote the distribution function of Q_n, and use the same symbol for the signed measure $P_r(-\Phi_{0,V}: \{\chi_\nu\})$ and its distribution function to get

$$\sup_{x \in R^1} (1 + |x|^s) \left| F_n(x) - \sum_{r=0}^{s-2} n^{-r/2} P_r(-\Phi_{0,V}: \{\chi_\nu\})(x) \right|$$

$$= o(n^{-(s-2)/2}) \qquad (n \to \infty). \tag{20.53}$$

It is possible to extend Theorem 20.1 to the non-i.i.d. case as follows.

THEOREM 20.6 *Let* $\{\mathbf{X}_n : n \geqslant 1\}$ *be a sequence of independent random vectors with values in* R^k, *having zero means, positive-definite covariance matrices, and finite* s*th absolute moments for some integer* $s \geqslant 3$. *Assume that (i) the smallest eigenvalues* λ_n *of* $V_n = n^{-1}\sum_{j=1}^n \mathrm{Cov}(\mathbf{X}_j)$ *are bounded away from zero, (ii) the average* s*th absolute moments* $n^{-1}\sum_{j=1}^n E\|\mathbf{X}_j\|^s$ *are bounded away from infinity and*

$$\lim_{n \to \infty} n^{-1} \sum_{j=1}^{n} \int_{\{\|\mathbf{X}_j\| > \varepsilon n^{1/2}\}} \|\mathbf{X}_j\|^s = 0 \tag{20.54}$$

for every positive ε, *and (iii) the characteristic functions* g_n *of* $\mathbf{X}_n$ *satisfy*

$$\overline{\lim_{n \to \infty}} \sup_{\|t\| > b} |g_n(t)| < 1 \tag{20.55}$$

for every positive b. *Then for every real-valued, Borel-measurable function* f *on* R^k *satisfying* (20.4) *for some* s', $0 \leqslant s' \leqslant s$, *one has*

$$\left| \int f\, d\left(Q_n - \sum_{r=0}^{s-2} n^{-r/2} P_r(-\Phi: \{\bar{\chi}_{\nu,n}\}) \right) \right|$$

$$\leqslant M_{s'}(f)\delta_1(n) + c(s,k)\bar{\omega}_f(2e^{-dn}: \Phi), \tag{20.56}$$

where Q_n *is the distribution of* $n^{-1/2}B_n(\mathbf{X}_1 + \cdots + \mathbf{X}_n)$, *with* $B_n^2 = V_n^{-1}$. *Also* $\delta_1(n)$ *and* d *are as in Theorem 20.1, and* $\bar{\chi}_{\nu,n} =$ *average* ν*th cumulant of* $B_n\mathbf{X}_j$ $(1 \leqslant j \leqslant n)$.

The proof of Theorem 20.6 is entirely analogous to that of Theorem 20.1 and is therefore omitted.

As indicated in the introduction to the present chapter, there are special functions f, for example, trigonometric polynomials, for which the expansion of $\int f\,dQ_n$ is valid whatever may be the type of distribution of $\mathbf{X}_1$. This follows from Theorems 9.10–9.12. Our next theorem provides a class of functions of this type. For the sake of simplicity we state it for the i.i.d. case.

THEOREM 20.7 *Let $\{\mathbf{X}_n: n \geqslant 1\}$ be an i.i.d. sequence of random vectors with values in R^k. Assume that the common distribution has mean zero, positive-definite covariance matrix* I *and a finite* s*th absolute moment ρ_s for some integer $s \geqslant 3$. Let f be a (real or complex-valued) Borel-measurable function on R^k that is the Fourier–Stieltjes transform of a finite signed measure μ satisfying*

$$\int \|x\|^{s-2} |\mu|(dx) < \infty. \tag{20.57}$$

Then

$$\left| \int f\,d\left(Q_n - \sum_{r=0}^{s-2} n^{-r/2} P_r(-\Phi: \{\chi_\nu\}) \right) \right| = o(n^{-(s-2)/2}) \qquad (n \to \infty). \tag{20.58}$$

Here Q_n is the distribution of $n^{-1/2}(\mathbf{X}_1 + \cdots + \mathbf{X}_n)$, $\chi_\nu = \nu$th cumulant of $\mathbf{X}_1$.

Proof. By Parseval's relation [Theorem 5.1(viii)] and Theorem 9.12,

$$\begin{aligned}
&\left| \int f\,d\left(Q_n - \sum_{r=0}^{s-2} n^{-r/2} P_r(-\Phi: \{\chi_\nu\}) \right) \right| \\
&\quad = \left| \int \left[\hat{Q}_n(t) - \sum_{r=0}^{s-2} n^{-r/2} \tilde{P}_r(it: \{\chi_\nu\}) \exp\{-\tfrac{1}{2}\|t\|^2\} \right] \mu(dt) \right| \\
&\quad \leqslant \int_{\{\|t\| \leqslant c_{20}(s,k) n^{1/2} \rho_s^{-1/(s-2)}\}} \delta(n) n^{-(s-2)/2} \left[\|t\|^2 + \|t\|^{3(s-2)} \right] \\
&\qquad \times \exp\left\{ -\frac{\|t\|^2}{4} \right\} |\mu|(dt) \\
&\qquad + \int_{\{\|t\| > c_{20}(s,k) n^{1/2} \rho_s^{-1/(s-2)}\}} \left[1 + \sum_{r=0}^{s-2} n^{-r/2} |\tilde{P}_r(it: \{\chi_\nu\})| \right. \\
&\qquad \left. \times \exp\{-\tfrac{1}{2}\|t\|^2\} \right] |\mu|(dt) \\
&\quad \leqslant c'\delta(n) n^{-(s-2)/2} \|\mu\| \\
&\qquad + c n^{-(s-2)/2} \int_{\{\|t\| > c_{20}(s,k) n^{1/2} \rho_s^{-1/(s-2)}\}} \|t\|^{s-2} |\mu|(dt)
\end{aligned} \tag{20.59}$$

where c depends only on the distribution of $\mathbf{X}_1$, and

$$c' = \sup_{t \in R^k} \left[\left(\|t\|^s + \|t\|^{3(s-2)} \right) \exp\left\{ \frac{-\|t\|^2}{4} \right\} \right].$$

Note that we have used a Chebyshev-type inequality

$$\int_{\{\|t\| > c_{20}(s,k) n^{1/2} \rho_s^{-1/(s-2)}\}} |\mu|(dt) \leqslant \left(c_{20}(s,k) n^{1/2} \rho_s^{-1/(s-2)} \right)^{-(s-2)}$$

$$\times \int_{\{\|t\| > c_{20}(s,k) n^{1/2} \rho_s^{-1/(s-2)}\}} \|t\|^{s-2} |\mu|(dt).$$

Q.E.D.

We point out that if μ is discrete, assigning its entire mass to a finite number of points of R^k, then the above theorem applies, and thus f may be taken to be an arbitrary trigonometric polynomial. However the result applies to a much larger class of functions f, including the class of all Schwartz functions (see Section A.2 for a definition of this class).

Finally, for strongly nonlattice distributions we have the following result.

THEOREM 20.8 *Let* $\{\mathbf{X}_n : n \geqslant 1\}$ *be a sequence of i.i.d. strongly nonlattice random vectors with values in* R^k. *If* $E\mathbf{X}_1 = 0$, $\mathrm{Cov}(\mathbf{X}_1) = \mathrm{I}$, *and* $\rho_3 \equiv E\|\mathbf{X}_1\|^3 < \infty$, *then for every real-valued, bounded, Borel-measurable function* f *on* R^k *one has*

$$\left| \int f d\left(Q_n - \Phi - n^{-1/2} P_1(-\Phi : \{\chi_\nu\}) \right) \right| = \omega_f(R^k) \cdot o(n^{-1/2}) + 0(\omega_f^*(\delta_n : \Phi))$$

$$(n \to \infty), \qquad (20.60)$$

where Q_n *is the distribution of* $n^{-1/2}(\mathbf{X}_1 + \cdots + \mathbf{X}_n)$, $\chi_\nu = \nu$*th cumulant of* $\mathbf{X}_1$, *and* $\delta_n = o(n^{-1/2})$; δ_n *does not depend on* f.

Proof. Given $\eta > 0$, we show that there exists $n(\eta)$ such that for all $n \geqslant n(\eta)$ the left side of (20.60) is less than

$$\omega_f(R^k) \cdot o(n^{-1/2}) + c(k) \omega_f^*(\eta n^{-1/2} : \Phi). \qquad (20.61)$$

Introduce truncated random vectors $\mathbf{Y}_{j,n}(1 \leqslant j \leqslant n)$ as in the proof of Theorem 20.6 and recall that Q'_n and Q''_n are the distributions of the normalized sums of these random vectors with and without centering, respectively. Then by (20.8), (20.11)–(20.13) (with $s=2$, $s'=0$)

$$\left|\int f\,d(Q_n - Q_n'')\right| = \omega_f(R^k)\cdot o(n^{-1/2}) \qquad (n\to\infty),$$

$$\int f\,dQ_n'' = \int f_{a_n}\,dQ_n', \tag{20.62}$$

$$\left|\int (f_{a_n} - f)\,d(\Phi + n^{-1/2}P_1(-\Phi:\{\chi_\nu\}))\right| = \omega_f(R^k)\cdot o(n^{-1/2}) \qquad (n\to\infty),$$

$$\left|\int f_{a_n}\,d(\Phi + n^{-1/2}P_1(-\Phi:\{\chi_\nu\}) - \Phi_{0,D_n} - n^{-1/2}P_1(-\Phi_{0,D_n}:\{\chi_{3,n}\}))\right|$$
$$= \omega_f(R^k)\cdot o(n^{-1/2}) \qquad (n\to\infty).$$

Also, by (20.42),

$$\left|\int f_{a_n}\,d\left(\sum_{r=2}^{k+1} n^{-r/2}P_r(-\Phi_{0,D_n}:\{\chi_{\nu,n}\})\right)\right|$$
$$= \omega_f(R^k)\cdot o(n^{-1/2}) \qquad (n\to\infty). \tag{20.63}$$

Write

$$H_n = Q_n' - \sum_{r=0}^{k+1} n^{-r/2}P_r(-\Phi_{0,D_n}:\{\chi_{\nu,n}\}). \tag{20.64}$$

In view of (20.62), (20.63), it is enough to estimate $\int f_{a_n}\,dH_n$. By Corollary 11.5,

$$\left|\int f_{a_n}\,dH_n\right| \leqslant (2\alpha-1)^{-1}\Bigg[\tfrac{1}{2}\omega_f(R^k)\|H_n * K_\varepsilon\|$$
$$+\,\omega^*_{f_{a_n}}\left(2\varepsilon: \left|\sum_{r=0}^{k+1} n^{-r/2}P_r(-\Phi_{0,D_n}:\{\chi_{\nu,n}\})\right|\right)\Bigg], \tag{20.65}$$

where the kernel probability measure K satisfies

$$\alpha \equiv K(x:\|x\|<1\}) > \tfrac{1}{2}, \qquad \int \|x\|^{k+4}K(dx) < \infty,$$
$$\hat{K}(t) = 0 \qquad \text{if } \|t\| > c_1(k). \tag{20.66}$$

Such a choice is possible by Theorem 10.1. Also,

$$K_\varepsilon(B)=K(\varepsilon^{-1}B) \qquad (B\in\mathfrak{B}^k),$$
$$\varepsilon=\tfrac{1}{2}\eta n^{-1/2}. \tag{20.67}$$

By Lemma 11.6,

$$\|H_n * K_\varepsilon\| \leqslant c_2(k)\max_{|\beta|\leqslant k+1}\int_{\{\|t\|\leqslant 2c_1(k)\eta^{-1}n^{1/2}\}}|D^\beta(\hat{H}_n\hat{K}_\varepsilon)(t)|\,dt. \tag{20.68}$$

Letting $c_n=n^{1/2}/(16\rho_3)$, one has, by (20.24), (20.26), and (20.27) with $s=3$,

$$\int_{\{\|t\|\leqslant c_n\}}|D^\beta(\hat{H}_n\hat{K}_\varepsilon)(t)|\,dt=o(n^{-1/2}) \qquad (n\to\infty). \tag{20.69}$$

Also,

$$|D^\beta(\hat{H}_n\hat{K}_\varepsilon)(t)|\leqslant c_3(k,\rho_3)\sum_{0\leqslant\alpha\leqslant\beta}|D^{\beta-\alpha}\hat{H}_n(t)|$$

and, since apart from $\hat{Q}_n'$ the remaining terms of $\hat{H}_n$ (and their derivatives) contain an exponential factor, it is enough to estimate

$$I\equiv\int_{\{n^{1/2}/(16\rho_3)<\|t\|<2c_1(k)\eta^{-1}n^{1/2}\}}|D^{\beta-\alpha}\hat{Q}_n'(t)|\,dt. \tag{20.70}$$

By (20.28),

$$|D^{\beta-\alpha}\hat{Q}_n'(t)|\leqslant n^{|\beta-\alpha|}|g_n(t)|^{n-|\beta-\alpha|}, \tag{20.71}$$

where [see (20.29) and (14.111)]

$$g_n(t)=E\left(\exp\{i\langle n^{-1/2}t,\mathbf{Y}_{1,n}-E\mathbf{Y}_{1,n}\rangle\}\right),$$
$$|g_n(t)|\leqslant\left|E\left(\exp\{i\langle n^{-1/2}t,\mathbf{X}_1\rangle\}\right)\right|+o(n^{-1/2}) \qquad (n\to\infty). \tag{20.72}$$

Since $\mathbf{X}_1$ is strongly nonlattice,

$$\sup\left\{|g_n(t)|:\ \frac{n^{1/2}}{16\rho_3}<\|t\|\leqslant 2c_1(k)\eta^{-1}n^{1/2}\right\}<\theta(\eta)<1 \tag{20.73}$$

for all $n \geqslant n(\eta)$, say. Using estimates (20.71), (20.73) in (20.70), we have

$$I = o(n^{-1/2}) \qquad (n \to \infty), \tag{20.74}$$

so that

$$\|H_n * K_\varepsilon\| = o(n^{-1/2}) \qquad (n \to \infty). \tag{20.75}$$

Finally, as in (20.36)–(20.41),

$$\omega^*_{f_{a_n}}\left(2\varepsilon : \left|\sum_{r=0}^{s-2} n^{-r/2} P_r(-\Phi_{0,D_n} : \{\chi_{\nu,n}\})\right|\right)$$
$$= \omega_f^*(2\varepsilon : \Phi) + \omega_f(R^k)\cdot o(n^{-1/2}) \qquad (n \to \infty). \tag{20.76}$$

Q.E.D.

Theorems 20.7, 20.8 hold with I replaced by an arbitrary symmetric, positive-definite matrix V. We chose I merely to simplify notation.

As an immediate consequence of (20.60) one has, in case $\mathbf{X}_1$ is a strongly nonlattice distribution with mean zero and covariance I,

$$\lim_{n \to \infty} \sup_{x \in R^k} n^{1/2} |F_n(x) - \Phi(x)| = \sup_{x \in R^k} |P_1(-\Phi : \{\chi_\nu\})(x)|, \tag{20.77}$$

where F_n is the distribution function of Q_n. For $k = 1$ (20.77) yields

$$\lim_{n \to \infty} \sup_{x \in R^1} n^{1/2} |F_n(x) - \Phi_{0,\sigma^2}(x)| = \frac{|\mu_3|}{6(2\pi)^{1/2}\sigma^3}, \tag{20.78}$$

provided that $\mathbf{X}_1$ is nonlattice and has zero mean, variance $\sigma^2 > 0$, and third moment μ_3. It may be noted that for $k = 1$ "strongly nonlattice" is the same as "nonlattice." One may also easily write down analogs of (20.77) for more general classes of sets (than rectangles), for example, the class of all Borel-measurable convex subsets of R^k, or the class $\mathcal{Q}_1^*(d : \Phi)$ introduced in (17.3).

NOTES

Although Chebyshev [1] and Edgeworth [1] had conceived of the formal expansions of this chapter, it was not until Cramér's important work [1, 3] (Chapter VII) that a proper foundation was laid and the first important results derived.

Section 19. For an early local limit theorem for densities in the non-i.i.d. case and its significant applications to statistical mechanics the reader is referred to Khinchin [3]. Theorem 19.1 is essentially proved in Gnedenko and Kolmogorov [1], pp. 224–227; in this book (pp. 228–230) one also finds the following result of Gnedenko: *in one dimension under the hypothesis of Theorem* 19.2 *one has*

$$\sup_{x\in R^1}\left|q_n(x)-\sum_{j=0}^{s-2} n^{-j/2}P_j(-\phi:\{\chi_\nu\})(x)\right|=o(n^{-(s-2)/2}). \tag{N.1}$$

For $k=1$ and $s\geqslant 3$ the relation (19.17) in Theorem 19.2 was proved by Petrov [1] assuming boundedness of q_n for some n; however this assumption has been shown to be equivalent to ours in Theorem 19.1. Theorems 19.2, 19.3, and Corollary 19.4 appear here for the first time in their present forms. The assumptions (19.47), (19.49) may be considered too restrictive for the non-i.i.d. case; however it is not difficult to weaken them and get somewhat weaker results; we have avoided this on the ground that the conditions would look messier. Theorem 19.5 is due to Bhattacharya [8]; it strengthens Corollary 19.6, which was proved earlier by Bikjalis [4] for $s\geqslant 3$.

Section 20. Cramér [1,3] (Chapter VII) proved that

$$\sup_{x\in R^1}\left|F_n(x)-\sum_{j=0}^{s-3} n^{-j/2}P_j(-\Phi_{0,\sigma^2}:\{\chi_\nu\})(x)\right|=O(n^{-(s-2)/2}) \tag{N.2}$$

in one dimension under the hypothesis of Theorem 20.1; here F_n is the distribution function of $n^{-1/2}(\mathbf{X}_1+\cdots+\mathbf{X}_n)$ and $\operatorname{var}(\mathbf{X}_1)=\sigma^2$. This was sharpened by Esseen [1], who obtained a remainder $o(n^{-(s-2)/2})$ by adding one more term to the expansion; Esseen's result is equivalent to (20.49) when specialized to $k=1$. R. Rao [1,2] was the first to obtain multidimensional expansions under Cramér's condition (20.1) and prove that in the i.i.d. case one can expand probabilities of Borel-measurable convex sets with an error term $O(n^{-(s-2)/2}(\log n)^{(k-1)/2})$ uniformly over the class $\mathcal{C}$, provided that the hypothesis of Theorem 20.1 holds. This was extended to more general classes of sets independently by von Bahr [3] and Bhattacharya [1,2]. Esseen's result on the expansion of the distribution function (mentioned above) was extended to R^k independently by Bikjalis [4] and von Bahr [3]. Corollaries 20.4, 20.5 as well as the relation (20.49), which refine earlier results of Bhattacharya [1,2] and von Bahr [3], were obtained in Bhattacharya [4,5]. The very general Theorem 20.1 is new; this extends Corollaries 20.2, 20.3 proved earlier by Bhattacharya [5]. Theorems 20.6, 20.7 are due to Bhattacharya [4,5]. There is a result in Osipov [1] that yields $O(n^{-(s-2)/2})$ in place of $o(n^{-(s-2)/2})$ as the right side of (20.53). Some analogs of Theorem 20.8 have been obtained independently by Bikjalis [6]. Earlier Esseen [1] had proved (20.60) for the distribution function of Q_n (i.e., for the class of functions $\{f=I_{(-\infty,x]}: x\in R^1\}$) in one dimension and derived (20.78).

CHAPTER 5

Asymptotic Expansions–Lattice Distributions

The Cramér–Edgeworth expansions of Chapter 4 are not valid for purely discrete distributions. For example, if $\{\mathbf{X}_n\colon n \geqslant 1\}$ is a sequence of i.i.d. lattice random variables ($k=1$), then the distribution Q_n of the nth normalized partial sum is easily shown to have point masses each of order $n^{-1/2}$ (if variance of $\mathbf{X}_1$ is finite and nonzero). Thus the distribution function of Q_n cannot possibly be expanded in terms of the absolutely continuous distribution functions of $P_r(-\Phi)$, $0 \leqslant r \leqslant s-2$, with a remainder term $o(n^{-(s-2)/2})$, when $\mathbf{X}_1$ has a finite sth moment for some integer s not smaller than 3. However the situation may be salvaged in the following manner. The multiple Fourier series $\hat{Q}_n$ is easily inverted to yield the point masses of Q_n. Making use of the approximation of $\hat{Q}_n$ by $\exp\{-\frac{1}{2}\langle t, Vt\rangle\}\Sigma_{r=0}^{s-2} n^{-r/2}\tilde{P}_r(it)$ as provided by Chapter 2, Section 9, one obtains an asymptotic expansion of the point masses of Q_n in terms of $\Sigma_{r=0}^{s-2} n^{-r/2} P_r(-\phi)$. To obtain an expansion of $Q_n(B)$ for a Borel set B, one has to add up the asymptotic expansions of the point masses in B. For $B=(-\infty, x]$, $x \in R^k$, this sum may be expressed in a simple closed form. A multidimensional extension of the classical Euler–Maclaurin summation formula is used for this purpose.

21. LATTICE DISTRIBUTIONS

Consider R^k as a group under vector addition. A subgroup L of R^k is said to be a *discrete subgroup* if there is a ball $B(0:d)$, $d>0$, around the origin such that $L \cap B(0:d) = \{0\}$. Equivalently, a subgroup L is discrete if every ball in R^k has only a finite number of points of L in it. In particular, a discrete subgroup is a closed subset of R^k. The following theorem gives the structure of discrete subgroups.

THEOREM 21.1 *Let* L *be a discrete subgroup of* R^{k} *and let* r *be the number of elements contained in a maximal set of linearly independent vectors in* L. *Then there exist* r *linearly independent vectors* $\xi_1,\ldots,\xi_{\mathrm{r}}$ *in* L *such that*

$$L=\mathbf{Z}\cdot\xi_1+\cdots+\mathbf{Z}\cdot\xi_r\equiv\{m_1\xi_1+\cdots+m_r\xi_r\colon m_1,\ldots,m_r \text{ integers}\}$$

$$(\mathbf{Z}=\{0,\pm1,\pm2,\ldots\}). \qquad (21.1)$$

Proof. First consider the case $k=1$. If L is a discrete subgroup, $L\neq\{0\}$, then $t_0=\min\{t\colon t\in L, t>0\}$ is positive. If $t\in L$ is arbitrary, let n be an integer such that $nt_0\leqslant t<(n+1)t_0$. Then $0\leqslant t-nt_0<t_0$ and $t-nt_0\in L$, so that the minimality of t_0 implies that $t=nt_0$ or $L=\mathbf{Z}\cdot t_0$.

Now consider the case $k>1$. The theorem will be proved if we can construct linearly independent vectors $\xi_1,\ldots,\xi_r$ in L such that $L\cap(R\cdot\xi_1+\cdots+R\cdot\xi_r)=\mathbf{Z}\cdot\xi_1+\cdots+\mathbf{Z}\cdot\xi_r$, since it follows from the definition of the integer r that $L\subset R\cdot\xi_1+\cdots+R\cdot\xi_r$. Here R is the field of reals. We construct these vectors inductively. Assume that linearly independent vectors $\xi_1,\ldots,\xi_s$ have been found for some s, $s<r$, such that (i) $\xi_j\in L$, $j=1,2,\ldots,s$ and (ii) $L\cap(R\cdot\xi_1+\cdots+R\cdot\xi_s)=\mathbf{Z}\cdot\xi_1+\cdots+\mathbf{Z}\cdot\xi_s$. Now we describe a method for choosing ξ_{s+1}. Let $\alpha\in L\backslash(R\cdot\xi_1+\cdots+R\cdot\xi_s)$. Then α is linearly independent of $\xi_1,\ldots,\xi_s$. Let $M=\{t_0\alpha\colon t_0\alpha+t_1\xi_1+\cdots+t_s\xi_s\in L$ for some choice of real numbers $t_1,\ldots,t_s\}$. Since $\xi_1,\ldots,\xi_s\in L$, it follows that $M=\{t_0\alpha\colon t_0\alpha+a_1\xi_1+\cdots+a_s\xi_s\in L$ for some choice of a_j with $0\leqslant a_j<1$, $1\leqslant j\leqslant s\}$. It is clear that the number of points in $M\cap B(0:d)$ is less than or equal to the number of points in $L\cap B(0:d')$ where $d'=d+\|\xi_1\|+\cdots+\|\xi_s\|$. Thus $M\cap B(0:d)$ is finite for each d, and M is thus a discrete subgroup of $R\cdot\alpha$. Thus there exists $a_0>0$ such that $M=\mathbf{Z}\cdot a_0\alpha$. Choose constants a_j, $1\leqslant j\leqslant s$, with $0\leqslant a_j<1$ such that $\xi_{s+1}\equiv a_0\alpha+a_1\xi_1+\cdots+a_s\xi_s\in L$. If $\eta=t_0\alpha+t_1\xi_1+\cdots+t_s\xi_s\in L$, then $t_0=na_0$ for some integer n, so that $\eta-n\xi_{s+1}\in L\cap(R\cdot\xi_1+\cdots+R\cdot\xi_s)=\mathbf{Z}\cdot\xi_1+\cdots+\mathbf{Z}\cdot\xi_s$. Thus $\xi_1,\ldots,\xi_{s+1}$ are linearly independent, and $L\cap(R\cdot\xi_1+\cdots+R\cdot\xi_{s+1})=\mathbf{Z}\cdot\xi_1+\cdots+\mathbf{Z}\cdot\xi_{s+1}$. Since the argument works for $s=0$, the proof is complete. Q.E.D.

The set of vectors $\{\xi_1,\ldots,\xi_r\}$ in the representation (1.1) is called a *basis* of L, and the number r is called the *rank* of L. A discrete subgroup of R^k having rank k is called a *lattice*.

The next theorem describes the structure of closed subgroups of R^k.

THEOREM 21.2 *Let* L *be a closed subgroup of* R^k. *Let* r *be the maximum of dimensions of linear subspaces of* L, *and let* r+s *be the number of elements contained in a maximal set of linearly independent vectors in* L. *Then there exist linearly independent vectors* $\eta_1,\dots,\eta_r$, $\xi_1,\dots,\xi_s$ *in* L *such that*

$$L=R\cdot\eta_1+\cdots+R\cdot\eta_r+\mathbf{Z}\cdot\xi_1+\cdots+\mathbf{Z}\cdot\xi_s. \tag{21.2}$$

Proof. Let $r(d)$ denote the number of elements in a maximal set of linearly independent vectors in $L\cap B(0:d)$. As $d\downarrow 0$, $r(d)\downarrow$. Let $r_0=\lim_{d\to 0}r(d)$. Clearly, $r_0\geq r$. We shall show that $r_0=r$. Since $r(d)$ is integer-valued, there exists $d_0>0$ such that $r(d)=r_0$ for $0<d\leq d_0$. Let $\eta_1,\dots,\eta_{r_0}$ be linearly independent vectors in $L\cap B(0:d_0)$. Then

$$L\cap B(0:d_0)\subset R\cdot\eta_1+\cdots+R\cdot\eta_{r_0}. \tag{21.3}$$

We claim

$$R\cdot\eta_1+\cdots+R\cdot\eta_{r_0}\subset L. \tag{21.4}$$

To prove this let $\varepsilon>0$ be arbitrary and let $d_1=\min\{\varepsilon/k,d_0\}$. Then $r(d_1)=d_0$, and there exist linearly independent vectors $\beta_1,\dots,\beta_{r_0}$ in $L\cap B(0:d_1)$. It follows that $\beta_j\in R\cdot\eta_1+\cdots+R\cdot\eta_{r_0}$ and, therefore, $R\cdot\beta_1+\cdots+R\cdot\beta_{r_0}\subset R\cdot\eta_1+\cdots+R\cdot\eta_{r_0}$, which implies

$$R\cdot\beta_1+\cdots+R\cdot\beta_{r_0}=R\cdot\eta_1+\cdots+R\cdot\eta_{r_0}. \tag{21.5}$$

Now let $\xi\in R\cdot\eta_1+\cdots+R\cdot\eta_{r_0}$ be arbitrary. Then $\xi=t_1\beta_1+\cdots+t_{r_0}\beta_{r_0}$. Write $t_j=m_j+t_j$ with m_j integral and $|t_j|\leq 1$, $1\leq j\leq r_0$. Thus there exists $\beta\in L$, $\beta=m_1\beta_1+\cdots+m_{r_0}\beta_{r_0}$, such that $\|\xi-\beta\|\leq kd_1\leq\varepsilon$. Since ε is arbitrary, this implies that $\xi\in$ closure of $L=L$, that is, $R\cdot\eta_1+\cdots+R\cdot\eta_{r_0}\subset L$, that is, $r=r_0$. Let $L'=R\cdot\eta_1+\cdots+R\cdot\eta_r$. Then $L'\subset L$. Extend $\eta_1,\dots,\eta_r$ to a basis $\eta_1,\dots,\eta_k$ of R^k. Let $L''=L\cap(R\cdot\eta_{r+1}+\cdots+R\cdot\eta_k)$. Then $L=L'+L''$. We claim that L'' is a discrete subgroup of R^k, since $L\cap B(0:d_0)\subset L'$, which implies that $L''\cap B(0:d_0)=\{0\}$. The present theorem follows on applying Theorem 21.1 to L''. Q.E.D.

COROLLARY 21.3 *The following two statements are equivalent.*

(i) L *is a lattice.*
(ii) *The quotient group* R^k/L *is compact.*

Proof. Let $\{\xi_1,\dots,\xi_k\}$ be a basis of L. Then $R^k=U+L$, where

$$U=\{t_1\xi_1+\cdots+t_k\xi_k:0\leq t_j\leq 1,\ 1\leq j\leq k\}. \tag{21.6}$$

Now $R^k/L=\{\dot{x}\equiv x+L\colon\ x\in R^k\}$ with group operation $\dot{x}+\dot{y}=(x+y)^{\cdot}$, and (quotient) topology defined as the strongest topology on R^k/L that makes the map $x\to\dot{x}$ (on R^k into R^k/L) continuous. The restriction of this map on U (into R^k/L) is continuous and has R^k/L as its image. Since U is compact, its continuous image R^k/L is compact. Conversely, suppose that R^k/L is compact. If possible, let $r\equiv\text{rank } L<k$. Then there exists $u\in R^k$, $u\neq 0$, such that $\langle u,\xi\rangle=0$ for all $\xi\in L$, so that the function $\dot{x}\to\langle u,x\rangle$ is well defined and continuous on R^k/L. Since R^k/L is compact, this function is bounded, implying that $x\to\langle u,x\rangle$ is bounded on R^k. This is impossible since $u\neq 0$. Q.E.D.

A random vector $\mathbf{X}$ on a probability space $(\Omega, \mathcal{B}, P)$ into R^k is a *lattice random vector* if there exist $x_0\in R^k$ and a lattice L such that

$$P(\mathbf{X}\in x_0+L)=1. \tag{21.7}$$

The distribution of $\mathbf{X}$ is then said to be a *lattice distribution*. A random vector $\mathbf{X}$ (or its distribution) is said to be *degenerate* if there exists a hyperplane H (i.e., a set of the form $\{x\colon\ \langle a,x\rangle=c\}$ for some nonzero vector a and some real number c) such that

$$P(\mathbf{X}\in H)=1. \tag{21.8}$$

LEMMA 21.4 *Let* $\mathbf{X}$ *be a lattice random vector. Then there exists a unique discrete subgroup* L_0 *with the following two properties*: (*i*) $P(\mathbf{X}\in x+L_0)=1$ *for every* x *such that* $P(\mathbf{X}=x)>0$; (*ii*) *if* M *is any closed subgroup such that* $P(\mathbf{X}\in y_0+M)=1$ *for some* $y_0\in R^k$, *then* $L_0\subset M$. *This discrete subgroup* L_0 *is generated by* $\{\xi\colon\ P(\mathbf{X}=x_0+\xi)>0\}$, *where* x_0 *is any given vector satisfying* $P(\mathbf{X}=x_0)>0$.

Proof. Since $\mathbf{X}$ is a lattice random vector, there exist $x_0\in R^k$ and a lattice L such that $P(\mathbf{X}\in x_0+L)=1$. We may, and shall, also assume that $P(\mathbf{X}=x_0)>0$. Let L_0 be the subgroup generated by the set (contained in L) $\{\xi\colon\ P(\mathbf{X}=x_0+\xi)>0\}$. Then $L_0\subset L$, so L_0 is a discrete subgroup and $P(\mathbf{X}\in x_0+L_0)=1$. If $P(\mathbf{X}=x_1)>0$, then $x_1\in x_0+L_0$, so that $x_1+L_0=x_0+L_0$ and $P(\mathbf{X}\in x_1+L_0)=1$, proving property (i). To prove (ii) suppose $P(\mathbf{X}\in y_0+M)=1$. Since $P(\mathbf{X}=x_0)>0$, it follows that $x_0\in y_0+M$ and so $x_0+M=y_0+M$, so that $P(\mathbf{X}\in x_0+M)=1$. The definition of L_0 now implies that $L_0\subset M$. Thus L_0 has all the properties stated in the lemma. The uniqueness is clear from property (ii). Q.E.D.

For a lattice random vector $\mathbf{X}$, the unique discrete subgroup L_0 with properties (i) and (ii) of Lemma 21.4 is called the *minimal group associated with* **X**. The rank r of L_0 is called the *rank of* **X**. When L_0 is of rank k, L_0 is called the *minimal lattice for* **X**.

LEMMA 21.5 *Let* **X** *be a lattice random vector. Then the following two statements are equivalent.*

(i) **X** *is nondegenerate.*
(ii) *The minimal subgroup associated with* **X** *is a lattice.*

Proof. Fix a vector $x_0 \in R^k$ such that $P(\mathbf{X}=x_0)>0$. Suppose that **X** is nondegenerate. If the minimal subgroup L_0 associated with **X** is not a lattice, then there exist a basis $\{\xi_1,\ldots,\xi_k\}$ of R^k and $r<k$ such that $L_0=\mathbf{Z}\cdot\xi_1+\cdots+\mathbf{Z}\cdot\xi_r$. In particular, this implies $P(\mathbf{X}\in H)=1$, where $H=x_0+R\cdot\xi_1+\cdots+R\cdot\xi_{k-1}$, contradicting (i). To show that (ii) implies (i), suppose that **X** is degenerate. Then there exists a linear subspace of dimension $(k-1)$, for example, M, such that $P(\mathbf{X}\in x_0+M)=1$. Minimality of the subgroup L_0 associated with **X** implies $L_0 \subset M$, showing that L_0 is not a lattice. Q.E.D.

Let f be the characteristic function of a random vector **X**. A vector $t_0 \in R^k$ is said to be a *period* of $|f|$ if $|f(t+t_0)|=|f(t)|$ for all $t\in R^k$.

LEMMA 21.6 *Let* **X** *be a lattice random vector and* L *the minimal subgroup associated with* **X**. *Let* f *be the characteristic function of* **X** *and* L* *the set of periods of* |f|. *Then*

$$\begin{aligned} &\text{(i) } L^*=\{t\colon |f(t)|=1\} \\ &\text{(ii) } L^*=\{t\colon \langle t,\xi\rangle \in 2\pi\mathbf{Z} \text{ for all } \xi\in L\} \\ &\text{(iii) } L=\{\xi\colon \langle t,\xi\rangle\in 2\pi\mathbf{Z} \text{ for all } t\in L^*\}. \end{aligned} \tag{21.9}$$

In particular, L* *is a closed subgroup of* R^k.

Proof. Let $t_0\in L^*$. Then $|f(t_0)|=|f(0)|=1$. Conversely, suppose that $|f(t_0)|=1$. Then there exists a real number α such that $f(t_0)=\exp\{i\alpha\}$. Therefore

$$E(1-\cos(\langle t_0,\mathbf{X}\rangle-\alpha))=0, \tag{21.10}$$

or $P(\cos(\langle t_0,\mathbf{X}\rangle-\alpha)=1)=1$. Equivalently,

$$P(\langle t_0,\mathbf{X}\rangle\in\alpha+2\pi\mathbf{Z})=1, \tag{21.11}$$

so that

$$f(t+t_0)=E(\exp\{i\langle t+t_0,\mathbf{X}\rangle\})=e^{i\alpha}f(t), \tag{21.12}$$

and $|f(t+t_0)|=|f(t)|$ for all t, proving (i). To prove (ii), choose x_0 such that $P(\mathbf{X}=x_0)>0$ and let $t_0 \in L^*$. Then (21.11) implies that $\langle t_0, x_0\rangle \in \alpha + 2\pi\mathbf{Z}$, so that

$$P(\langle t_0, \mathbf{X}-x_0\rangle \in 2\pi\mathbf{Z})=1. \tag{21.13}$$

If $S=\{\xi: P(\mathbf{X}=x_0+\xi)>0\}$, then (21.13) is equivalent to

$$\langle t_0, \xi\rangle \in 2\pi\mathbf{Z} \qquad \text{for all } \xi\in S. \tag{21.14}$$

Since S generates L, (21.13) is equivalent to

$$\langle t_0, \xi\rangle \in 2\pi\mathbf{Z} \qquad \text{for all } \xi\in L. \tag{21.15}$$

Thus t_0 belongs to the right side of (ii). Conversely, if (21.15) holds for some t_0, (21.13) holds and

$$|f(t+t_0)|=|E(\exp\{i\langle t+t_0,\mathbf{X}\rangle\})|=|E(\exp\{i\langle t+t_0,\mathbf{X}\rangle - i\langle t_0, \mathbf{X}-x_0\rangle\})|=|E(\exp\{i\langle t,\mathbf{X}\rangle+i\langle t_0,x_0\rangle\})|=|f(t)| \qquad (t\in R^k). \tag{21.16}$$

Thus $t_0\in L^*$, and (ii) is proved. It remains to prove (iii). By Theorem 21.1, there exists a basis $\{\xi_1,\ldots,\xi_k\}$ of R^k such that $L=\mathbf{Z}\cdot\xi_1+\cdots+\mathbf{Z}\cdot\xi_r$, where r is the rank of L. Let $\{\eta_1,\ldots,\eta_k\}$ be the dual basis, that is, $\langle\xi_j,\eta_{j'}\rangle=\delta_{jj'}$, $1\leqslant j,j'\leqslant k$, where $\delta_{jj'}$ is Kronecker's delta. Then (ii) implies

$$L^*=2\pi(\mathbf{Z}\cdot\eta_1+\cdots+\mathbf{Z}\cdot\eta_r)+R\cdot\eta_{r+1}+\cdots+R\cdot\eta_k. \tag{21.17}$$

The relation (iii) follows immediately from (21.17). The last assertion is an immediate consequence of (i) and (ii) [or (21.17)]. Q.E.D.

COROLLARY 21.7 *Let* $\mathbf{X}$ *be a lattice random vector with characteristic function* f. *The set* L* *of periods of* |f| *is a lattice if and only if* $\mathbf{X}$ *is nondegenerate.*

Proof. This follows immediately from representation (21.17) and Lemma 21.5. Q.E.D.

Let L be a lattice and $\{\xi_1,\ldots,\xi_k\}$ a basis of L; that is,

$$L=\mathbf{Z}\cdot\xi_1+\cdots+\mathbf{Z}\cdot\xi_k. \tag{21.18}$$

Let $\{\eta_1,\ldots,\eta_k\}$ be a dual basis, that is, $\langle\xi_j,\eta_{j'}\rangle=\delta_{jj'}$, and let L^* be the lattice defined as

$$L^*=2\pi(\mathbf{Z}\cdot\eta_1+\cdots+\mathbf{Z}\cdot\eta_k). \tag{21.19}$$

Write $\operatorname{Det}(\xi_1,\ldots,\xi_k)$ for the determinant of the matrix whose jth row is $\xi_j=(\xi_{j1},\ldots,\xi_{jk})$, so that $\operatorname{Det}(\xi_1,\ldots,\xi_k)=\operatorname{Det}(\xi_{jj'})$. If $\{\xi'_1,\ldots,\xi'_k\}$ is another basis of L, then there exist integral matrices $A=(a_{jj'})$ and $B=(b_{jj'})$ such that $\xi_j=\sum a_{j'j}\xi'_j$ and $\xi_{j'}=\sum b_{jj'}\xi_j$. Then $\operatorname{Det} A=\pm 1$, so that

$$\operatorname{Det}(\xi_1,\ldots,\xi_k)=\pm\operatorname{Det}(\xi'_1,\ldots,\xi'_k).$$

Thus the quantity defined by

$$\det L=|\operatorname{Det}(\xi_1,\ldots,\xi_k)| \tag{21.20}$$

is independent of the basis and depends only on the lattice L. Also, with this definition,

$$\det L^*=(2\pi)^k|\operatorname{Det}(\eta_1,\ldots,\eta_k)|=\frac{(2\pi)^k}{\det L}. \tag{21.21}$$

Consider a domain $\mathscr{F}^*$ defined as follows:

$$\mathscr{F}^*=\{t_1\eta_1+\cdots+t_k\eta_k:\ |t_j|<\pi \text{ for all } j\}. \tag{21.22}$$

Then $\mathscr{F}^*$ has the following properties:

(i) $\mathscr{F}^*\cap L^*=\{0\}$;
(ii) For any two $\eta,\ \eta'\in L^*$, $\eta\neq\eta'$ the sets $\mathscr{F}^*+\eta$ and $\mathscr{F}^*+\eta'$ are disjoint;
(iii) For any $x\in R^k$ there exists $\eta\in L^*$ such that $x-\eta\in\operatorname{Cl}\mathscr{F}^*$; that is,

$$R^k=\cup_{\eta\in L^*}\operatorname{Cl}(\mathscr{F}^*+\eta). \tag{21.23}$$

In view of these properties, $\mathscr{F}^*$ is also called *a fundamental domain for* L*. Note that $\mathscr{F}^*=\{x:\ |\langle\xi_j,x\rangle|<\pi \text{ for all } j\}$, so that

$$\begin{aligned}\operatorname{vol}\mathscr{F}^* &= \int_{\mathscr{F}^*}dx=\int_{C(\pi)}\frac{\partial(x_1,\ldots,x_k)}{\partial(y_1,\ldots,y_k)}\,dy\\ &= \frac{(2\pi)^k}{|\operatorname{Det}(\xi_1,\ldots,\xi_k)|},\end{aligned} \tag{21.24}$$

where $y_j=\langle\xi_j,x\rangle$, and $\dfrac{\partial(y_1,\ldots,y_k)}{\partial(x_1,\ldots,x_k)}$, the Jacobian of y with respect to x, is $|\operatorname{Det}(\xi_1,\ldots,\xi_k)|$, and $C(\pi)$ is the cube equal to $\{(y_1,\ldots,y_k):\ |y_j|<\pi$ for all

$j\}$. From (21.21) we get

$$\operatorname{vol} \mathcal{F}^* = \det L^* = \frac{(2\pi)^k}{\det L}. \tag{21.25}$$

An evaluation similar to (21.24) gives

$$\int_{\mathcal{F}^*} e^{i\langle \xi, x\rangle} dx = 0 \qquad \text{for all } \xi \in L,\ \xi \neq 0. \tag{21.26}$$

Now let $\mathbf{X}$ be a nondegenerate, lattice random vector having characteristic function f. Let L be a lattice and $x_0 \in R^k$ be such that $P(\mathbf{X} \in x_0 + L) = 1$. Then the characteristic function f is given by

$$f(t) = \sum_{\xi \in L} P(\mathbf{X} = x_0 + \xi) e^{i\langle t, x_0 + \xi\rangle}. \tag{21.27}$$

Multiplying (21.27) by $\exp\{-i\langle t, x_0 + \xi\rangle\}$ and integrating over $\mathcal{F}^*$ we get, from (21.25) and (21.26), for all $\xi \in L$,

$$P(\mathbf{X} = x_0 + \xi) = \frac{\det L}{(2\pi)^k} \int_{\mathcal{F}^*} e^{-i\langle t, x_0 + \xi\rangle} f(t)\, dt. \tag{21.28}$$

The above formula is called the *inversion formula in the lattice case.* If $E\|\mathbf{X}\|^s$ is finite for some positive integer s, then on differentiating both sides of (21.27), multiplying the result by $\exp\{-i\langle t, x_0 + \xi\rangle\}$, and finally integrating over $\mathcal{F}^*$, one obtains, for $|\nu| \leqslant s$,

$$(x_0 + \xi)^\nu P(\mathbf{X} = x_0 + \xi) = (2\pi)^{-k} \det L (-i)^{|\nu|} \int_{\mathcal{F}^*} \exp\{-i\langle t, x_0 + \xi\rangle\} D^\nu f(t)\, dt$$
$$(\xi \in L). \tag{21.29}$$

22. LOCAL EXPANSIONS

Throughout this section $\{\mathbf{X}_j : j \geqslant 1\}$ is a sequence of independent and identically distributed, lattice random vectors defined on a probability space $(\Omega, \mathcal{B}, P)$ with values in R^k; the common distribution is assumed to be nondegenerate having a *minimal lattice L*; also,

$$E\mathbf{X}_1 = \mu, \qquad \operatorname{Cov}(\mathbf{X}_1) = I, \qquad P(\mathbf{X}_1 \in L) = 1. \tag{22.1}$$

Note that for every nondegenerate, lattice random vector $\mathbf{X}$ having finite second moments, there exists $x_0 \in R^k$ and a nonsingular matrix T such that $\mathbf{X}_1 \equiv T(\mathbf{X} - x_0)$ satisfies the above hypothesis.

For each n define sequences of truncated random vectors $\{\mathbf{Y}_{j,n}: j \geqslant 1\}$, $\{\mathbf{Z}_{j,n}: j \geqslant 1\}$ by

$$\mathbf{Y}_{j,n} = \begin{cases} \mathbf{X}_j & \text{if} \quad \|\mathbf{X}_j - \mu\| \leqslant n^{1/2} \\ 0 & \text{if} \quad \|\mathbf{X}_j - \mu\| > n^{1/2}, \end{cases}$$

$$\mathbf{Z}_{j,n} = \mathbf{Y}_{j,n} - E\mathbf{Y}_{j,n} \qquad (j \geqslant 1, \quad n \geqslant 1). \tag{22.2}$$

We shall assume that $\rho_s \equiv E\|\mathbf{X}_1 - \mu\|^s$ is finite for some integer $s \geqslant 2$, and write

$$D_n = \operatorname{Cov}(\mathbf{Z}_{1,n}), \qquad \chi_\nu = \nu\text{th cumulant of } \mathbf{X}_1, \qquad l = \det L,$$

$$\chi_{\nu,n} = \nu\text{th cumulant of } \mathbf{Z}_{1,n} \qquad (|\nu| \leqslant s),$$

$$y_{\alpha,n} = n^{-1/2}(\alpha - n\mu), \qquad y'_{\alpha,n} = n^{-1/2}(\alpha - nE\mathbf{Y}_{1,n}) \qquad (\alpha \in L),$$

$$p_n(y_{\alpha,n}) = P(\mathbf{X}_1 + \cdots + \mathbf{X}_n = \alpha) = P\left(n^{-1/2} \sum_{j=1}^{n} (\mathbf{X}_j - \mu) = y_{\alpha,n}\right),$$

$$p'_n(y'_{\alpha,n}) = P(\mathbf{Y}_{1,n} + \cdots + \mathbf{Y}_{n,n} = \alpha) = P\left(n^{-1/2} \sum_{j=1}^{n} \mathbf{Z}_{j,n} = y'_{\alpha,n}\right),$$

$$q_{n,m} = l n^{-k/2} \sum_{r=0}^{m-2} n^{-r/2} P_r(-\phi: \{\chi_\nu\}) \qquad (2 \leqslant m \leqslant s),$$

$$q'_{n,m} = l n^{-k/2} \sum_{r=0}^{m-2} n^{-r/2} P_r(-\phi_{0,D_n}: \{\chi_{\nu,n}\}) \qquad (2 \leqslant m < \infty). \tag{22.3}$$

The following theorem constitutes the main result of this section.

THEOREM 22.1 *If* $\rho_s \equiv \mathrm{E}\|\mathbf{X}_1 - \mu\|^s$ *is finite for some integer* $s \geqslant 2$, *then*

$$\sup_{\alpha \in L} (1 + \|y_{\alpha,n}\|^s) |p_n(y_{\alpha,n}) - q_{n,s}(y_{\alpha,n})| = o(n^{-(k+s-2)/2}) \qquad (n \to \infty). \tag{22.4}$$

Also,

$$\sum_{\alpha \in L} |p_n(y_{\alpha,n}) - q_{n,s}(y_{\alpha,n})| = o(n^{-(s-2)/2}) \qquad (n \to \infty). \tag{22.5}$$

Proof. Let g_1 denote the characteristic function of $\mathbf{X}_1$ and f_n that of $n^{-1/2}(\mathbf{X}_1+\cdots+\mathbf{X}_n-n\mu)$. Then

$$f_n(t)=\left(g_1(n^{-1/2}t)\exp\{-i\langle n^{-1/2}t,\mu\rangle\}\right)^n \qquad (t\in R^k). \tag{22.6}$$

The inversion formulas (21.28), (21.29) yield

$$\begin{aligned} p_n(y_{\alpha,n}) &= l(2\pi)^{-k}\int_{\mathfrak{F}^*} g_1^n(t)\exp\{-i\langle t,\alpha\rangle\}\,dt \\ &= l(2\pi)^{-k}n^{-k/2}\int_{n^{1/2}\mathfrak{F}^*} f_n(t)\exp\{-i\langle t,y_{\alpha,n}\rangle\}\,dt, \end{aligned} \tag{22.7}$$

$$y^\beta_{\alpha,n}p_n(y_{\alpha,n}) = l(2\pi)^{-k}n^{-k/2}(-i)^{|\beta|}\int_{n^{1/2}\mathfrak{F}^*}\exp\{-i\langle t,y_{\alpha,n}\rangle\}D^\beta f_n(t)\,dt$$

where β is a nonnegative integral vector satisfying $|\beta|\leqslant s$, and

$$n^{1/2}A=\{n^{1/2}x\colon x\in A\} \qquad (A\subset R^k). \tag{22.8}$$

Also, clearly,

$$\begin{aligned} y^\beta_{\alpha,n}q_{n,s}(y_{\alpha,n}) &= l(2\pi)^{-k}n^{-k/2}(-i)^{|\beta|}\int_{R^k}\exp\{-i\langle t,y_{\alpha,n}\rangle\} \\ &\quad\times D^\beta\left[\sum_{r=0}^{s-2}n^{-r/2}\tilde{P}_r(it\colon\{\chi_\nu\})\exp\left\{\frac{-\|t\|^2}{2}\right\}\right]dt \qquad (|\beta|\leqslant s). \end{aligned} \tag{22.9}$$

Hence

$$|y^\beta_{\alpha,n}(p_n(y_{\alpha,n})-q_{n,s}(y_{\alpha,n}))| \leqslant l(2\pi)^{-k}n^{-k/2}(I_1+I_2+I_3), \tag{22.10}$$

where, writing (the constant c_{20} being the same as in Theorem 9.10)

$$E_1=\{t\colon \|t\|\leqslant c_{20}(s,k)\rho_s^{-1/(s-2)}\}, \tag{22.11}$$

one has, using Theorem 9.12,

$$\begin{aligned} I_1 &\equiv \int_{n^{1/2}E_1}\delta(n)n^{-(s-2)/2}(\|t\|^{s-|\beta|}+\|t\|^{3(s-2)+|\beta|})\exp\left\{\frac{-\|t\|^2}{4}\right\}dt \\ &= o(n^{-(s-2)/2}) \qquad (n\to\infty), \\ I_2 &\equiv \int_{n^{1/2}\mathfrak{F}^*\backslash n^{1/2}E_1}|D^\beta f_n(t)|\,dt, \\ I_3 &\equiv \int_{R^k\backslash n^{1/2}E_1}\left|D^\beta\left[\sum_{r=0}^{s-2}n^{-r/2}\tilde{P}_r(it\colon\{\chi_\nu\})\exp\left\{\frac{-\|t\|^2}{2}\right\}\right]\right|dt \\ &= o(n^{-(s-2)/2}) \qquad (n\to\infty). \end{aligned} \tag{22.12}$$

To estimate I_2, note that

$$\delta \equiv \sup\{|g_1(t)|: t \in \mathfrak{F}^* \backslash E_1\} < 1, \qquad (22.13)$$

since the closure of $\mathfrak{F}^* \backslash E_1$ contains no period of $|g_1|$. In view of (22.13) and (22.6), we get (using Leibniz' formula for differentiating a product)

$$I_2 \leqslant \rho_{|\beta|} n^{|\beta|/2} \delta^{n-|\beta|} \qquad (|\beta| \leqslant s). \qquad (22.14)$$

Thus

$$\sup_{\alpha \in L} \left| y_{\alpha,n}^{\beta} (p_n(y_{\alpha,n}) - q_{n,s}(y_{\alpha,n})) \right| = o(n^{-(k+s-2)/2}) \qquad (n \to \infty), \qquad (22.15)$$

from which (22.4) easily follows.

If $s \geqslant k+1$, then (22.5) follows from (22.4). For the general case, we shall use truncation. * N.B.

First note that [see (14.111)]

$$\sum_{\alpha \in L} |p_n(y_{\alpha,n}) - p_n'(y_{\alpha,n}')| \leqslant 2 \sum_{j=1}^{n} P(X_j \neq Y_{j,n})$$

$$= 2nP(\|\mathbf{X}_1 - \mu\| \geqslant n^{1/2})$$

$$= o(n^{-(s-2)/2}) \qquad (n \to \infty). \qquad (22.16)$$

Next, by Lemma 14.6,

$$|q_{n,s}(y_{\alpha,n}) - q_{n,s}'(y_{\alpha,n}')| \leqslant n^{-k/2} \delta_1(n) \sum_{r=0}^{s-2} (1 + \|y_{\alpha,n}\|^{3r+2})$$

$$\times \exp\left\{ \frac{-\|y_{\alpha,n}\|^2}{6} + \frac{\|y_{\alpha,n}\|}{8k^{1/2}} \right\} \qquad (\alpha \in L), \qquad (22.17)$$

where $\delta_1(n) = o(n^{-(s-2)/2})$. Now

$$n^{-k/2} \sum_{\alpha \in L} (1 + \|y_{\alpha,n}\|^{3r+2}) \exp\left\{ \frac{-\|y_{\alpha,n}\|^2}{6} + \frac{\|y_{\alpha,n}\|}{8k^{1/2}} \right\}$$

$$= n^{-k/2} \sum_{\alpha \in L - n\mu} (1 + \|n^{-1/2}\alpha\|^{3r+2}) \exp\left\{ \frac{-\|n^{-1/2}\alpha\|^2}{6} + \frac{\|n^{-1/2}\alpha\|}{8k^{1/2}} \right\}$$

$$\leqslant n^{-k/2} \sup_{\|x\| \leqslant \Delta n^{-1/2}} \sum_{\alpha \in L} h(n^{-1/2}\alpha + x) \qquad (\Delta = \text{diameter of } \mathfrak{F}^*). \qquad (22.18)$$

where

$$h(y)=(1+\|y\|^{3r+2})\exp\left\{\frac{-\|y\|^2}{6}+\frac{\|y\|}{8k^{1/2}}\right\} \qquad (y\in R^k). \quad (22.19)$$

Let T be a nonsingular linear transformation on R^k such that $TL=\mathbf{Z}^k$. Then

$$n^{-k/2}\sup_{\|x\|\leqslant \Delta n^{-1/2}}\sum_{\alpha\in L} h(n^{-1/2}\alpha+x)$$

$$=n^{-k/2}\sup_{\|x\|\leqslant \Delta n^{-1/2}}\sum_{\alpha'\in \mathbf{Z}^k} h(n^{-1/2}T^{-1}\alpha'+x)\to\int_{R^k}h(T^{-1}y)\,dy<\infty$$

$$(n\to\infty). \quad (22.20)$$

Thus (22.17) yields

$$\sum_{\alpha\in L}|q_{n,s}(y_{\alpha,n})-q'_{n,s}(y'_{\alpha,n})|=o(n^{-(s-2/2)}) \qquad (n\to\infty). \quad (22.21)$$

Let f'_n denote the characteristic function of $n^{-1/2}\Sigma_{j=1}^n\mathbf{Z}_{j,n}$. Then writing

$$s_0=\max\{s,k\}+1, \quad (22.22)$$

one has, for all $\alpha\in L$ [since $P(\mathbf{Y}_{1,n}\in L)=1$],

$$(y'_{\alpha,n,j})^{s'}p'_n(y'_{\alpha,n})=l(2\pi)^{-k}n^{-k/2}(-i)^{s'}$$

$$\times\int_{n^{1/2}\mathcal{F}^*}\exp\{-i\langle t,y'_{\alpha,n}\rangle\}\left(\frac{\partial}{\partial t_j}\right)^{s'}f'_n(t)\,dt,$$

$$(y'_{\alpha,n,j})^{s'}q'_{n,s_0-1}(y'_{\alpha,n})=l(2\pi)^{-k}n^{-k/2}(-i)^{s'}\int_{R^k}\exp\{-i\langle t,y'_{\alpha,n}\rangle\}$$

$$\times\left(\frac{\partial}{\partial t_j}\right)^{s'}\left(\sum_{r=0}^{s_0-3}n^{-r/2}\tilde{P}_r(it:\{\chi_{\nu,n}\})\exp\{-\tfrac{1}{2}\langle t,D_nt\rangle\}\right)dt$$

$$(s'=0,1,\ldots;\ y_{\alpha,n,j}=j\text{th coordinate of }y_{\alpha,n}). \quad (22.23)$$

For all sufficiently large n, let $B_n=D_n^{-1/2}$ and write

$$E_{1,n}=\left\{t:\|t\|\leqslant\frac{c_{20}(s,k)n^{1/2}}{(E\|B_n\mathbf{Z}_{1,n}\|^{s_0})^{1/(s_0-2)}\Lambda_n^{1/2}}\right\}, \quad (22.24)$$

where Λ_n is the largest eigenvalue of D_n. By Lemmas 14.1, 14.2,

$$E_{1,n} \supset E_{2,n} \equiv \left\{ t:\ \|t\| \leqslant \frac{c'_{20}(s,k) n^{(s-2)/2(s_0-2)}}{\rho_s^{1/(s_0-2)}} \right\} \tag{22.25}$$

for some positive constant $c'_{20}(s,k)$. By Theorem 9.10, for $s'=0,1,\dots,s_0$, one has

$$\left| \int_{E_{2,n}} \exp\{-i\langle t, y'_{\alpha,n}\rangle\} \left(\frac{\partial}{\partial t_j}\right)^{s'} \left[f'_n(t) - \sum_{r=0}^{s_0-3} n^{-r/2} \tilde{P}_r(it:\ \{\chi_{\nu,n}\}) \times \exp\{-\tfrac{1}{2}\langle t, D_n t\rangle\} \right] dt \right|$$

$$\leqslant c'_{21}(s_0,k) E\|B_n \mathbf{Z}_{1,n}\|^{s_0} n^{-(s_0-2)/2} = o(n^{-(s-2)/2}), \tag{22.26}$$

using Lemma 14.1(v) for the last step. By Lemma 14.3,

$$\int_{\{\|t\| \leqslant n^{1/2}/(16\rho_3)\} \setminus E_{2,n}} \left| \left(\frac{\partial}{\partial t_j}\right)^{s'} f'_n(t) \right| dt \leqslant \int_{R^k \setminus E_{2,n}} c'(s',s,k)(1+\|t\|^s) \times \exp\{-\tfrac{5}{24}\|t\|^2\} dt = o(n^{-(s-2)/2}) \qquad (s'=0,1,\dots). \tag{22.27}$$

Next, because of the presence of an exponential factor,

$$\int_{R^k \setminus E_{2,n}} \left| \left(\frac{\partial}{\partial t_j}\right)^{s'} \left(\sum_{r=0}^{s_0-3} n^{-r/2} \tilde{P}_r(it:\ \{\chi_{\nu,n}\}) \exp\{-\tfrac{1}{2}\langle t, D_n t\rangle\} \right) \right| dt$$

$$= o(n^{-(s-2)/2}) \qquad (s'=0,1,\dots), \tag{22.28}$$

using Lemma 14.2.

Finally,

$$\int_{n^{1/2}\mathfrak{F}^* \setminus \{\|t\| > n^{1/2}/(16\rho_3)\}} \left| \left(\frac{\partial}{\partial t_j}\right)^{s'} f'_n(t) \right| dt$$

$$\leqslant n^{s'/2} E\|\mathbf{Z}_{1,n}\|^{s'} (\delta' + 2\rho_s n^{-(s-2)/2})^{n-s'}$$

$$= o(n^{-(s-2)/2}) \qquad (s'=0,1,\dots), \tag{22.29}$$

where δ' is defined by

$$\delta' \equiv \sup\{|g_1(t)|:\ t \in \mathfrak{F}^* \setminus \{t':\ \|t'\| < (16\rho_3)^{-1}\}\} < 1. \tag{22.30}$$

Note that in (22.29) we have used the inequality

$$|E(\exp\{i\langle t,\mathbf{Z}_{1,n}\rangle\})|=|E(\exp\{i\langle t,\mathbf{Y}_{1,n}\rangle\})|\leqslant|g_1(t)|+2\rho_s n^{-(s-2)/2}. \tag{22.31}$$

Using (22.26)–(22.29) in (22.23), one obtains

$$\sup_{\alpha\in L}\left(1+\sum_{j=1}^{k}|y'_{\alpha,n,j}|^{k+1}\right)|p'_n(y'_{\alpha,n})-q'_{n,s_0-1}(y'_{\alpha,n})|$$

$$=o(n^{-(k+s-2)/2})\qquad (n\to\infty), \tag{22.32}$$

from which follows (Lemma A.4.6 may be used to do the summation)

$$\sum_{\alpha\in L}|p'_n(y'_{\alpha,n})-q'_{n,s_0-1}(y'_{\alpha,n})|=o(n^{-(s-2)/2})\qquad (n\to\infty). \tag{22.33}$$

Using Lemmas 9.5 and 14.1(v) (as well as Lemma A.4.6),

$$\sum_{\alpha\in L}|q'_{n,s_0-1}(y'_{\alpha,n})-q'_{n,s}(y'_{\alpha,n})|$$

$$=n^{-k/2}\sum_{\alpha\in L}\left|\sum_{r=s-1}^{s_0-3}n^{-r/2}P_r(-\phi_{0,D_n}:\{\chi_{\nu,n}\})(y'_{\alpha,n})\right|$$

$$=o(n^{-(s-2)/2})\qquad (n\to\infty), \tag{22.34}$$

and one has, combining (22.16), (22.21), (22.33), and (22.34), the desired result (22.5). Q.E.D.

The following corollary is immediate from (22.5).

COROLLARY 22.2 *Under the hypothesis of Theorem* 22.1,

$$\sup_{B\in\mathfrak{B}^k}\left|Q_n(B)-\sum_{\alpha\in B\cap L}q_{n,s}(y_{\alpha,n})\right|=o(n^{-(s-2)/2})\qquad (n\to\infty), \tag{22.35}$$

where Q_n *is the distribution of* $n^{-1/2}\sum_{j=1}^{n}(\mathbf{X}_j-\mu)$.

The next section provides a generalized Euler–Maclaurin summation formula for expressing $\sum_{\alpha\in B\cap L}q_{n,s}(y_{\alpha,n})$ in a more convenient integral form.

For notational convenience we later take $L=\mathbf{Z}^k$. Again, this does not involve any essential loss of generality provided that one works with a general nonsingular covariance matrix V, instead of the identity matrix I. For example, if $\mathbf{X}_1$ satisfies (22.1) and has minimal lattice L, and if T is a nonsingular linear transformation such that $TL=\mathbf{Z}^k$, then $\mathbf{Y}_1 \equiv T\mathbf{X}_1$ is a lattice random vector having minimal lattice $\mathbf{Z}^k$ and satisfying

$$\operatorname{Cov}(\mathbf{Y}_1)=TT'=V, \qquad P\left(\mathbf{Y}_1 \in \mathbf{Z}^k\right)=1. \tag{22.36}$$

We arrive at the following corollary:

COROLLARY 22.3 *Let $\{\mathbf{Y}_j\colon j \geqslant 1\}$ be a sequence of i.i.d. lattice random vectors with values in R^k. Assume that $\mathrm{E}\mathbf{Y}_1=\mu$ and that $\mathbf{Y}_1$ satisfies* (22.36) *and has the minimal lattice $\mathbf{Z}^k$. If $\rho_s \equiv \mathrm{E}\|\mathbf{Y}_1-\mu\|^s<\infty$ for some integer $s \geqslant 2$, then*

$$\sup_{\alpha \in \mathbf{Z}^k}\left(1+\|y_{\alpha,n}\|^s\right)\left|p_n(y_{\alpha,n})-q_{n,s}(y_{\alpha,n})\right|=o\left(n^{-(k+s-2)/2}\right),$$
$$\sum_{\alpha \in \mathbf{Z}^k}\left|p_n(y_{\alpha,n})-q_{n,s}(y_{\alpha,n})\right|=o\left(n^{-(s-2)/2}\right) \qquad (n \to \infty), \tag{22.37}$$

where

$$y_{\alpha,n}=n^{-1/2}(\alpha-n\mu), \qquad p_n(y_{\alpha,n})=P\left(\mathbf{Y}_1+\cdots+\mathbf{Y}_n=\alpha\right) \qquad (\alpha \in \mathbf{Z}^k),$$
$$=P\left(n^{-1/2}\sum_{j=1}^{n}(\mathbf{Y}_j-\mu)=y_{\alpha,n}\right),$$

$$q_{n,s}=n^{-k/2}\sum_{r=0}^{s-2} n^{-r/2}P_r\left(-\phi_{0,V}\colon\{\chi_\nu\}\right), \tag{22.38}$$

χ_ν being the νth cumulant of $\mathbf{Y}_1$.

23. ASYMPTOTIC EXPANSIONS OF DISTRIBUTION FUNCTIONS

In this section we apply the generalized Euler–Maclaurin expansion obtained in Appendix A.4 to the local expansion of Section 22.

As noted toward the end of Section 22, we may assume, without any essential loss of generality, that the sequence of *i.i.d. lattice* random vectors $\{\mathbf{X}_n\colon n \geqslant 1\}$ satisfies

$$E\mathbf{X}_1=\mu, \qquad \operatorname{Cov}(\mathbf{X}_1)=V, \qquad P\left(\mathbf{X}_1 \in \mathbf{Z}^k\right)=1, \qquad E\|\mathbf{X}_1-\mu\|^s<\infty,$$
$$\text{minimal lattice for } \mathbf{X}_1=\mathbf{Z}^k. \tag{23.1}$$

Here μ is an arbitrary vector in R^k, V is an arbitrary symmetric, positive-definite matrix, and s is an integer $\geqslant 2$. We also write

$$x_{\alpha,n} = n^{-1/2}(\alpha - n\mu) = -n^{1/2}\mu + n^{-1/2}\alpha \qquad (\alpha \in \mathbf{Z}^k),$$
$$p_n(x_{\alpha,n}) = \mathrm{Prob}(\mathbf{X}_1 + \cdots + \mathbf{X}_n = \alpha)$$
$$= \mathrm{Prob}\big(n^{-1/2}(\mathbf{X}_1 + \cdots + \mathbf{X}_n - n\mu) = x_{\alpha,n}\big),$$
$$q_{n,s} = n^{-k/2} \sum_{r=0}^{s-2} n^{-r/2} P_r(-\phi_{0,V}: \{\chi_\nu\}). \tag{23.2}$$

As before, χ_ν denotes the νth cumulant of $\mathbf{X}_1$.

THEOREM 23.1 *Let a sequence of i.i.d. lattice random vectors* $\{\mathbf{X}_n: n \geqslant 1\}$ *satisfy* (23.1). *Let* F_n *denote the distribution function of* $n^{-1/2} \times (\mathbf{X}_1 + \cdots + \mathbf{X}_n - n\mu)$. *Then*

$$\sup_{x \in R^k} \Big| F_n(x) - \sum_{|\alpha| \leqslant s-2} n^{-|\alpha|/2}(-1)^{|\alpha|} S_\alpha(n\mu + n^{1/2}x)(D^\alpha \Phi_{0,V})(x)$$
$$- n^{-1/2} \sum_{|\alpha| \leqslant s-3} n^{-|\alpha|/2}(-1)^{|\alpha|} S_\alpha(n\mu + n^{1/2}x)(D^\alpha P_1(-\Phi_{0,V}: \{\chi_\nu\}))(x)$$
$$- \cdots - n^{-(s-2)/2} P_{s-2}(-\Phi_{0,V}: \{\chi_\nu\})(x) \Big|$$
$$= o(n^{-(s-2)/2}) \qquad (n \to \infty), \tag{23.3}$$

where S_α *are defined by* $(A.4.14)$ *and* $(A.4.2)$.

Proof. First, by Corollary 22.3,

$$\sum_{\alpha \in \mathbf{Z}^k} |p_n(x_{\alpha,n}) - q_{n,s}(x_{\alpha,n})| = o(n^{-(s-2)/2}) \qquad (n \to \infty). \tag{23.4}$$

Hence it is enough to prove (23.3) with $F_n(x)$ replaced by

$$Q_{n,s}(x) \equiv \sum_{\{\alpha:\, x_{\alpha,n} \leqslant x\}} q_{n,s}(x_{\alpha,n}). \tag{23.5}$$

But by Theorem A.4.3 (taking $r = s-1$),

$$\sup_{x \in R^k} \Big| Q_{n,s}(x) - \sum_{j(\alpha) \leqslant s-2} n^{-|\alpha|/2}(-1)^{|\alpha|} S_\alpha(n\mu + n^{1/2}x)$$
$$\times D^\alpha \Big(\sum_{j=0}^{s-2} n^{-j/2} P_j(-\Phi_{0,V}: \{\chi_\nu\}) \Big)(x) \Big|$$
$$= O(n^{-(s-1)/2}) \qquad (n \to \infty). \tag{23.6}$$

The relation (23.3) is now obtained by omitting from the expansion in (23.6) all terms of order $n^{-j/2}, j \geqslant s-1$. Q.E.D.

As an immediate consequence one obtains

COROLLARY 23.2 *If* s=3 *in the hypothesis of Theorem* 23.1, *then*

$$\sup_{x \in R^k}\left| F_n(x) - \Phi_{0,V}(x) + n^{-1/2} \sum_{j=1}^{k} S_1(n\mu_j + n^{1/2}x_j)(D_j\Phi_{0,V})(x) - n^{-1/2}P_1(-\Phi_{0,V}: \{\chi_\nu\})(x)\right| = o(n^{-1/2}); \quad (23.7)$$

and if $s \geqslant 4$, *then the left side of* (23.7) *is* $O(n^{-1})$.

Recall that for every nonnegative integral vector $\alpha = (\alpha_1,\ldots,\alpha_k)$ the integer $j(\alpha)$ is defined by (A.4.21). To state the next result it would be convenient to write $\Lambda_{r,n}(F)$ for the finite signed measure whose distribution function is given by

$$\Lambda_{r,n}(F)(x) \equiv \sum_{j(\alpha)<r} (-1)^{|\alpha|} n^{-|\alpha|/2} S_\alpha(n\mu + n^{1/2}x)(D^\alpha F)(x). \quad (23.8)$$

THEOREM 23.3 *Under the hypothesis of Theorem* 23.1, *one has*

$$\left\| Q_n - \sum_{j=0}^{s-2} n^{-j/2}\Lambda_{s-j-1,n}(P_j(-\Phi_{0,V}: \{\chi_\nu\}))\right\| = o(n^{-(s-2)/2}), \quad (23.9)$$

where Q_n *is the distribution of* $n^{-1/2}(\mathbf{X}_1 + \cdots + \mathbf{X}_n - n\mu)$.

Proof. This theorem is an immediate consequence of Theorem A.4.3 and relation (23.4) if one omits all terms with variation norms $O(n^{-j/2})$, $j \geqslant s-1$. Q.E.D.

Specializing Corollary 23.2 to $k=1$, one has

$$n^{1/2} \sup_{x \in R^1} |F_n(x) - \Phi_{0,\sigma^2}(x)| = \sup_{x \in R^1}\left| S_1(n\mu_1 + n^{1/2}x)\phi_{0,\sigma^2}(x) - \frac{\mu_3}{6\sqrt{2\pi}\,\sigma^3}\left(1 - \frac{x^2}{\sigma^2}\right)e^{-x^2/2\sigma^2}\right| + o(1), \quad (23.10)$$

where $\mu_1 = E\mathbf{X}_1$, $\mu_3 = E\mathbf{X}_1^3$, and $\sigma^2 = \mathrm{var}(\mathbf{X}_1)$. Since the function $x \to S_1(n\mu_1 + n^{1/2}x)$ is periodic with a period $n^{-1/2}$, for sufficiently large n

the function $x \to S_1(n\mu_1 + n^{1/2}x)\phi_{0,\sigma^2}(x)$ takes values arbitrarily close to both $\frac{1}{2}\phi_{0,\sigma^2}(0)$ and $-\frac{1}{2}\phi_{0,\sigma^2}(0)$ near zero. Also, the supremum

$$\sup_{x \in R^1} \left| \frac{\mu_3}{6\sqrt{2\pi}\,\sigma^3}\left(1 - \frac{x^2}{\sigma^2}\right)e^{-x^2/2\sigma^2} \right| = \frac{|\mu_3|}{6\sqrt{2\pi}\,\sigma^3} \tag{23.11}$$

is attained at zero. It follows that the left side of (23.10) has a limit (as $n \to \infty$) given by

$$\begin{aligned}\lim_{n\to\infty} n^{1/2} \sup_{x \in R^1} |F_n(x) - \Phi_{0,\sigma^2}(x)| &= \frac{1}{2\sqrt{2\pi}\,\sigma} + \frac{|\mu_3|}{6\sqrt{2\pi}\,\sigma^3} \\ &= \frac{3\sigma^2 + |\mu_3|}{6\sqrt{2\pi}\,\sigma^3}.\end{aligned} \tag{23.12}$$

It has been shown by Esseen [2] that (if $\mathbf{X}_1$ has a lattice distribution)

$$3\sigma^2 + |\mu_3| \leqslant (\sqrt{10} + 3)\rho_3, \tag{23.13}$$

where $\rho_3 = E|\mathbf{X}_1|^3$. Hence

$$\lim_{n\to\infty} n^{1/2} \sup_{x \in R^1} |F_n(x) - \Phi_{0,\sigma^2}(x)| \leqslant \frac{\sqrt{10} + 3}{6\sqrt{2\pi}}\left(\frac{\rho_3}{\sigma^3}\right). \tag{23.14}$$

This bound is actually attained by a Bernoulli random variable $\mathbf{X}_1$ with probabilities

$$P\left(\mathbf{X}_1 = -\frac{\sqrt{10} - 3}{2}\right) = \frac{\sqrt{10} - 2}{2}, \qquad P\left(\mathbf{X}_1 = \frac{\sqrt{10} - 3}{2}\right) = \frac{4 - \sqrt{10}}{2}. \tag{23.15}$$

Recall also [see (20.78)] that if $\mathbf{X}_1$ is nonlattice, then

$$\lim_{n\to\infty} n^{1/2} \sup_{x \in R^1} |F_n(x) - \Phi_{0,\sigma^2}(x)| = \frac{|\mu_3|}{6\sqrt{2\pi}\,\sigma^3}. \tag{23.16}$$

Therefore the best asymptotic constant in the Berry–Esseen theorem (Theorem 12.4) is $\frac{1}{6}(\sqrt{10} + 3)/\sqrt{2\pi}$. *

Going through the proof of Theorem 22.1, one can easily obtain some sufficient conditions for the validity of analogs of Theorems 22.1, 23.1, and 23.3 for a sequence of nonidentically distributed, lattice random vectors.

* = .409722.

For example, one may assume that the random vectors have a common minimal lattice, say $\mathbf{Z}^k$, that their average covariance matrices and average sth absolute moments satisfy (19.47), and that their characteristic functions satisfy (22.13) uniformly, to obtain the above theorems in the non-i.i.d. case with o replaced by O in the remainders. These conditions can be somewhat relaxed.

NOTES

Section 21. The structure of lattice random vectors was investigated by Esseen [1]; our treatment is more detailed.

Section 22. The first local limit theorem (for lattice random variables) is that of DeMoivre [1] and Laplace [1] for Bernoulli random variables. A much sharpened form of this appears in Uspensky [1]. Among the earlier local limit theorems for general lattice random variables are those of Esseen [1] and Gnedenko [1]. Esseen [1] also obtained the expansion

$$\sup_{\alpha \in L} |p_n(y_{\alpha,n}) - q_{n,s}(y_{\alpha,n})| = o(n^{-(k+s-2)/2}) \tag{N.1}$$

in one dimension under the hypothesis of Theorem 22.1. R. Rao [1,2] extended this to R^k and proved a version of Theorem 22.1 having a factor $(\log n)^{k/2}$ with the remainder in (22.4). For $k=1$ and $s \geqslant 3$ Theorem 22.1 was proved by Petrov [1], whereas the result of Rao was refined in the multidimensional case by Bikjalis [5]. It follows from a result of Gnedenko (see Gnedenko and Kolmogorov [1], pp. 233–235) that for $s=2$ the hypothesis in Theorem 22.1 is also necessary.

Section 23. Esseen, in his important basic work [1], used the Euler–Maclaurin summation formula to prove Theorem 23.1 in one dimension. Theorem A.4.2, which is a multidimensional extension of the Euler–Maclaurin sum formula, is due to R. Rao [1,2]; using this, Rao was able to prove versions of Theorems 23.1 and 23.3, the remainders having an additional factor $(\log n)^{k/2}$. Bikjalis [5] removed this factor from the remainders and proved versions of Theorems 23.1, 23.3 (having considerably more terms in the expansion than we have). We would like to note here that Theorem 4 in R. Rao [2] contains an error in its first assertion. Comparing it with our Theorem 23.3 with $s=4$, we get (assuming $\rho_4 < \infty$)

$$\|Q_n - \Lambda_{3,n}(\Phi_{0,V}) - n^{-1/2}\Lambda_{2,n}(P_1(-\Phi_{0,V}: \{\chi_\nu\}))\| = O(n^{-1}), \tag{N.2}$$

while the corresponding expansion in Rao's Theorem 4 omits some of the above terms. This omission is permissible for expanding F_n (see, for example, Corollary 23.2) since the functions S_j are bounded; it is not permissible when one is computing the variation norm. The complicated nature of the signed measures corresponding to the functions F_r and Λ_r [see (A.4.19) and (A.4.20)] has compelled us to give a fairly long account of these objects. For example, even the fact that F_r is of bounded variation turns out to be nontrivial to prove (see Theorem A.2.3 in the Appendix). Also, there remains the outstanding problem of finding computable expressions for $\Lambda_{r,n}(P_j(-\Phi_{0,V}: \{\chi_\nu\}))(A)$ for a large enough class of "nice" sets A (not just rectangles properly aligned with the lattice). For some results in this direction see *Yarnold [1].

There is one important theorem of Esseen [1] that we have not discussed in this monograph: *if* $\{\mathbf{X}_n: n \geqslant 1\}$ *is a sequence of i.i.d. random vectors with (common) mean zero, nonsingular covariance matrix* V, *and a finite fourth absolute moment, then*

$$\sup_{A \in \mathcal{E}} |Q_n(A) - \Phi_{0,V}(A)| = O(n^{-k/(k+1)}), \tag{N.3}$$

where $\mathcal{E}$ *is the class of all ellipsoids of the form*

$$A = \{x \in R^k : \langle x, V^{-1}x \rangle \leqslant c\} \qquad (c > 0), \tag{N.4}$$

and Q_n *is the distribution of* $n^{-1/2}(\mathbf{X}_1 + \cdots + \mathbf{X}_n)$. Note that if the common distribution of $\{\mathbf{X}_n : n \geqslant 1\}$ satisfies Cramér's condition (20.1), then Corollary 20.4 yields a smaller error, namely $O(n^{-1})$, in (N.3), since $P_1(-\Phi_{0,V} : \{\chi_\nu\})(A) = 0$ for $A \in \mathcal{E}$. The strength of Esseen's result lies in its validity for discrete distributions. Indeed, he showed that when specialized to lattice random vectors (N.3) is equivalent to a deep result of Landau in analytic number theory: *the number* B(c) *of lattice points, that is, points with integer coordinates, in the ellipsoid* (N.4) *is of the order*

$$B(c) = (\text{volume of } A) + O(c^{k/2 - k/k+1}) \qquad (c \to \infty). \tag{N.5}$$

A significant extension of (N.3) to a class of smooth convex bodies has been recently obtained by Matthes [1]. Since in this case the contribution from $P_1(-\Phi_{0,V} : \{\chi_\nu\})$ is nonzero, one must include this term in (N.3).

Appendix

A.1 RANDOM VECTORS AND INDEPENDENCE

A *measure space* is a triple $(\Omega, \mathcal{B}, \mu)$, where Ω is a nonempty set, $\mathcal{B}$ is a sigma-field of subsets of Ω, and μ is a measure defined on $\mathcal{B}$. A measure space $(\Omega, \mathcal{B}, P)$ is called a *probability space* if the measure P is a probability measure, that is, if $P(\Omega)=1$.

Let $(\Omega, \mathcal{B}, P)$ be a probability space. A *random vector* $\mathbf{X}$ *with values in* R^k is a map on Ω into R^k satisfying

$$\mathbf{X}^{-1}(A) \equiv \{\omega : \mathbf{X}(\omega) \in A\} \in \mathcal{B} \tag{A.1.1}$$

for all $A \in \mathcal{B}^k$, where $\mathcal{B}^k$ is the Borel sigma-field of R^k. When $k=1$, such an $\mathbf{X}$ is also called a *random variable*. If $\mathbf{X}$ is an integrable random variable, the *mean*, or *expectation*, of $\mathbf{X}$, denoted by $E\mathbf{X}$ [or $E(\mathbf{X})$], is defined by

$$E\mathbf{X} \equiv \int_{\Omega} \mathbf{X}\, dP. \tag{A.1.2}$$

If $\mathbf{X}=(\mathbf{X}_1, \ldots, \mathbf{X}_k)$ is a random vector (with values in R^k) each of whose coordinates is integrable, then the *mean*, or *expectation*, $E\mathbf{X}$ of $\mathbf{X}$ is defined by

$$E\mathbf{X} \equiv (E\mathbf{X}_1, \ldots, E\mathbf{X}_k). \tag{A.1.3}$$

If $\mathbf{X}$ is an integrable random variable, then the *variance* of $\mathbf{X}$, denoted by var $\mathbf{X}$ [or var($\mathbf{X}$)], is defined by

$$\operatorname{var} \mathbf{X} \equiv E(\mathbf{X} - E\mathbf{X})^2. \tag{A.1.4}$$

Let $\mathbf{X}$, $\mathbf{Y}$ be two random variables defined on $(\Omega, \mathcal{B}, P)$. If $\mathbf{X}$, $\mathbf{Y}$, and $\mathbf{XY}$

are all integrable, one defines the *covariance between* **X** *and* **Y**, denoted $\text{cov}(\mathbf{X},\mathbf{Y})$, by

$$\text{cov}(\mathbf{X},\mathbf{Y}) \equiv E(\mathbf{X}-E\mathbf{X})(\mathbf{Y}-E\mathbf{Y}) = E\mathbf{X}\mathbf{Y} - (E\mathbf{X})(E\mathbf{Y}). \qquad \text{(A.1.5)}$$

If $\mathbf{X} = (\mathbf{X}_1,\dots,\mathbf{X}_k)$ is a random vector (with values in R^k), such that $\text{cov}(\mathbf{X}_i,\mathbf{X}_j)$ is defined for every pair of coordinates $(\mathbf{X}_i,\mathbf{X}_j)$, then one defines the *covariance matrix* $\text{Cov}(\mathbf{X})$ of **X** as the $k \times k$ matrix whose (i,j) element is $\text{cov}(\mathbf{X}_i,\mathbf{X}_j)$. The *distribution* $P_{\mathbf{X}}$ of a random vector **X** (with values in R^k) is the induced probability measure $P \circ \mathbf{X}^{-1}$ on R^k, that is,

$$P_{\mathbf{X}}(A) \equiv P\left(\mathbf{X}^{-1}(A)\right) \qquad (A \in \mathfrak{B}^k). \qquad \text{(A.1.6)}$$

Since the mean and the covariance matrix of a random vector **X** depend only on its distribution, one also defines the mean and the covariance matrix of a probability measure Q on R^k as those of a (any) random vector having distribution Q.

Random vectors $\mathbf{X}_1,\dots,\mathbf{X}_m$ (with values in R^k) defined on $(\Omega, \mathfrak{B}, P)$ are *independent* if

$$P(\mathbf{X}_1 \in A_1, \mathbf{X}_2 \in A_2, \dots, \mathbf{X}_m \in A_m)$$
$$= P(\mathbf{X}_1 \in A_1) P(\mathbf{X}_2 \in A_2) \cdots P(\mathbf{X}_m \in A_m) \qquad \text{(A.1.7)}$$

for every m-tuple $(A_1,\dots,A_m)$ of Borel subsets of R^k. In other words, $\mathbf{X}_1,\dots,\mathbf{X}_m$ are independent if the induced measure $P \circ (\mathbf{X}_1,\dots,\mathbf{X}_m)^{-1}$ is a product measure. A *sequence* $\{\mathbf{X}_n : n \geq 1\}$ *of random vectors* [defined on $(\Omega, \mathfrak{B}, P)$] *are independent* if every finite subfamily is so.

A.2 FUNCTIONS OF BOUNDED VARIATION AND DISTRIBUTION FUNCTIONS

Let μ be a finite signed measure on R^k. The *distribution function* F_μ *of* μ is the real-valued function on R^k defined by

$$F_\mu(x) = \mu((-\infty, x]) \qquad (x \in R^k), \qquad \text{(A.2.1)}$$

where

$$(-\infty, x] = (-\infty, x_1] \times (-\infty, x_2] \times \cdots \times (-\infty, x_k]$$
$$\left[x = (x_1,\dots,x_k) \in R^k\right]. \qquad \text{(A.2.2)}$$

It is simple to check that F_μ is right continuous. For a random vector $\mathbf{X}$ defined on some probability space $(\Omega, \mathcal{B}, P)$, the *distribution function of* $\mathbf{X}$ is merely the distribution function of its distribution $P_{\mathbf{X}}$. The distribution function F_μ completely determines the (finite) signed measure μ. To see this consider the class $\mathcal{F}$ of all *rectangles* of the form

$$(a,b]=(a_1,b_1]\times\cdots\times(a_k,b_k] \tag{A.2.3}$$

$$\left(a_i<b_i \text{ for all } i; a, b\in R^k\right).$$

Also define the *difference operators* $\Delta_1^{h_1},\ldots,\Delta_k^{h_k}$ by

$$\Delta_1^{h_1}F(x)=F(x_1+h_1,x_2,\ldots,x_k)-F(x_1-h_1,x_2,\ldots,x_k),$$

$$\Delta_1^{h_2}F(x)=F(x_1,x_2+h_2,x_3,\ldots,x_k)-F(x_1,x_2-h_2,x_3,\ldots,x_k), \tag{A.2.4}$$

$$\vdots$$

$$\Delta_k^{h_k}F(x)=F(x_1,x_2,\ldots,x_{k-1},x_k+h_k)$$

$$-F(x_1,x_2,\ldots,x_{k-1},x_k-h_k) \qquad \left[x=(x_1,\ldots,x_k)\in R^k\right],$$

where $h_1, h_2,\ldots,h_k$ are positive numbers, and F is an arbitrary real-valued function on R^k. The difference operators are associative and commutative, and one can define the operator Δ^h by

$$\Delta^h=\Delta_1^{h_1}\cdots\Delta_k^{h_k} \qquad \left[h=(h_1,\ldots,h_k),\quad h_i>0 \text{ for } i=1,\ldots,k\right]. \tag{A.2.5}$$

If $k=1$, we shall write Δ^h for the difference operator. One can also show† that for every $(a,b]\in\mathcal{F}$

$$\mu((a,b])=\Delta^h F_\mu(x), \tag{A.2.6}$$

where

$$h=\tfrac{1}{2}(b-a), \qquad x=\tfrac{1}{2}(a+b). \tag{A.2.7}$$

The class $\mathcal{R}$ of all finite disjoint unions of sets in $\mathcal{F}$ is a ring over which μ is determined by (A.2.6). Since the sigma-ring generated by $\mathcal{R}$ is $\mathcal{B}^k$, the

---

†See Cramér [4], pp. 78–80.

uniqueness of the Caratheodory extension† implies that μ on $\mathcal{B}^k$ is determined by μ on $\mathcal{R}$ (and, hence by the distribution function F_μ). One may also show by an induction argument‡ that

$$\Delta^h F(x) = \sum \pm F(x_1 + \epsilon_1 h_1, x_2 + \epsilon_2 h_2, \ldots, x_k + \epsilon_k h_k), \qquad \text{(A.2.8)}$$

where the summation is over all k-tuples $(\epsilon_1, \epsilon_2, \ldots, \epsilon_k)$, each ϵ_i being either $+1$ or -1. The sign of a summand in (A.2.8) is plus or minus depending on whether the number of negative ϵ's is even or odd.

Now let F be an arbitrary real-valued function on an open set U. Define a set function μ_F on the class $\mathcal{F}_U$ of all those sets in $\mathcal{F}$ that are contained in U by

$$\mu_F((a,b]) = \Delta^h F(x), \qquad \text{(A.2.9)}$$

where x and h are given by (A.2.7). One can check that μ_F is finitely additive in $\mathcal{F}$. The function F is said to be *of bounded variation on an open set* U *if*

$$\sup \sum_j |\mu_F(I_j)| \qquad \text{(A.2.10)}$$

is finite, where the supremum is over all finite collections $\{I_1, I_2, \ldots\}$ of pairwise disjoint sets in $\mathcal{F}$ such that $I_j \subset U$ for all j. The expression (A.2.10) is called the *variation of* F *on* U. The following theorem is proved in Saks [1] (Theorem 6.2, p. 68).

THEOREM A.2.1. *Let* F *be a right continuous function of bounded variation on a nonempty open set* U. *There exists a unique finite signed measure on* U *that agrees with* μ_F *on the class* $\mathcal{F}_U$ *of all sets in* $\mathcal{F}$ *contained in* U.

It may be checked that the variation on U of a right continuous function F of bounded variation (on U) coincides with the variation norm of the signed measure whose existence is asserted in Theorem A.2.1.

A function F is said to be *absolutely continuous on an open set* U if given $\epsilon > 0$ there exists $\delta > 0$ such that

$$\sum_j |\mu_F(I_j)| < \epsilon \qquad \text{(A.2.11)}$$

for all finite collections $\{I_1, \ldots\}$ of pairwise disjoint rectangles $I_j \in \mathcal{F}_U$ satisfying

$$\sum_j \lambda_k(I_j) < \delta, \qquad \text{(A.2.12)}$$

†See Halmos [1], p. 54.

‡See Cramér [4], pp. 78–80.

where λ_k denotes the Lebesgue measure on R^k. If F is absolutely continuous on a bounded open set U, then it may be shown that F is of bounded variation on U.†

THEOREM A.2.2. *Let* F *be a right continuous function of bounded variation on an open set* $U \subset R^k$. *Let* μ_F *be the measure on* R^k *defined by* (A.2.6) (*and Theorem* A.2.1). *Suppose that on* U *the successive derivatives* $D_k F$, $D_{k-1}D_k F, \ldots, D_1 \cdots D_k F$ *exist and are continuous. Then* F *is absolutely continuous on* U *and one has*

$$\mu_F(A) = \int_A (D_1 \cdots D_k F)(x)\,dx \tag{A.2.13}$$

for every Borel subset A *of* U. *Also,*

$$\lim_{h_1 \downarrow 0, \ldots, h_k \downarrow 0} (2^k h_1 \cdots h_k)^{-1} \Delta^h F = D_1 D_2 \cdots D_k F \tag{A.2.14}$$

on U.

Proof. Let the closed rectangle $[a,b]$ be contained in U. Let h and x be defined by (A.2.7). Then

$$\begin{aligned}
\mu_F((a,b]) &= \Delta^h F(x) = \Delta_1^{h_1} \cdots \Delta_k^{h_k} F(x) \\
&= \Delta_1^{h_1} \cdots \Delta_{k-1}^{h_{k-1}} (F(x_1, \ldots, x_{k-1}, x_k + h_k) - F(x_1, \ldots, x_{k-1}, x_k - h_k)) \\
&= \Delta_1^{h_1} \cdots \Delta_{k-1}^{h_{k-1}} \int_{x_k - h_k}^{x_k + h_k} (D_k F)(x_1, \ldots, x_{k-1}, y_k)\,dy_k
\end{aligned} \tag{A.2.15}$$

by the fundamental theorem of integral calculus. Since the integrand has a continuous derivative with respect to x_{k-1},

$$\begin{aligned}
\mu_F((a,b]) = \Delta_1^{h_1} \cdots \Delta_{k-2}^{h_{k-2}} \int_{x_k - h_k}^{x_k + h_k} [(D_k F)(x_1, \ldots, x_{k-1} + h_{k-1}, y_k) \\
- (D_k F)(x_1, \ldots, x_{k-1} - h_{k-1}, y_k)]\,dy_k
\end{aligned}$$

$$= \Delta_1^{h_1} \cdots \Delta_{k-2}^{h_{k-2}} \int_{x_k - h_k}^{x_k + h_k} \left[\int_{x_{k-1} - h_{k-1}}^{x_{k-1} + h_{k-1}} (D_{k-1} D_k F)(x_1, \ldots, y_{k-1}, y_k)\,dy_{k-1} \right] dy_k. \tag{A.2.16}$$

†Saks [1], p. 93.

Proceeding in this manner, we arrive at (A.2.13) for $A=(a,b]$, remembering that by Fubini's theorem the iterated integral as obtained by the above procedure is equal to the integral on $(a,b]$ with respect to Lebesgue measure on R^k. We next show that $D_1 \cdots D_k F$ is integrable on U. For if this is false, then for every integer $n \geqslant 1$ there exists an integer m_n and pairwise disjoint rectangles $(a^1,b^1],\ldots,(a^{m_n},b^{m_n}]$ such that $[a^i,b^i]\subset U$, $i=1,\ldots,m_n$, and

$$\left|\sum_{i=1}^{m_n}\int_{(a^i,b^i]}(D_1\cdots D_k F)(x)\,dx\right|>n.$$

By (A.2.13), which we have proved for sets like $(a^i,b^i]$, one then has

$$\sum_{i=1}^{m_n}|\mu_F((a_i,b_i])|>n$$

for all n, contradicting the hypothesis that F is of bounded variation on U. Thus we have two finite signed measures on U, defined by

$$A\to\mu_F(A),\qquad A\to\int_A(D_1\cdots D_k F)(x)\,dx,$$

that coincide on the class of all rectangles $(a,b]$ such that $[a,b]\subset U$. Therefore the two signed measures on U are equal, and (A.2.13) is established. To prove (A.2.14), let $x\in U$. Choose $h=(h_1,\ldots,h_k)$ such that $h_i>0$ for all i and $[x-h,x+h]\subset U$. Then by (A.2.13) one has

$$\Delta^h F(x)=\mu_F((x-h,x+h])=\int_{(x-h,x+h]}(D_1\cdots D_k F)(y)\,dy.$$

From this and continuity of $D_1\cdots D_k F$ on U the relation (A.2.14) follows. Q.E.D.

It follows from definition that the sum of a finite number of functions of bounded variation, or absolutely continuous, on an open set U is itself of bounded variation, or absolutely continuous, on U. Our next result establishes the bounded variation of a product of a special set of functions of bounded variation. We say that a function g on R^k (into R^1) is *Schwartz* if it is infinitely differentiable and if for every nonnegative integral vector α and every positive integer m one has

$$\sup_{x\in R^k}\|x\|^m|(D^\alpha g)(x)|<\infty. \tag{A.2.17}$$

THEOREM A.2.3. *Let* $F_1,\ldots,F_p$ *be right continuous, real-valued functions on* R^1, *each periodic with period one*; *in* $(0,1)$ *each* F_i *is differentiable* $(1\leqslant i\leqslant p)$ *and has a bounded derivative. Suppose that* ψ *is a real-valued, bounded, measurable function on* R^{k-p} *and that* g *is a Schwartz function on* R^k, $k \geqslant p$, *and let*

$$G(x)=\int_{-\infty}^{x_{p+1}} \cdots \int_{-\infty}^{x_k} \psi(y_{p+1},\ldots,y_k)\, g(x_1,\ldots,x_p,y_{p+1},\ldots,y_k)\, dy_k \cdots dy_{p+1}$$

$$\left[x=(x_1,\ldots,x_k)\in R^k\right], \qquad \text{(A.2.18)}$$

if $k>p$, *and* $G=g$ *if* $k=p$. *Then the function*

$$F(x)\equiv F_1(x_1)\cdots F_p(x_p)G(x) \qquad (x\in R^k), \qquad \text{(A.2.19)}$$

is of bounded variation on R^k.

Proof. Consider an arbitrary function H_0 on R^p. We first show that

$$\begin{aligned}\Delta_1^{h_1}\cdots\Delta_p^{h_p}&F_1(x_1)\cdots F_p(x_p)H_0(x_1,\ldots,x_p)\\ &=\sum F_{i_1}(x_{i_1}-h_{i_1})\cdots F_{i_s}(x_{i_s}-h_{i_s})\left[\Delta_{i_1}^{h_{i_1}}\cdots\Delta_{i_s}^{h_{i_s}}H_0(x')\right]\\ &\quad\times\left[\Delta^{h_{j_1}}F_{j_1}(x_{j_1})\right]\cdots\left[\Delta^{h_{j_{p-s}}}F_{j_{p-s}}(x_{j_{p-s}})\right],\end{aligned} \qquad \text{(A.2.20)}$$

where $x'=(x_1',\ldots,x_p')$, $x_{i_1}'=x_{i_1},\ldots,x_{i_s}'=x_{i_s}$, $x_{j_1}'=x_{j_1}+h_{j_1},\ldots,$ $x_{j_{p-s}}'=x_{j_{p-s}}+h_{j_{p-s}}$, and the summation is over all partitions of $\{1,2,\ldots,p\}$ into two disjoint subsets $\{i_1,\ldots,i_s\}$, $\{j_1,\ldots,j_{p-s}\}$, $0\leqslant s\leqslant p$; when one of these subsets is empty, the corresponding factor drops out. If $p=1$, then

$$\begin{aligned}\Delta^h F_1(x)H_0(x)&=F_1(x+h)H_0(x+h)-F_1(x-h)H_0(x-h)\\ &=F_1(x-h)\left[H_0(x+h)-H_0(x-h)\right]\\ &\quad+H_0(x+h)\left[F_1(x+h)-F_1(x-h)\right]\\ &=F_1(x-h)\Delta^h H_0(x)+H_0(x+h)\Delta^h F_1(x),\end{aligned}$$

which proves (A.2.20) for $p=1$. Assume, as an induction hypothesis, that (A.2.20) holds for some p. Then

$$\begin{aligned}\Delta_1^{h_1}\cdots&\Delta_{p+1}^{h_{p+1}}F_1(x_1)\cdots F_{p+1}(x_{p+1})H_0(x_1,\ldots,x_{p+1})\\ &=\Delta_1^{h_1}\cdots\Delta_p^{h_p}\left[F_1(x_1)\cdots F_p(x_p)\cdot\Delta_{p+1}^{h_{p+1}}(F_{p+1}(x_{p+1})H_0(x_1,\ldots,x_{p+1}))\right]\\ &=\Delta_1^{h_1}\cdots\Delta_p^{h_p}\left[F_1(x_1)\cdots F_p(x_p)\left\{F_{p+1}(x_{p+1}-h_{p+1})\Delta_{p+1}^{h_{p+1}}H_0(x_1,\ldots,x_{p+1})\right.\right.\\ &\qquad\left.\left.+H_0(x_1,\ldots,x_p,x_{p+1}+h_{p+1})\Delta^{h_{p+1}}F_{p+1}(x_{p+1})\right\}\right]\\ &=\Delta_1^{h_1}\cdots\Delta_p^{h_p}F_1(x_1)\cdots F_p(x_p)H_0'(x_1,\ldots,x_{p+1})\\ &\quad+\Delta_1^{h_1}\cdots\Delta_p^{h_p}F_1(x_1)\cdots F_p(x_p)H_0''(x_1,\ldots,x_{p+1}),\end{aligned} \qquad \text{(A.2.21)}$$

where

$$H_0'(x_1,\dots,x_{p+1}) = F_{p+1}(x_{p+1}-h_{p+1})\Delta_{p+1}^{h_{p+1}}H_0(x_1,\dots,x_{p+1}),$$
$$H_0''(x_1,\dots,x_{p+1}) = H_0(x_1,\dots,x_p,x_{p+1}+h_{p+1})\Delta^{h_{p+1}}F_{p+1}(x_{p+1}). \quad \text{(A.2.22)}$$

Now apply the induction hypothesis to each of the two summands of the last expression in (A.2.21) and then substitute from (A.2.22) to see that (A.2.20) holds with p replaced by $p+1$. This completes the proof of (A.2.20) for all p.

Looking at F given by (A.2.19), one sees that

$$\Delta_1^{h_1}\cdots\Delta_k^{h_k}F(x) = \Delta_1^{h_1}\cdots\Delta_p^{h_p}F_1(x_1)\cdots F_p(x_p)H_0(x_1,\dots,x_k), \quad \text{(A.2.23)}$$

where

$$H_0(x_1,\dots,x_k) = \Delta_{p+1}^{h_{p+1}}\cdots\Delta_k^{h_k}G(x_1,\dots,x_k). \quad \text{(A.2.24)}$$

By (A.2.20) we obtain

$$\Delta_1^{h_1}\cdots\Delta_k^{h_k}F(x) = \sum F_{i_1}(x_{i_1}-h_{i_1})\cdots F_{i_s}(x_{i_s}-h_{i_s})$$
$$\times\left[\Delta_{i_1}^{h_{i_1}}\cdots\Delta_{i_s}^{h_{i_s}}\Delta_{p+1}^{h_{p+1}}\cdots\Delta_k^{h_k}G(x')\right]$$
$$\times\left[\Delta^{h_{j_1}}F_{j_1}(x_{j_1})\right]\cdots\left[\Delta^{h_{j_{p-s}}}F_{j_{p-s}}(x_{j_{p-s}})\right], \quad \text{(A.2.25)}$$

where $x' = (x_1',\dots,x_k')$, $x_{i_1}' = x_{i_1},\dots,x_{i_s}' = x_{i_s}$, $x_{p+1}' = x_{p+1},\dots,x_k' = x_k$, $x_{j_1}' = x_{j_1}+h_{j_1},\dots,x_{j_{p-s}}' = x_{j_{p-s}}+h_{j_{p-s}}$, and the summation is over all partitions of $\{1,2,\dots,p\}$ into two disjoint subsets $\{i_1,\dots,i_s\}$, $\{j_1,\dots,j_{p-s}\}$, $0 \leqslant s \leqslant p$. For the sake of simplicity consider the summand on the right side of (A.2.25) corresponding to $i_1 = 1,\dots,i_s = s$. Then by the definition (A.2.18) of G, one has

$$\Delta_1^{h_1}\cdots\Delta_s^{h_s}\Delta_{p+1}^{h_{p+1}}\cdots\Delta_k^{h_k}G(x')$$
$$= \int_{x_{p+1}-h_{p+1}}^{x_{p+1}+h_{p+1}}\cdots\int_{x_k-h_k}^{x_k+h_k}\psi(y_{p+1},\dots,y_k)\Delta_1^{h_1}$$
$$\cdots\Delta_s^{h_s}g(x_1,\dots,x_s,x_{s+1}+h_{s+1},\dots,x_p+h_p,y_{p+1},\dots,y_k)\,dy_k\cdots dy_{p+1}$$
$$= \int_{x_1-h_1}^{x_1+h_1}\cdots\int_{x_s-h_s}^{x_s+h_s}\int_{x_{p+1}-h_{p+1}}^{x_{p+1}+h_{p+1}}$$
$$\cdots\int_{x_k-h_k}^{x_k+h_k}\psi(y_{p+1},\dots,y_k)(D_1\cdots D_s g)(y_1,\dots,y_s,x_{s+1}$$
$$+h_{s+1},\dots,x_p+h_p,y_{p+1},\dots,y_k)\,dy_k\cdots dy_{p+1}\,dy_s\cdots dy_1. \quad \text{(A.2.26)}$$

Let the derivative of F_i on $(0,1)$ be bounded above in magnitude by b_i, and let c_i denote the magnitude of the jump of F_i at 0 (c_i may be zero), $1 \leqslant i \leqslant p$. Assume that $2h_i < \epsilon / b_i$, $s+1 \leqslant i \leqslant q$, and that the intervals $(x_{s+1}-h_{s+1}, x_{s+1}+h_{s+1}], \ldots, (x_q - h_q, x_q + h_q]$ each contain exactly one integer point, whereas $(x_j - h_j, x_j + h_j]$, $q+1 \leqslant j \leqslant p$, contain no integer point. Here q is an integer, $s+1 \leqslant q \leqslant p$. Then

$$\begin{aligned}
&\Big|F_1(x_1-h_1)\cdots F_s(x_s-h_s)\big[\Delta_1^{h_1}\cdots\Delta_s^{h_s}\Delta_{p+1}^{h_{p+1}}\cdots\Delta_k^{h_k}G(x')\big] \\
&\qquad\times\big[\Delta^{h_{s+1}}F_{s+1}(x_{s+1})\big]\cdots\big[\Delta^{h_p}F_p(x_p)\big]\Big| \\
&\qquad\leqslant \|F_1\|_\infty\cdots\|F_s\|_\infty(\epsilon+c_{s+1})\cdots(\epsilon+c_q)b_{q+1}\cdots b_p h_{q+1}\cdots h_p \\
&\quad\times 2^{p-q}\|\psi\|_\infty\cdot\int_{\{x_i-h_i<y_i\leqslant x_i+h_i;\, i=1,\ldots,s,p+1,\ldots,k\}}|(D_1\cdots D_s g)(y_1,\ldots \\
&\ldots,y_s,x_{s+1}+h_{s+1},\ldots,x_p+h_p,y_{p+1},\ldots,y_k)|\,dy_k\cdots dy_{p+1}\,dy_s\cdots dy_1. \qquad \text{(A.2.27)}
\end{aligned}$$

Here $\|f\|_\infty$ denotes the supremum of $|f|$. The sum of the left side of (A.2.27) over an arbitrary partitioning of the $(x_1,\ldots,x_s,x_{p+1},\ldots,x_p)$-space into rectangles of the form $x_i - h_i < y_i \leqslant x_i + h_i$ for $i-1,\ldots,s$, $p+1,\ldots,k$, is therefore bounded above by

$$\begin{aligned}
&\|F_1\|_\infty\cdots\|F_s\|_\infty(\epsilon+c_{s+1})\cdots(\epsilon+c_q)b_{q+1}\cdots b_p h_{q+1}\cdots h_p 2^{p-q}\|\psi\|_\infty \\
&\qquad\times\int_{R^{s+k-p}}|(D_1\cdots D_s g)(y_1,\ldots,y_s,x_{s+1}+h_{s+1},\ldots,x_p+h_p, \\
&\qquad\qquad y_{p+1},\ldots,y_k)|\,dy_k\cdots dy_{p+1}\,dy_s\cdots dy_1 \\
&\qquad\leqslant c' h_{q+1}\cdots h_p\big(1+|x_{s+1}|^{k+1}+\cdots+|x_p|^{k+1}\big)^{-1} \qquad (\epsilon\leqslant 1),
\end{aligned}$$

since g is a Schwartz function. Now the lim sup (as $h_i \downarrow 0$) of the sum of the last expression over all pairwise disjoint intervals $(x_i - h_i, x_i + h_i]$ in each variable $i = s+1,\ldots,p$, as prescribed preceding the inequality (A.2.27), is

$$\begin{aligned}
&c'\sum_{n_{s+1}=-\infty}^{\infty}\cdots\sum_{n_q=-\infty}^{\infty}\int_{R^{p-q}}\big(1+|n_{s+1}|^{k+1} \\
&\quad+\cdots+|n_q|^{k+1}+|x_{q+1}|^{k+1}+\cdots+|x_p|^{k+1}\big)^{-1}dx_{q+1}\cdots dx_p<\infty, \qquad \text{(A.2.28)}
\end{aligned}$$

where each summation is over the set of all integers. The variation of F on R^k is given by the expression (A.2.10) (with $U = R^k$) and is also obtained in the limit by computing this sum over successively finer partitions

$\{I_j : j \geqslant 1\}$ of R^k into rectangles in $\mathcal{F}$. For a typical rectangle I, $\mu_F(I)$ is given by (A.2.23) and (A.2.25). There are a finite (bounded) number of terms on the right side of (A.2.25), and we have shown above that the limit of the sum of the absolute values of a typical term over successively finer partitions is finite. This has been accomplished by further splitting a partition into subpartitions; in a subpartition $(x_i - h_i, x_i + h_i]$ contains one integer point for a specified set of i's, but none for a disjoint set of i's. This was necessary to take care of the possible jumps of the F_i's at integer points. Q.E.D.

A.3 ABSOLUTELY CONTINUOUS, SINGULAR, AND DISCRETE PROBABILITY MEASURES

We recall that a finite signed measure μ on a sigma-field $\mathcal{B}$ of subsets of a set Ω is said to be *absolutely continuous with respect to a sigma-finite measure* ν (also defined on $\mathcal{B}$) if every ν-null set is also a μ-null set. The Radon–Nikodym theorem asserts that in this case there exists a ν-integrable function f, called the *Radon–Nikodym derivative of* μ *with respect to* ν, such that

$$\mu(B) = \int_B f \, d\nu \qquad (B \in \mathcal{B}). \tag{A.3.1}$$

A well-known characterization of absolute continuity is the following:

THEOREM A.3.1. *A finite signed measure* μ *(on* $\mathcal{B}$*) is absolutely continuous with respect to a sigma-finite measure* ν *(on* $\mathcal{B}$*) if and only if for every positive* ϵ *there exists a* $\delta > 0$ *such that* $|\mu(\mathrm{B})| < \epsilon$ *for every* B *(in* $\mathcal{B}$*) for which* $\nu(\mathrm{B}) < \delta$.

For a proof see Halmos [1], Theorem B, pp. 125, 126. The opposite of absolute continuity is singularity. A finite signed measure μ *is singular with respect to a sigma-finite measure* ν *if there exists* a set $B_0 \in \mathcal{B}$ such that $|\mu|(B_0) = 0$, $\nu(\Omega \setminus B_0) = 0$. The so-called Lebesgue decomposition theorem says that if μ is a finite signed measure and ν a sigma-finite measure (on $\mathcal{B}$), then there exists a unique decomposition

$$\mu = \mu_1 + \mu_2, \tag{A.3.2}$$

where μ_1 is absolutely continuous and μ_2 is singular (both with respect to ν).

We now specialize to (signed) measures on R^k. Let μ be a finite (signed) measure on R^k. Let (A.3.2) denote the Lebesgue decomposition of μ with respect to *Lebesgue measure*. It is easy to see that μ_2 may be decomposed uniquely as

$$\mu_2 = \mu_3 + \mu_4, \tag{A.3.3}$$

where μ_4 is (purely) *discrete* and μ_3 has no point masses; that is, there exists a countable set $A = \{x_1, x_2, \ldots\}$ such that

$$|\mu_4|(A) = |\mu_4|(R^k), \qquad \mu_3(\{x\}) = 0 \qquad \text{for all } x \in R^k.$$

The (signed) measures μ_3 and μ_4 are called the *singular* and *discrete components* of μ, respectively. Thus μ is decomposed uniquely as

$$\mu = \mu_1 + \mu_3 + \mu_4, \tag{A.3.4}$$

the sum of its absolutely continuous, singular, and discrete components. The (signed) measure μ is simply called *absolutely continuous or singular or discrete* if the other components vanish. When μ is absolutely continuous, its Radon–Nikodym derivative (with respect to Lebesgue measure) is simply referred to as its *density*. Theorem A.3.1 and the definition of absolute continuity of a *function* on R^k lead to the following proposition: *a finite signed measure* μ *on* R^k *is absolutely continuous if and only if its distribution function* F_μ *is absolutely continuous.*

Let P be a probability measure on R^k. If P is absolutely continuous, then it follows from the Riemann–Lebesgue lemma [Theorem 4.1(iii)] that its characteristic function $\hat{P}$ satisfies

$$\lim_{\|t\|\to\infty} |\hat{P}(t)| = 0. \tag{A.3.5}$$

It follows that if P has a nonzero absolutely continuous component, then

$$\lim_{\|t\|\to\infty} |\hat{P}(t)| < 1. \tag{A.3.6}$$

If P is discrete, then $\hat{P}$ is a uniformly convergent trigonometric series. A fundamental theorem of Bohr (see Bohr [1], pp. 80, 81) implies that in this case

$$\overline{\lim_{\|t\|\to\infty}} |\hat{P}(t)| = 1. \tag{A.3.7}$$

For singular probability measures the situation is a little more complex. It is known (see Esseen [1], pp. 28, 29) that there are singular probability measures P_1, P_2 such that

$$\lim_{\|t\|\to\infty} |\hat{P}_1(t)| = 0, \qquad \overline{\lim_{\|t\|\to\infty}} |\hat{P}_2(t)| = 1. \tag{A.3.8}$$

By taking convex combinations of P_1, P_2, one shows that for each $\alpha \in [0,1]$ there exists a singular probability measure P such that

$$\overline{\lim_{\|t\|\to\infty}} |\hat{P}(t)| = \alpha. \tag{A.3.9}$$

A.4 THE EULER–MACLAURIN SUM FORMULA FOR FUNCTIONS OF SEVERAL VARIABLES

d=1

Let f be a function of the real variable x on an interval $0 \leqslant x \leqslant n$ (n integral), with continuous derivative f'. Then a closed expression for the sum $\sum_{j=0}^{n} f(j)$ may be given as follows:

$$\sum_{j=0}^{n} f(j) = \int_0^n f(x)\,dx + \tfrac{1}{2}(f(0)+f(n)) + \int_0^n (x-[x]-\tfrac{1}{2})f'(x)\,dx$$

See H. Rademacher Topics in An. Number Theory Springer #169 (1973) pp. 1–30

where $[x]$ denotes the largest integer less than or equal to x. This is the Euler–Maclaurin sum formula in its simplest form. In this section we obtain a general summation formula for functions of several variables. The results of this section are used in Section 23 to get asymptotic expansions for distribution functions of normalized sums of independent, lattice random vectors.

The Functions S_j $(j \geqslant 0)$.

Let S_j $(j=0,1,2,\ldots)$ be a sequence of real-valued periodic functions on R^1 of period one, possessing the following properties:

(i) For $j \geqslant 0$, S_j is differentiable at all nonintegral points and $S'_{j+1}(x) = S_j(x)$ (at all nonintegral x),

(ii) $S_0(x) \equiv 1$, S_1 is right continuous and S_j is continuous for $j \geqslant 2$. (A.4.1)

Such a sequence is uniquely determined by the above properties and plays a fundamental role in the summation formula. To see this, write $S_j(0) = B_j/(j!)$ and observe that (i) leads to

$$S_1(x) = x + B_1, \qquad S_2(x) = \tfrac{1}{2}x^2 + B_1 x + B_2/2!, \ldots$$

$$S_j(x) = \frac{1}{j!}x^j + \frac{B_1}{1!}\frac{x^{j-1}}{(j-1)!} + \cdots + \frac{B_j}{j!} = \frac{1}{j!}\sum_{r=0}^{j} B_r \binom{j}{r} x^{j-r}$$

$$(0 < x < 1,\ j \geqslant 1) \tag{A.4.2}$$

The constants B_j's are determined by the property (A.4.1). In fact, $S_j(0) = S_j(1)$ for $j \geqslant 2$, which yields

$$1+\binom{j}{1}B_1+\binom{j}{2}B_2+\cdots+\binom{j}{j-1}B_{j-1}=0 \qquad (j=2,3,\ldots). \quad \text{(A.4.3)}$$

The sequence of constants B_j is recursively defined by the relation (A.4.3), thus completely determining the sequence of functions S_j in the interval $0<x<1$. The continuity assumption determines their values at integral points. The numbers B_j defined by (A.4.3) are called *Bernoulli numbers*, and the polynomial

$$B_j(x)=\sum_{r=0}^{j} B_r\binom{j}{r}x^{j-r} \qquad \text{(A.4.4)}$$

is called the jth *Bernoulli polynomial*. Clearly, $S_j(x)=B_j(x)/(j!)$ for $0 \leqslant x < 1$. Since the sequence $\{(-1)^j S_j(-x) : j \geqslant 0\}$ has the properties (A.4.1), excepting right continuity of $(-1)S_1(-x)$, it follows from uniqueness that

$$S_j(-x)=(-1)^j S_j(x) \qquad (\text{for all } x \text{ if } j\neq 1, \quad \text{for nonintegral } x \text{ if } j=1). \qquad \text{(A.4.5)}$$

The functions S_j are thus even or odd depending on whether j is even or odd. In particular,

$$S_j(0)=\frac{B_j}{j!}=0 \qquad \text{for } j \text{ odd}, j \geqslant 3. \qquad \text{(A.4.6)}$$

The first few Bernoulli numbers are

$$B_0=1, \quad B_1=-\tfrac{1}{2}, \quad B_2=\tfrac{1}{6}, \quad B_3=0, \quad B_4=-\tfrac{1}{30}, \quad B_5=0. \qquad \text{(A.4.7)}$$

Therefore

$$S_1(x)=x-\tfrac{1}{2}, \qquad S_2(x)=\tfrac{1}{2}\left(x^2-x+\tfrac{1}{6}\right),$$

$$S_3(x)=\tfrac{1}{6}\left(x^3-\tfrac{3}{2}x^2+\tfrac{1}{2}x\right),\ldots \qquad (0\leqslant x<1), \qquad \text{(A.4.8)}$$

and so on. The periodic functions S_j have the following Fourier series expansions when x is not an integer:

$$S_j(x)=\begin{cases}(-1)^{j/2-1}\sum_{n=1}^{\infty}\dfrac{2\cos(2n\pi x)}{(2n\pi)^j}, & j \text{ even}, j>0,\\ (-1)^{(j-1)/2}\sum_{n=1}^{\infty}\dfrac{2\sin(2n\pi x)}{(2n\pi)^j}, & j \text{ odd}.\end{cases} \tag{A.4.9}$$

This may be seen as follows. Let u_j denote the function represented by the Fourier series ($j \geq 1$). It can be checked directly that u_1 is the Fourier series of S_1 and that $u'_{j+1}=u_j$ for $j \geq 1$. Thus $S_j=u_j$ for all $j \geq 2$, and $S_1(x)=u_1(x)$ for all nonintegral x.

THEOREM A.4.1. *Let* f *be a real-valued function on* R^1 *having* r *continuous derivatives*, $r \geq 1$, *and let*

$$\int_{R^1}|D^jf|\,dx<\infty \tag{A.4.10}$$

for $j \leq r$. *Then for every Borel set* A

$$\sum_{n\in A} f(n)=\int_A dF_r, \tag{A.4.11}$$

where

$$F(x)=\int_{-\infty}^{x} f(t)\,dt,$$

$$F_r(x)=\sum_{j=0}^{r}(-1)^jS_j(x)(D^jF)(x)+(-1)^{r+1}\int_{-\infty}^{x}S_r(t)(D^{r+1}F)(t)\,dt. \tag{A.4.12}$$

Proof. It is clear from the definition of F_r that it is a function of bounded variation. Also, since each one of the functions $S_j(x)$ is absolutely continuous with a continuous differential coefficient in the interval $n<x<n+1$ for every integer n, it follows that F_r is absolutely continuous in $R^1\backslash\mathbf{Z}$. By using property (A.4.1i) it is easily verified that $(d/dx)F_r(x)=0$ for $x\in R^1\backslash\mathbf{Z}$; that is, the signed measure dF_r has no mass on $R^1\backslash\mathbf{Z}$. Because of the presence of the term $-S_1D^1F=-S_1f$, which is the only summand in the expression for F_r that has discontinuities, F_r has a jump of magnitude $-S_1(n)f(n)+S_1(n-0)f(n)=f(n)$ at the integer point n. This proves (A.4.11). Q.E.D.

Generalization to Functions on R^{k}

Let $\mathcal{S}$ denote the *Schwartz space* on R^k; that is, $f \in \mathcal{S}$ if and only if f is infinitely differentiable and

$$\sup_{x \in R^k} |x^{\beta}(D^{\alpha}f)(x)| < \infty \tag{A.4.13}$$

for all pairs of nonnegative integral vectors α, β.

Our first step is to construct functions on R^k analogous to (A.4.12). Write

$$S_{\alpha}(x) = S_{\alpha_1}(x_1) S_{\alpha_2}(x_2) \cdots S_{\alpha_k}(x_k)$$

$$\left[\alpha = (\alpha_1, \ldots, \alpha_k) \text{ nonnegative integral vector, } x = (x_1, \ldots, x_k)\right], \tag{A.4.14}$$

and for an integrable f write

$$F(x) = \int_{-\infty}^{x_1} \cdots \int_{-\infty}^{x_k} f(y)\,dy \quad \left[f \in \mathcal{S},\ x = (x_1, \ldots, x_k) \in R^k\right]. \tag{A.4.15}$$

For a function f that is Schwartz in the jth coordinate x_j, let the operator $I_{r,j}(F)$ be defined by

$$\begin{aligned} I_{r,j}(F)(x) &= \int_{-\infty}^{x_j} S_r(y_j)\left(D_j^{r+1}F\right)(x_1, \ldots, x_{j-1}, y_j, x_{j+1}, \ldots, x_k)\,dy \\ &= \int_{-\infty}^{x_1} \cdots \int_{-\infty}^{x_k} S_r(y_j)(D_j^r f)(y)\,dy \qquad (x \in R^k), \end{aligned} \tag{A.4.16}$$

where $D_j = (\partial / \partial x_j)$. By Fubini's theorem, the operators $I_{r,j_1}, \ldots, I_{r,j_p}$ commute if f is Schwartz in $x_{j_1}, \ldots, x_{j_p}$, and

$$\begin{aligned} &I_{r,j_1} \cdots I_{r,j_p}(F)(x) \\ &= \int_{-\infty}^{x_1} \cdots \int_{-\infty}^{x_k} S_r(y_{j_1}) \cdots S_r(y_{j_p})\left(D_{j_1}^r \cdots D_{j_p}^r f\right)(y)\,dy. \end{aligned} \tag{A.4.17}$$

From Lemma A.4.4 it follows that if f is Schwartz in $x_{s_1}, \ldots, x_{s_q}$, $x_{j_1}, \ldots, x_{j_p}$,

$$D_{s_1}^{\beta_1} \cdots D_{s_q}^{\beta_q} I_{r,j_1} \cdots I_{r,j_p}(F)(x)$$

$$= \int_{-\infty}^{x_1} \cdots \int_{-\infty}^{x_k} S_r(y_{j_1}) \cdots S_r(y_{j_p})\left(D_{s_1}^{\beta_1} \cdots D_{s_q}^{\beta_q} D_{j_1}^r \cdots D_{j_p}^r f\right)(y)\,dy, \tag{A.4.18}$$

where the sets $\{s_1, \ldots, s_q\}$ and $\{j_1, \ldots, j_p\}$ are disjoint.

Theorems A.4.2 and A.4.3 below are the main results of this appendix.

THEOREM A.4.2. *Let* f *be a Schwartz function on* R^k *and let*

$$F_r(x)$$

$$= \prod_{j=1}^{k} \{1 - S_1(x_j)D_j + \cdots + (-1)^r S_r(x_j)D_j^r + (-1)^{r+1} I_{r,j}\}(F)(x),$$

(A.4.19)

where F *is defined by* (A.4.15). *Then* F_r *is a well-defined function of bounded variation, and for any Borel set* A

$$\sum_{n \in A \cap \mathbf{Z}^k} f(n) = \int_A dF_r.$$

In the next theorem we obtain an error estimate when we expand the product defining F_r and replace it by a partial sum.

THEOREM A.4.3. *Let* $f \in \mathcal{S}$, $v \in R^k$, *and* $h > 0$, *and let* r *be a positive integer. Define*

$$\Lambda_r(x) = \sum_{j(\alpha) < r} (-1)^{|\alpha|} h^{|\alpha|} S_\alpha\left(\frac{x - v}{h}\right)(D^\alpha F)(x) \qquad \text{(A.4.20)}$$

where F *is defined by* (A.4.15), *and for any nonnegative integral vector* $\alpha = (\alpha_1, \ldots, \alpha_k)$

$$j(\alpha) = \sum_{a_j \geq 2} (\alpha_j - 1), \qquad j(\alpha) = 0 \qquad \text{if } \alpha_j < 2 \text{ for all } j. \qquad \text{(A.4.21)}$$

For every $m > k/2$ *there exists a constant* c(r, m, k) *such that for all Borel sets* A

$$\left| h^k \sum_{v + hn \in A} f(v + hn) - \int_A d\Lambda_r \right|$$

$$\leq c(r,m,k) \sum_{r \leq |\gamma| \leq kr} h^{|\gamma|} \nu_m(D^\gamma f), \qquad \text{(A.4.22)}$$

where ν_m *is defined by*

$$\nu_m(\phi) = \sup\left\{ (1 + \|x\|^2)^{m/2} |\phi(x)| : x \in R^k \right\}. \qquad \text{(A.4.23)}$$

The rest of this section is devoted to proofs of these two theorems.

Proof of Theorem A.4.2

The first step is to show that the various operations occuring in the definition of F_r commute and that F_r is a well-defined function of bounded variation. This is done by using Lemma A.4.4. The proof of the theorem is then carried out by induction on k.

The following lemma on differentiation under the integral sign is well known,† and its proof is omitted.

LEMMA A.4.4. *Let* ψ *be a function on* R^k *having continuous partial derivatives of all orders* $\alpha=(\alpha_1,\ldots,\alpha_p,0,\ldots,0)$ *with* $|\alpha|\leqslant s$. *Assume that there exists an integrable function* H *on* R^{k-p} *such that*

$$|(D^\alpha\psi)(x_1,\ldots,x_k)|\leqslant H(x_{p+1},\ldots,x_k) \qquad (x=(x_1,\ldots,x_k)\in R^k).$$

Then the function

$$\phi=\int\psi\,dx_{p+1}\cdots dx_k$$

has continuous derivatives of all orders α, $|\alpha|\leqslant s$, *in the variables* $x_1,\ldots,x_p$, *and*

$$D^\alpha\phi=\int D^\alpha\psi\,dx_{p+1}\cdots dx_k$$

for all $(x_1,\ldots,x_p)\in R^p$.

Write

$$T_{r,j}=1-S_1D_j+\cdots+(-1)^rS_rD_j^r+(-1)^{r+1}I_{r,j}. \qquad \text{(A.4.24)}$$

One can regard $T_{r,j}$ as an operator acting on functions G

$$G(x)=\int_{-\infty}^{x_1}\cdots\int_{-\infty}^{x_k}g(y)\,dy,$$

where g is an integrable function that is Schwartz in the variable x_j. Also note that if $g\in\mathcal{S}$, then $T_{r,j}(G)$ as a function of the remaining variables is the indefinite integral of a Schwartz function in these variables. Thus if $g\in\mathcal{S}$, it makes sense to apply $T_{r,j_1},\ldots,T_{r,j_p}$ successively to G.

†See Loève [1], p. 126.

It is then clear that $F_r = \prod_{j=1}^{k} T_{r,j}(F)$ is a well-defined function, independent of the order of application of the operators $T_{r,j}$. By Theorem A.2.3 of the Appendix, it is a function of bounded variation. Moreover

$$D_s T_{r,j_1} T_{r,j_2} \cdots T_{r,j_p}(F) = T_{r,j_1} T_{r,j_2} \cdots T_{r,j_p} D_s F$$

if $s \not\in \{j_1, \ldots, j_p\}$.

We now complete the proof of Theorem A.4.2 by induction on k.

If $x = (x_1, \ldots, x_{k+1})$, write $x' = (x_1, \ldots, x_k)$ so that $x = (x', x_{k+1})$. Then by the induction hypothesis,

$$\sum_{n' \leqslant x'} f(n', y) = G_r(x', y), \tag{A.4.25}$$

where

$$G_r(x', y) = \prod_{j=1}^{k} T_{r,j}(G)(x', y)$$

and

$$G(x', y) = G(x_1, \ldots, x_k, y) = \int_{-\infty < z' \leqslant x'} f(z', y)\, dz'.$$

Now

$$\sum_{n \leqslant x} f(n_1, \ldots, n_{k+1}) = \sum_{n_{k+1} \leqslant x_{k+1}} \sum_{n' \leqslant x'} f(n', n_{k+1}) = \sum_{n_{k+1} \leqslant x_{k+1}} h(x', n_{k+1}),$$

where

$$h(x', y) = \sum_{n' \leqslant x'} f(n', y) = G_r(x', y)$$

by (A.4.25). For each x', $h(x', y)$ is a Schwartz function of y. Consequently, from Theorem A.4.1 one has

$$\sum_{n_{k+1} \leqslant x_{k+1}} h(x', n_{k+1}) = T_{r,k+1}(H)(x', x_{k+1}),$$

where

$$H(x', x_{k+1}) = \int_{-\infty}^{x_{k+1}} h(x', y)\, dy = \int_{-\infty}^{x_{k+1}} G_r(x', y)\, dy$$

$$= \left(\prod_{j=1}^{k} T_{r,j} \right) \int_{-\infty}^{x_{k+1}} G(x', y)\, dy = \prod_{j=1}^{k} T_{r,j}(F)(x).$$

Thus

$$\sum_{n \leqslant x} f(n) = T_{r,k+1}(H)(x) = F_r(x),$$

which proves the theorem for $k+1$. Since the theorem follows for $k=1$ from Theorem A.4.1, the proof is complete. Q.E.D.

Proof of Theorem A.4.3

We begin with some preparation. The main problem is to estimate the mass of $\int_A dF_r - \int_A d\Lambda_r$ on the various planes defined by the restriction that some of the coordinates are integers.

LEMMA A.4.5. *Let* $h>0$, $v \in R^1$. *Then for* $m > \frac{1}{2}$

$$\sum_{n=-\infty}^{\infty} \left[1+(hn+v)^2+a^2\right]^{-m} \leqslant c_m h^{-1}(1+a^2)^{-m+1/2} + 2(1+a^2)^{-m},$$

where

$$c_m = \int_{-\infty}^{\infty} (1+x^2)^{-m}\,dx = B\left(\tfrac{1}{2}, m-\tfrac{1}{2}\right);$$

here $B(m,n)$ *denotes the standard beta function.*

Proof. The above summation can be split up as

$$\sum_{h(n-1)+v \geqslant 0} + \sum_{h(n+1)+v \leqslant 0} + \sum_{|hn+v|<h} = \sum_1 + \sum_2 + \sum_3, \qquad \text{(A.4.26)}$$

say. Then

$$\sum_1 \leqslant \sum_{(n-1) \geqslant -v/h} \int_{n-1}^{n} \left[1+(hx+v)^2+a^2\right]^{-m} dx$$

$$\leqslant \int_{-v/h}^{\infty} \left[1+(hx+v)^2+a^2\right]^{-m} dx = h^{-1}\int_0^{\infty} (1+x^2+a^2)^{-m}\,dx,$$

$$\sum_2 \leqslant \sum_{h(n+1)+v \leqslant 0} \int_n^{n+1} \left[1+(hx+v)^2+a^2\right]^{-m} dx \qquad \text{(A.4.27)}$$

$$\leqslant \int_{-\infty}^{-v/h} \left[1+(hx+v)^2+a^2\right]^{-m} dx = h^{-1}\int_{-\infty}^{0} (1+x^2+a^2)^{-m}\,dx.$$

Also, there can be no more than two integers satisfying the inequality

$$-1-\frac{v}{h}<n<1-\frac{v}{h},$$

so that

$$\sum_{3} \leqslant 2(1+a^2)^{-m}. \qquad \text{(A.4.28)}$$

The lemma follows from (A.4.27) and (A.4.28). Q.E.D.

LEMMA A.4.6. *Assume that* $0<h<1$, $v\in R^p$ *and* $m>p/2$. *Then*

$$h^p\sum\left[1+\|h\cdot n+v\|^2+a^2\right]^{-m}\leqslant(c_m+2)^p(1+a^2)^{-m+p/2},$$

where the summation is over all integral vectors $n=(n_1,\dots,n_p)$ *in* R^p.

Proof. Fix a and h. Put

$$\psi_p(v:m)=\sum\left[1+\|h\cdot n+v\|^2+a^2\right]^{-m}.$$

Then by Lemma A.4.5,

$$\psi_1(v:m)\leqslant(c_m+2)h^{-1}(1+a^2)^{-m+1/2}.$$

So, writing $v=(v_1,\dots,v_p)$ and $v'=(v_1,\dots,v_{p-1})$,

$$\psi_p(v:m)\leqslant(c_m+2)h^{-1}\psi_{p-1}(v':m-\tfrac{1}{2}),$$

and the Lemma follows by induction on p. Q.E.D.

LEMMA A.4.7. *Let* $H(x)=S_1(x_1)\cdots S_1(x_s)G(x)$, *where* G *is a function of bounded variation on* R^k, *continuous in the variables* $x_1,\dots,x_s$. *Let* H *also denote the signed measure with distribution function* H. *If* $L(n_1,\dots,n_s)=\{x=(x_1,\dots,x_k)\colon x_1=n_1,\dots,x_s=n_s\}$, *then the signed measure* $H|L(n_1,\dots,n_s)$ [*i.e., the signed measure* H *restricted to* $L(n_1,\dots,n_s)$] *has distribution function* $(-1)^sG(n_1,\dots,n_s,x_{s+1},\dots,x_k)$.

Proof. We shall prove it for the case $s=1$. The general case follows by induction on s. Let the difference operator Δ_j^h be defined as $\Delta_j^h\phi(x)=\phi(x+he_j)-\phi(x-he_j)$, where e_j is the vector with 1 at the jth coordinate and 0 at others, $h>0$. Then if $h=(h_1,\dots,h_k)$, $h_i>0$ for all i,

$$\int_{x-h<y\leqslant x}dH=\Delta_1^{h_1}\cdots\Delta_k^{h_k}(H)=\Delta_1^{h_1}(S_1\Delta_2^{h_2}\cdots\Delta_k^{h_k}G). \qquad \text{(A.4.29)}$$

The result follows by putting $x_1 = n_1$ and taking the limit as $h_1 \downarrow 0$. Q.E.D.

LEMMA A.4.8. *Let* $H(x) = S_1(x_1) \cdots S_1(x_p)\psi_1(x_{p+1},\ldots,x_q)G(x)$, *where*

$$G(x) = \int_{-\infty}^{x_{q+1}} \cdots \int_{-\infty}^{x_k} \psi_2(y_{q+1},\ldots,y_k)\, g(x_1,\ldots,x_q,y_{q+1},\ldots,y_k)\, dy_{q+1} \cdots dy_k \tag{A.4.30}$$

and $g \in \mathcal{S}$. *Suppose that* ψ_1 *is absolutely continuous in its variables and*

$$\sup|D_{p+1}^{\beta_{p+1}} \cdots D_q^{\beta_q}\psi_1| \leqslant c(\psi), \qquad \sup|\psi_2| \leqslant c(\psi), \tag{A.4.31}$$

where the constant $c(\psi)$ *depends only on* $\psi = (\psi_1, \psi_2)$, *and the first supremum is over all* $\beta_j = 0$ *or* 1, $p+1 \leqslant j \leqslant q$ *and all* $x_{p+1},\ldots,x_q$. *If* $\lambda_{(n_1,\ldots,n_s)}$ *denotes the absolutely continuous component of* $H|L(n_1,\ldots,n_s)$, *then for* $s \leqslant p$

$$\|\lambda_{(n_1,\ldots,n_s)}\| \leqslant c(\psi) \sum_\alpha \int_{R^{k-s}} |(D_{s+1}^{\alpha_{s+1}} \cdots D_q^{\alpha_q} g)(n_1,\ldots,n_s,x_{s+1},\ldots,x_k)|\, dx_{s+1} \cdots dx_k \tag{A.4.32}$$

where the sum is over all $\alpha = (\alpha_{s+1},\ldots,\alpha_q)$ *with* $\alpha_j = 0$ *or* 1, $s+1 \leqslant j \leqslant q$.

Proof. If $H_{(n_1,\ldots,n_s)}$ is the distribution function of $H|L(n_1,\ldots,n_s)$ for some integers $n_1,\ldots,n_s$, then by Lemma A.4.7, $H_{(n_1,\ldots,n_s)}(x_{s+1},\ldots,x_k) = (-1)^s \times S_1(x_{s+1}) \cdots S_1(x_p)\psi_1(x_{p+1},\ldots,x_q) \cdot G(n_1,\ldots,n_s, x_{s+1},\ldots,x_k)$. Since the density function of $\lambda_{(n_1,\ldots,n_s)}$ is $D_{s+1} \cdots D_k H_{(n_1,\ldots,n_s)}$, the estimate (A.4.32) follows if one uses (A.4.31). Q.E.D.

Let ν_m be the seminorm defined by (A.4.23). We then have

LEMMA A.4.9. *With the same notation and assumptions as in Lemma* A.4.8, *let* $g(x) = \phi(v + hx)$, *where* $v \in R^k$, $h > 0$, *and* $\phi \in \mathcal{S}$. *Then for all* $m > k/2$,

$$h^k \|H\| \leqslant c(m,k,\psi) \sum_\alpha h^{|\alpha|} \nu_m(D^\alpha \phi), \tag{A.4.33}$$

where $c(m,k,\psi)$ *is a constant depending only on its arguments, and the summation is over all* α *such that* $\alpha_j = 0$ *or* 1 *and* $\alpha_j = 0$ *for* $j > q$.

Proof. Let $\{j_1,\ldots,j_s\} \subset \{1,2,\ldots,p\}$ and $L(j_1,\ldots,j_s\colon n_1,\ldots,n_s) = \{x \in R^k\colon x_{j_1} = n_1,\ldots,x_{j_s} = n_s\}$ for some integers $n_1,\ldots,n_s$. Let $\lambda(j_1,\ldots,j_s\colon n_1,\ldots,n_s)$

denote the absolutely continuous component of $H|L(j_1,\ldots,j_s\colon n_1,\ldots,n_s)$ with respect to the Lebesgue measure on the $(k-s)$-dimensional "plane" L. Then from the nature of the function H, it follows that

$$H=\lambda_0+\sum_{(j_1,\ldots,j_s)}\sum_{(n_1,\ldots,n_s)}\lambda(j_1,\ldots,j_s\colon n_1,\ldots,n_s), \tag{A.4.34}$$

where λ_0 is the absolutely continuous component of the signed measure H. We shall estimate

$$\sum_{(n_1,\ldots,n_s)}\|\lambda(j_1,\ldots,j_s\colon n_1,\ldots,n_s)\|$$

assuming $j_1=1,\ldots,j_s=s$. By Lemma A.4.8, this is less than or equal to

$$\begin{aligned}
&c(\psi)\sum_{\alpha}\sum_{n_1,\ldots,n_s}h^{|\alpha|}\nu_m(D^\alpha\phi)\int_{R^{k-s}}\left[1+\sum_{i=1}^{s}(hn_i+v_i)^2\right.\\
&\qquad\qquad\left.+\sum_{i=s+1}^{k}(hx_i+v_i)^2\right]^{-m}dx_{s+1}\cdots dx_k\\
&\leqslant c(\psi)\left[\sum_{\alpha}h^{|\alpha|}\nu_m(D^\alpha\phi)\right]\left\{\int_{R^{k-s}}\left[1+\|hx+v\|^2\right]^{-m+s/2}dx\right\}c'(m,s)h^{-s}\\
&\leqslant c(\psi)\left[\sum_{\alpha}h^{|\alpha|}\nu_m(D^\alpha\phi)\right]c(m,k)h^{-k}.
\end{aligned} \tag{A.4.35}$$

We have also used Lemma A.4.6 in this estimation. Lemma A.4.9 now follows. Q.E.D.

We are now ready to prove Theorem A.4.3. Let

$$w(x)=f(v+hx). \tag{A.4.36}$$

Then

$$W(x)=\int_{-\infty<y\leqslant x}w(y)\,dy=h^{-k}F(v+hx). \tag{A.4.37}$$

Then by Theorem A.4.2,

$$\sum_{v+hn\in A}f(v+hn)=\int_{v+hx\in A}dW_r(x), \tag{A.4.38}$$

where

$$W_r = \prod_{j=1}^{k} T_{r,j}(W). \tag{A.4.39}$$

If we now expand the product formally (as we may), then W_r is expressed as a sum of terms of the form

$$\pm S_{\gamma_1}(x_{\sigma_1}) \cdots S_{\gamma_t}(x_{\sigma_t}) D_{\sigma_1}^{\gamma_1} \cdots D_{\sigma_t}^{\gamma_t} I_{r,j_1} \cdots I_{r,j_p}(W), \tag{A.4.40}$$

where $\{\sigma_1, \ldots, \sigma_t, j_1, \ldots, j_p\}$ is a permutation of $\{1, 2, \ldots, k\}$, and $\gamma_j = 0, 1$, or larger than 1. We now obtain an estimate of the variation norm of the signed measure whose distribution function is given by (A.4.40). For this purpose the function (A.4.40) may be written in the form

$$\pm S_1(x_{k_1}) \cdots S_1(x_{k_q}) S_{\beta_1}(x_{m_1}) \cdots S_{\beta_s}(x_{m_s}) G(x), \tag{A.4.41}$$

where $\beta_j \geqslant 2$, $1 \leqslant j \leqslant s$, and

$$G(x) = \int_{-\infty}^{x_{i_1}} \cdots \int_{-\infty}^{x_{i_l}} \int_{-\infty}^{x_{j_1}} \cdots \int_{-\infty}^{x_{j_p}} S_r(x'_{j_1}) \cdots S_r(x'_{j_p}) g(x') \, dx'_{i_1} \cdots dx'_{j_p}, \tag{A.4.42}$$

where $(x')_a = x'_a$ for all $a \in \{i_1, \ldots, i_l, j_1, \ldots, j_p\}$ and $(x')_a = x_a$ for all other indices a. Also, writing $|\beta| = \beta_1 + \cdots + \beta_s$, we define the function g by

$$\begin{aligned} g(x) &= \left(D_{m_1}^{\beta_1 - 1} \cdots D_{m_s}^{\beta_s - 1} D_{j_1}^r \cdots D_{j_p}^r w\right)(x) \\ &= h^{pr + |\beta| - s} \left(D_{m_1}^{\beta_1 - 1} \cdots D_{m_s}^{\beta_s - 1} D_{j_1}^r \cdots D_{j_p}^r f\right)(v + hx) \quad (x \in R^k). \end{aligned} \tag{A.4.43}$$

Here $\{i_1, \ldots, i_l,\ k_1, \ldots, k_q,\ m_1, \ldots, m_s,\ j_1, \ldots, j_p\}$ is a permutation of $\{1, 2, \ldots, k\}$. Since $S_j (j \geqslant 2)$ and its derivative is bounded, it follows that Lemma A.4.9 can be applied to the function (A.4.41) and its variation is less than or equal to

$$c(r, m, k) h^{-k + pr + |\beta| - s} \sum_{\alpha} h^{|\alpha|} v_m\left(D^{\alpha} D_{m_1}^{\beta_1 - 1} \cdots D_{j_p}^r f\right), \tag{A.4.44}$$

where the summation is over all α with $\alpha_j = 0$ or 1 and $\alpha_j = 0$ unless $j \in \{k_1, \ldots, k_q,\ m_1, \ldots, m_s\}$. It follows that $|\alpha| \leqslant q + s$, and if $D^{\gamma} = D^{\alpha} D_{m_1}^{\beta_1 - 1} \cdots D_{m_s}^{\beta_s - 1} D_{j_1}^r \cdots D_{j_p}^r$, then $|\gamma| = |\alpha| + |\beta| - s + pr \leqslant |\beta| + q + pr \leqslant kr$. Also $|\gamma| \geqslant r$ if $p > 0$.

Let

$$W_r = W_r' + W_r'' + W_r''', \tag{A.4.45}$$

where

$$\begin{aligned} W_r'(x) &= \sum_{j(\alpha)<r} (-1)^{|\alpha|} S_\alpha(x)(D^\alpha W)(x), \\ W_r''(x) &= \sum_{j(\alpha)\geqslant r} (-1)^{|\alpha|} S_\alpha(x)(D^\alpha W)(x), \end{aligned} \tag{A.4.46}$$

and W'''_r is defined by the identity (A.4.45). Then W_r''' is the sum of terms of the form (A.4.41), where $p>0$. Thus the variation of W_r''' is not larger than

$$h^{-k}c(r,m,k) \sum_{r\leqslant|\gamma|\leqslant kr} h^{|\gamma|}\nu_m(D^\gamma f). \tag{A.4.47}$$

Now consider W_r''. If the general term occurring in W_r'' is written in the form (A.4.41), then $p=0$, $j(\alpha)=|\beta|-s\geqslant r$. Then from (A.4.44) it follows that

$$h^k\|W_r''\| \leqslant c(r,m,k) \sum_{r\leqslant|\gamma|\leqslant kr} h^{|\gamma|}\nu_m(D^\gamma f), \tag{A.4.48}$$

where $\|W_r''\|$ is the variation norm of the signed measure having distribution function W_r''. Finally, it is easy to check that

$$h^{-k}\Lambda_r(x) = W_r'\left(\frac{x-v}{h}\right) \tag{A.4.49}$$

so that

$$\int_{v+hx\in A} dW_r' = h^{-k}\int_A d\Lambda_r. \tag{A.4.50}$$

The theorem follows from (A.4.45)–(A.4.50). Q.E.D.

References

Agnew, R. P.

[1] Global versions of the central limit theorem. *Proc. Natl. Acad. Sci.* **40** (1954) 800–804.

Bahr, B. von

[1] On the convergence of moments in the central limit theorem. *Ann. Math. Stat.* **36** (1965) 808–818.

[2] On the central limit theorem in R_k. *Ark. Mat.* **7** (1967) 61–69.

[3] Multi-dimensional integral limit theorems. *Ark. Mat.* **7** (1967) 71–88.

Beek, P. Van

[1] An application of the Fourier method to the problem of sharpening the Berry–Esseen inequality. *Z. Wahrscheinlichkeitstheor. Verw. Geb.* **23** (1972) 187–197.

Bergström, H.

[1] On the central limit theorem in the space R_k, $k>1$. *Skand. Aktuarietidskr.* **28** (1945) 106–127.

[2] On the central limit theorem in the case of not equally distributed random variables. *Skand. Aktuarietidskr.* **33** (1949) 37–62.

[3] On the central limit theorem in R^k. *Z. Wahrscheinlichkeitstheor. Verw. Geb.* **14** (1969) 113–126.

Bernstein, S.

[1] Sur l'extension du théoremè limite du calcul des probabilités aux sommes de quantités dépendantes. *Math. Ann.* **97** (1927) 1–59.

Berry, A. C.

[1] The accuracy of the Gaussian approximation to the sum of independent variates. *Trans. Am. Math. Soc.* **48** (1941) 122–136.

Bhattacharya, R. N.

[1] Berry–Esseen bounds for the multi-dimensional central limit theorem. Ph.D. Dissertation, University of Chicago (1967).

[2] Berry–Esseen bounds for the multi-dimensional central limit theorem. *Bull. Am. Math. Soc.* **75** (1968) 285–287.

[3] Rates of weak convergence for the multi-dimensional central limit theorems. *Theory Probab. Appl.* **15** (1970) 68–86.

[4] Rates of weak convergence and asymptotic expansions for classical central limit theorems. *Ann. Math. Stat.* **42** (1971) 241–259.

[5] Recent results on refinements of the central limit theorem. *Proceedings of the Sixth Berkeley Symposium on Mathematical Statistics and Probability*, Vol. II, University of California Press (1972), pp. 453–484.

[6] Speed of convergence of the n-fold convolution of a probability measure on a compact group. *Z. Wahrscheinlichkeitstheor. Verw. Geb.* **23** (1972) 1–10.

[7] On errors of normal approximation. *Ann. Probab.* **3** (1975) 815–828.

Bickel, P. J.

[1] Edgeworth expansions in nonparametric statistics. *Ann. Stat.* **2** (1974) 1–20.

Bikjalis, A.

[1] On the refinement of the remainder term in the multidimensional central limit theorem. *Litov. Mat. Sb.* **4** (1964) 153–158 (in Russian).

[2] Estimates of the remainder term in the central limit theorem. *Litov. Mat. Sb.* **6** (1966) 321–346 (in Russian).

[3] On multivariate characteristic functions. *Litov. Mat. Sb.* **8** (1968) 21–39 (in Russian).

[4] Asymptotic expansions of distribution functions and the density functions of sums of independent identically distributed random vectors. *Litov. Mat. Sb.* **8** (1968) 405–422 (in Russian).

[5] Asymptotic expansions of distributions of sums of identically distributed independent lattice random variables. *Theory Probab. Appl.* **14** (1969) 481–489.

[6] On the central limit theorem in R^k, Parts I, II. *Litov. Mat. Sb.* **11** (1971) 27–58; **12** (1972) 53–84 (in Russian).

Billingsley, P.

[1] *Convergence of Probability Measures.* Wiley, New York (1968).

Billingsley, P. and Topsøe, F.

[1] Uniformity in weak convergence. *Z. Wahrscheinlichkeitstheor. Verw. Geb.* **7** (1967) 1–16.

Bohr, H.

[1] *Almost Periodic Functions.* Chelsea, New York (1947).

Brillinger, D.

[1] A note on the rate of convergence of a mean. *Biometrika* **49** (1962) 574–576.

Chebyshev, P. L.

[1] Sur deux théorèmes relatifs aux probabilitiés. *Acta Math.* **14** (1890) 305–315.

Chung, K. L.

[1] *A Course in Probability Theory*, 2nd ed. Academic Press, New York (1974).

Cramér, H.

[1] On the composition of elementary errors. *Skand. Aktuarietidskr.* **11** (1928) 13–74, 141–180.

[2] Sur un nouveau théorème-limite de la théorie des probabilités. *Act. Sci. Ind.* **736** (1938).

[3] *Random Variables and Probability Distributions.* Cambridge University Press, Cambridge (1937).

[4] *Mathematical Methods of Statistics.* Princeton University Press, Princeton, N. J. (1946).

Dieudonné, J.

[1] *Treatise on Analysis*, Vol. II (1970), Vol. III (1972). English translation by I. G. MacDonald, Academic Press, New York.

Doob, J. L.

[1] *Stochastic Processes.* Wiley, New York (1953).

Dudley, R. M.

[1] Convergence of Baire measures. *Stud. Math.* **27** (1966) 251–268.

[2] Distances of probability measures and random variables. *Ann. Math. Stat.* **39** (1968) 1563–1572.

Edgeworth, F. Y.
[1] The law of error. *Proc. Camb. Philos. Soc.* **20** (1905) 36–65.

Eggleston, H. G.
[1] *Convexity*. Cambridge University Press, Cambridge (1958).

Esseen, C. G.
[1] Fourier analysis of distribution functions. A mathematical study of the Laplace–Gaussian law. *Acta Math.* **77** (1945) 1–125.
[2] A moment inequality with an application to the central limit theorem. *Skand. Aktuarietidskr.* **3–4** (1956) 160–170.
[3] On mean central limit theorems. *Trans. R. Inst. Technol., Stockh.*, **121** (1958) 1–30.

Federer, H.
[1] *Geometric Measure Theory*, Die Grundlehren der Mathematischen Wissenschaften, Vol. 153. Springer, New York (1969).

Feller, W.
[1] Über den zentralen grenzwertsatz der wahrscheinlichkeitsrechnung. *Math. Z.* **40** (1935) 521–559.
[2] On the Berry-Esseen theorem. *Z. Wahrscheinlichkeitstheor. Verw. Geb.* **10** (1968) 261–268.
[3] *An Introduction to Probability Theory and Its Applications*, Vol. 2, 2nd ed. Wiley, New York (1971).

Gnedenko, B. V.
[1] On a local theorem for stable limit distributions. *Ukr. Mat. Zh.* **4** (1949) 3–15 (in Russian).

Gnedenko, B. V. and Kolmogorov, A. N.
[1] *Limit Distributions of Sums of Independent Random Variables*. English translation by K. L. Chung, Addison-Wesley, Reading, Mass. (1954).

Halmos, P.
[1] *Measure Theory*. Van Nostrand, Princeton, N. J. (1950).

Hardy, G. H.
[1] *Pure Mathematics*, 3rd ed. Cambridge University Press (1959).

Hardy, G. H., Littlewood, J. E., and Polya, G.
[1] *Inequalities*. Cambridge University Press (1934).

Heyde, C. C.
[1] On the influence of moments on the rate of convergence to the normal distribution. *Z. Wahrscheinlichkeitstheor. Verw. Geb.* **8** (1967) 12–18.

Ibragimov, I. A.
[1] On the accuracy of Gaussian approximation to the distribution functions of sums of independent variables. *Theory Probab. Appl.* **11** (1966) 559–579.

Ibragimov, I. A. and Linnik, Yu. V.
[1] *Independent and Stationary Sequences of Random Variables*. English translation, Wolters-Noordhoff, Gronigen (1971).

Ingham, A. E.
[1] A note on Fourier transforms. *J. Lond. Math. Soc.* **9** (1934) 29–32.

Katznelson, Y.
[1] *An Introduction to Harmonic Analysis*. Wiley, New York (1968).

Khinchin, A. Ya.
[1] Begrundung der normalkorrelation nach der Lindebergschen methode. *Izv. Assoc. Inst. Mosk. Univ.* **1** (1928) 37–45.

[2] *Asymptotische Gesetze der Wahrscheinlichkeitsrechnung*. Ergeb. der Mat. Springer, Berlin (1933).

[3] *Mathematical Foundations of Statistical Mechanics*. Gtti, Moscow–Leningrad (1938). English translation by G. Gamow, Dover, New York (1949).

Laplace, P. S.

[1] *Theorie Analytique des Probabilités*, 1st ed. (1812).

Lévy, P.

[1] *Calcul del Probabilités*. Paris (1925).

Liapounov, A. M.

[1] Sur une proposition de la théorie des probabilités. *Bull. Acad. Imp. Sci. St. Petersb.* (5) **13** (1900) 359–386.

[2] Nouvelle forme du théorème sur la limite de probabilité. *Mem. Acad. Sci. St. Petersb.*, (8) **12** (1901).

Lindeberg, J. W.

[1] Eine neue herleitung des exponential gesetzes in der Wahrscheinlichkeitsrechnung. *Math. Z.* **15** (1922) 211–225.

Loéve, M.

[1] *Probability Theory*, 3rd. ed. Van Nostrand, Princeton, N. J. (1963).

Markov, A. A.

[1] The law of large numbers and the method of least squares. *Izv. Fiz. Mat. Soc. Kazan. Univ.* (2) **8** (1898) 110–128 (in Russian).

Matthes, T. K.

[1] The multivariate central limit theorem for regular convex sets. *Ann. Probab.* **3** (1975) 503–515.

DeMoivre, A.

[1] *Miscellana Analytica Supplementum*. London (1730).

Nagaev, S. V.

[1] Some limit theorems for large derivations. *Theory Probab. Appl.* **10** (1965) 214–235.

Osipov, L. V.

[1] On asymptotic expansions of distribution functions of sums of independent random variables. *Vestn. Leningr. Univ.* **24** (1972) 51–59 (in Russian).

Osipov, L. V. and Petrov, V. V.

[1] On an estimate of the remainder in the central limit theorem. *Theory Probab. Appl.* **12** (1967) 281–286.

Paley, R. E. A. C. and Wiener, N.

[1] *Fourier Transforms in the Complex Domain*, Vol. XIX. A. M. S. Colloquium Publications (1934).

Paulauskas, V.

[1] On the multidimensional central limit theorem. *Litov. Mat. Sb.* **10** (1970) 783–789.

Petrov, V. V.

[1] On local limit theorems for sums of independent random variables. *Theory Probab. Appl.* **9** (1964) 312–320.

[2] *Sums of Independent Random Variables*. Nauka, Moscow (1972) (in Russian).

Rao, Ranga R.

[1] Some problems in probability theory. D. Phil, Thesis, Calcutta University (1960).

[2] On the central limit theorem in R_k. *Bull, Am. Math. Soc.* **67** (1961) 359–361.

[3] Relations between weak and uniform convergence of measures with applications. *Ann. Math. Stat.* **33** (1962) 659–680.

Rao, R. Ranga and Varadarajan, V. S.

[1] A limit theorem for densities. *Sankhya* **22** (1960) 261–266.

Rotar', V. I.

[1] A non-uniform estimate for the convergence speed in the multi-dimensional central limit theorem. *Theory Probab. Appl.* **15** (1970) 630–648.

Sazonov, V. V.

[1] On the multi-dimensional central limit theorem. *Sankhya*, Ser. A **30** (1968) 181–204.

[2] On a bound for the rate of convergence in the multidimensional central limit theorem. *Proceedings of the Sixth Berkeley Symposium on Mathematical Statistics and Probability*, Vol. II, University of California Press (1972), pp. 563–581.

Scheffé, H. Bounds on Moments : TPA 19 371–374 (1974)

[1] A useful convergence theorem for probability distributions. *Ann. Math. Stat.* **18** (1947) 434–438.

Stein, E. M. and Weiss, G.

[1] *Introduction to Fourier Analysis on Euclidean Spaces*. Princeton University Press, Princeton, N. J. (1971).

Takano, K.

[1] Multidimensional central limit criterion in the case of bounded variances. *Ann. Inst. Stat. Math.* Tokyo **7** (1956) 81–93.

Topsøe, F.

[1] On the Gilvenko–Cantelli theorem. *Z. Wahrscheinlichkeitstheor. Verw. Geb.* **14** (1969) 239–250.

Uspensky, J. V.

[1] *Introduction to Mathematical Probability*. McGraw-Hill, New York (1937).

Wallace, D. L.

[1] Asymptotic approximations to distributions. *Ann. Math. Stat.* **29** (1958) 635–654.

Wiener, N.

[1] *The Fourier Integral and Certain of Its Applications*. Dover Publications, New York (1933).

Yarnold, J. K.

[1] Asymptotic approximations for the probability that a sum of lattice random vectors lies in a convex set. *Ann. Math. Stat.* **43** (1972), 1566–1580.

Zahl, S.

[1] Bounds for the central limit theorem error. *SIAM Appl. Math.* **14** (1966) 1225–1245.

Zolotarev, V. M.

[1] On the closeness of the distributions of two sums of independent random variables. *Theor. Probab. Appl.* **10** (1965) 472–479.

[2] A sharpening of the inequality of Berry–Esseen. *Z. Wahrscheinlichkeitstheor. und Verw. Geb.* **8** (1967) 32–42.

Index

non uniform rates:
see paper of
[7] Bhat... A. Pr. 3
(1975) 815–828